# INTRODUCTION TO APPLIED STATISTICAL SIGNAL ANALYSIS

*Second Edition*

# INTRODUCTION TO APPLIED STATISTICAL SIGNAL ANALYSIS

## Second Edition

**Richard Shiavi**

*Vanderbilt University School of Engineering*
*Nashville, Tennessee*

**ACADEMIC PRESS**

*San Diego   San Francisco   New York   Boston   London   Sydney   Tokyo*

## Academic Press Series in Biomedical Engineering

The focus of this series is twofold. First, it will produce a series of core text/references for biomedical engineering undergraduate and graduate courses. With biomedical engineers coming from a variety of engineering and biomedical backgrounds, it is necessary to create new cross-disciplinary teaching and self-study books. Secondly, the Academic Press Series in Biomedical Engineering will also develop handbooks for each of the major subject areas of biomedical engineering.

The series editor, Joseph Bronzino, is one of the most renowned biomedical engineers in the world. Joseph Bronzino is the Vernon Roosa Professor of Applied Science at Trinity College in Hartford, Connecticut.

This book is printed on acid-free paper. ∞

Copyright © 1999 by Academic Press

ACADEMIC PRESS
*A Harcourt Science and Technology Company*
525 B. St. Suite 1900, San Diego, CA 92101-4495, USA
http://www.apnet.com

Academic Press
24–28 Oval Road, London NW1 7DX, UK
http://www.hbuk.co.uk/ap/

**Library of Congress Cataloging Number:** 98-22148

**International Standard Serial Number:** 0-12-640010-5

Printed in the United States of America
99  00  01  02  03  COB  9  8  7  6  5  4  3  2  1

# CONTENTS

# 1

# INTRODUCTION AND TERMINOLOGY

# II

# EMPIRICAL MODELING AND APPROXIMATION

# III

# FOURIER ANALYSIS

# IV
## PROBABILITY CONCEPTS AND SIGNAL CHARACTERISTICS

# V

# INTRODUCTION TO RANDOM PROCESSES AND SIGNAL CORRELATION

# VI
## RANDOM SIGNALS, LINEAR SYSTEMS, AND POWER SPECTRA

# VII
## SPECTRAL ANALYSIS FOR RANDOM SIGNALS: CLASSICAL ESTIMATION

# VIII
## RANDOM SIGNAL MODELING AND MODERN SPECTRAL ESTIMATION

# IX
## THEORY AND APPLICATION OF CROSS CORRELATION AND COHERENCE

# PREFACE

This book presents a practical introduction to signal analysis techniques that are commonly used in a broad range of engineering areas such as speech, biomedical signals, geophysics, communications, pattern recognition, etc. To emphasize the analytic approaches, a certain background is necessary. The book is designed for an individual who has the basic background in mathematics, science, and computer programming concepts required in an undergraduate engineering curriculum. In addition, one *should* have an *introductory level* background in probability, and statistics and in discrete time systems.

The sequence of material begins with a discussion of techniques for modeling and representing discrete data measurements, interpolation, and a definition of time series. Then the classical and some modern techniques for analyzing signals are treated. These topics include Fourier spectra, statistical properties of time series, correlation functions, spectral density functions, and time series modeling. Integrated into the treatment of these techniques are the methodologies for estimating them properly. The presentation style is designed for the individual who wants a theoretical introduction to the basic principles and then the knowledge necessary to implement them practically. The mode of presentation is to define a theoretical concept, show areas of engineering in which these concepts are useful, define the algorithms and assumptions needed to implement them, and then present detailed examples that have been implemented on a computer. It is hoped that the exposure to engineering applications will develop an appreciation for the utility and necessity of signal processing methodologies.

The exercises at the end of the chapters are designed with several goals. Some focus directly on the material presented, and some extend the material for applications that are less often encountered. The degree of difficulty ranges from simple pencil-and-paper problems to computer implementation of algorithms. All computer-oriented examples and problems are directly implementable on personal computers. One can use either signal processing environments such as MATLAB or a standard computing approach with a FORTRAN or C compiler integrated with suitable graphics hardware and software. For an introductory course, the environment and software should not be so sophisticated and complex that the student cannot comprehend the code or script. A good reference for software environments is the November, 1995 issue of the *IEEE Signal Processing Magazine*.

The two major additions to this edition of this book are found in the accompanying diskette. One is the number of real data sets. All of the data in the Appendices are available in data files, as are other signals from real-world applications in engineering. The other addition is the notebooks, which are written in the integrated environment of Microsoft Word and MATLAB. Each notebook presents a principle and demonstrates its implementation via script in MATLAB. The student is then asked to exercise other aspects of the principle interactively by making simple changes in the MATLAB script. The student then receives immediate feedback concerning what is happening and can relate theoretical concepts to real effects upon a signal. Finally, the student is asked to implement the learned procedure on a signal from a database of actual measurements.

When used as a course textbook, most of this book can be studied in one semester in a senior undergraduate or first-year graduate course. The topic selection is obviously the instructor's choice.

*This book is dedicated to my wife, Gloria,
and to my parents, who encouraged me and gave me
the opportunity to be where I am today.*

# ACKNOWLEDGMENTS

The author of a textbook is usually helped significantly by the institution by which he or she is employed and through surrounding circumstances. In particular, I am indebted to the Department of Biomedical Engineering and the School of Engineering at Vanderbilt University for giving me some released time and for placing a high priority on the publication of this book for academic purposes. The reviewers have been very helpful in their constructive criticism toward developing a presentation with balance of theory and application. In particular I would like to thank, in alphabetical order, K. S. Arun, University of Illinois; Edward Delp, Purdue University; O. K. Esroy, Purdue University; Alfred Hero, University of Michigan; and David Munson, University of Illinois for their detailed remarks and scholarly review of the manuscript.

# SYMBOLS

## English

| | |
|---|---|
| $a(i), b(i)$ | parameters of A, MA, and ARMA models |
| $A_m$ | polynomial coefficient |
| $B$ | bandwidth |
| $B_e$ | equivalent bandwidth |
| $c_x(k)$ | sample covariance function |
| $c_{yx}(k)$ | sample cross covariance function |
| $C_n$ | coefficients of trigonometric Fourier series |
| $C_x(k)$ | autocovariance function |
| $C_{xy}(k)$ | cross covariance function |
| COV [ ] | covariance operator |
| $d(n)$ | data window |
| $D(f)$ | data spectral window |
| $e_i$ | error in polynomial curve fitting |
| E[ ] | expectation operator |
| $E_M$ | sum of squared errors |
| $E_{\text{tot}}$ | total signal energy |
| $f$ | cyclic frequency |
| $f_d$ | frequency spacing |
| $f_N$ | folding frequency, highest frequency component |
| $f_s$ | sampling frequency |

| | |
|---|---|
| $f(t)$ | scaler function of variable **t** |
| $f_x(\alpha), f(x)$ | probability density function |
| $f_{xy}(\alpha, \beta), f(x, y)$ | bivariate probability density function |
| $F_x(\alpha), F(x)$ | probability distribution function |
| $F_{xy}(\alpha, \beta), F(x, y)$ | bivariate probability distribution function |
| $g$ | loss coefficient |
| $h(t), h(n)$ | impulse response |
| $H(f), H(\omega)$ | transfer function |
| $I(f), I(m)$ | periodogram |
| $\mathrm{Im}(\ )$ | imaginary part of a complex function |
| $\Im\ [\ ]$ | imaginary operator |
| $K^2(f), K^2(m)$ | magnitude squared coherence function |
| $L_i(x)$ | Lagrange coefficient function |
| $m$ | mean |
| $N$ | number of points in a discrete time signal |
| $p, q$ | order of AR, MA, and ARMA processes |
| $P$ | signal power, or signal duration |
| $P[\ ]$ | probability of [ ] |
| $P_m(x)$ | polynomial function |
| $\mathrm{Re}(\ )$ | real part of a complex function |
| $R_x(k)$ | autocorrelation function |
| $R_{yx}(k)$ | cross correlation function |
| $\Re[\ ]$ | real operator |
| $s_p^2$ | variance of linear prediction error |
| $S(f), S(m)$ | power spectral density function |
| $S_{yx}(f), s_{yx}(m)$ | cross-spectral density function |
| $T$ | sampling interval |
| $U(t)$ | unit step function |
| $\mathrm{Var}[\ ]$ | variance operator |
| $w(k)$ | lag window |
| $W(f)$ | lag spectral window |
| $x(t), x(n)$ | time function |
| $X(f), X(m), X(\omega)$ | Fourier transform |
| $z_m$ | coefficients of complex Fourier series |

## Greek

| | |
|---|---|
| $\alpha$ | significance level |
| $\gamma_x(t_0, t_1), \gamma_x(k)$ | ensemble autocovariance function |
| $\epsilon(n)$ | linear prediction error |
| $\delta(t)$ | impulse function, Dirac delta function |

| | |
|---|---|
| $\delta(n)$ | unit impulse, Kronecker delta function |
| $\lambda_i$ | energy in a function |
| $\Lambda_{yx}(f)$ | cospectrum |
| $\eta(n)$ | white-noise process |
| $\xi(\tau)$ | ensemble normalized autocovariance function |
| $\rho$ | correlation coefficient |
| $\rho_x(k)$ | normalized autocovariance function |
| $\rho_{yx}(k)$ | normalized cross covariance function |
| $\sigma^2$ | variance |
| $\sigma_e$ | standard error of the estimate |
| $\sigma_{xy}^2$ | covariance |
| $\phi(f)$ | phase response |
| $\phi_{yx}(f)$ , $\phi_{yx}(m)$ | cross phase spectrum |
| $\Phi_n(t)$ | orthogonal function set |
| $\Phi(z)$ | cdf for unit Gaussian function |
| $\varphi_x(t_0, t_1)$, $\varphi_x(k)$ | ensemble autocorrelation function |
| $\Psi_{yx}(f)$ | quadrature spectrum |
| $\omega$ | radian frequency |
| $\omega_d$ | radian frequency spacing |

## Acronyms

| | |
|---|---|
| ACF | autocorrelation function |
| ACVF | autocovariance function |
| AIC | Akaike's information criterion |
| AR | autoregressive |
| ARMA | autoregressive moving-average |
| BT | Blackman-Tukey |
| CCF | cross correlation function |
| CCVF | cross covariance function |
| cdf | cumulative distribution function |
| CF | correlation function |
| CSD | cross-spectral density |
| CTFT | continuous time Fourier transform |
| DFT | discrete Fourier transform |
| DTFT | discrete time Fourier transform |
| erf | error function |
| FPE | final prediction error |
| FS | Fourier series |
| IDFT | inverse discrete Fourier transform |
| IDTFT | inverse discrete time Fourier transform |

LPC    linear prediction coefficient
MA     moving average
MEM    maximum entropy method
MSC    magnitude squared coherence
MSE    mean squared error
NACF   normalized autocovariance function
NCCF   normalized cross covariance function
pdf    probability density function
PL     process loss
PSD    power spectral density
TSE    total squared error
VR     variance reduction
WN     white noise
YW     Yule-Walker

## Operators

$X(f)^*$              conjugation
$x(n) * y(n)$         convolution
$\hat{S}(m)$          sample estimate
$\tilde{S}(m)$        smoothing
$\bar{x}(n)$          periodic repetition

# 1

# INTRODUCTION
# AND TERMINOLOGY

## 1.1 INTRODUCTION

Historically, a *signal* meant any set of signs, symbols, or physical gesticulations that transmitted information or messages. The first electronic transmission of information was in the form of Morse code. In the most general sense, a signal can be embodied in two forms. In one form it is some measured or observed behavior or physical property of a phenomenon that contains information about that phenomenon. In the other form the signal can be generated by a man-made system and have the information encoded. Signals can vary over time or space. Our daily existence is replete with the presence of signals, and they occur not only in man-made systems but also in human and natural systems. A simple natural signal is the measurement of air temperature over time, as shown in Figure 1.1. Study of the fluctuations in temperature informs us about some characteristics of our environment. A much more complex phenomenon is speech. Speech is intelligence transmitted through a variation over time in the intensity of sound waves. Figure 1.2 shows an example of the intensity of a waveform associated with a typical sentence. Each sound has a different characteristic waveshape that conveys different information to the listener. In television systems, the signal is the variation in electromagnetic wave intensity that encodes the picture information. In human systems, measurements of heart and skeletal muscle

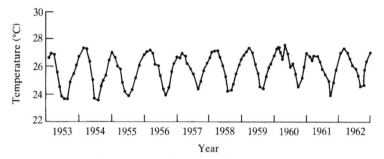

**FIGURE 1.1**   The average monthly air temperature at Recife, Brazil. [Adapted from Chatfield, fig. 1.2, with permission.]

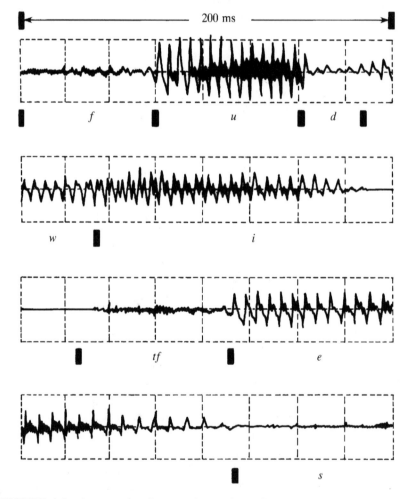

**FIGURE 1.2**   An example of a speech waveform illustrating different sounds. The utterance is "should we chase....." [Adapted from Oppenheim, fig. 3.3, with permission.]

activity in the form of electrocardiographic and electromyographic voltages are signals. With respect to these last three examples, the objective of signal analysis is to process these signals to extract information concerning the characteristics of the picture, cardiac function, and muscular function. Signal processing has been implemented for a wide variety of applications. Many of them will be mentioned throughout this textbook. Good sources for other applications are Chen (1988) and Cohen (1986).

A time-dependent signal measured at particular points in time is synonymously called a *time series*. The latter term arose within the field of applied mathematics and initially pertained to the application of probability and statistics to data varying over time. Some of the analyses were performed on economic or astronomic data such as the Beveridge wheat price index or Wolfer's sunspot numbers (Anderson, 1971). Many of the techniques that are used currently were devised by mathematicians decades ago. The invention of the computer and now the development of powerful and inexpensive computers has made the application of these techniques very feasible. In addition, the availability of inexpensive computing environments of good quality has made their implementation widespread. All of the examples in this textbook were implemented using inexpensive subroutine libraries or computing environments that are good for a broad variety of engineering and scientific applications (Ebel and Younan, 1995; Foster, 1992). These libraries are also good from a pedagogical perspective because the algorithms are explained in the accompanying books (Blahut, 1985; Press et al., 1992 & 1993, Stearns and David, 1988). Before beginning a detailed study of the techniques and capabilities of signal or time series analysis, an overview of terminology and basic properties of signal waveforms is necessary. As with any field, symbols and acronyms are a major component of the terminology, and standard definitions have been utilized as much as possible (Granger, 1982; Jenkins and Watts, 1968). Other textbooks that will provide complementary information of either an introductory or an advanced level are listed in the reference section.

## 1.2   SIGNAL TERMINOLOGY

### 1.2.1   Domain Types

The *domain* of a signal is the independent variable over which the phenomenon is considered. The domain encountered most often is the time domain. Figure 1.3 shows the electrocardiogram (ECG) measured from the abdomen of a pregnant woman. The ECG exists at every instant of time, thus the ECG evolves in the *continuous time domain*. Signals that have values at a finite set of time instants exist in the *discrete time domain*. The temperature plot in Figure 1.1 shows a discrete time signal with average temperatures given each month. There are two

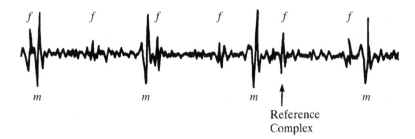

Reference
Complex

**FIGURE 1.3**   The abdominal ECG from a pregnant woman showing the maternal ECG waveform (m) and the fetal ECG waveform (f). [Adapted from Inbar, fig. 8, with permission.]

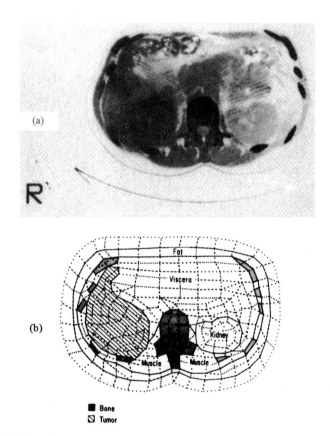

**FIGURE 1.4**   (a) A tomagraphic scan of a cross section of the abdomen. (b) A digitized model of the scan showing the calculated outlines of the major anatomical regions. [Adapted from Strohbehn and Douple, fig. 4, with permission.]

types of discrete time signals. If the dependent variable is processed in some way, it is an *aggregate* signal. Processing can be averaging, such as the temperature plot, or summing, such as a plot of daily rainfall. If the dependent variable is not processed but represents only an instantaneous measurement, it is simply called *instantaneous*. The time interval between points, called the *sampling interval*, is very important. The time intervals between successive points in a time series are usually equal. However, there are several applications that require this interval to change. The importance of the sampling interval will be discussed in Chapter 3.

Another domain is the *spatial* domain and usually this has two or three dimensions in the sense of having two or three independent variables. Images and moving objects have spatial components with these dimensions. Image analysis has become extremely important within the last decade. Applications are quite diverse and include medical imaging of body organs, robotic vision, remote sensing, and inspection of products on an assembly line. The signal is the amount of whiteness, called gray level, or color in the image. Figure 1.4 illustrates the task of locating a tumor in a tomagraphic scan. In an assembly line, the task may be to inspect objects for defects as in Figure 1.5. The spatial domain can also be discretized for computerized analyses. Image analysis is an extensive topic of study and will not be treated in this textbook.

### 1.2.2 Amplitude Types

The amplitude variable, like the time variable, also can have different forms. Most amplitude variables, such as temperature, are *continuous in magnitude*. The most pertinent *discrete-amplitude* variables involve counting. An example is the presentation of the number of monthly sales in Figure 1.6. Other phenomena that involve counting are radioactive decay and routing processes such as in telephone exchanges or other queueing processes. Another type of process has no amplitude value; these are called *point processes* and occur when one is only interested in the time or place of occurrence. The study of neural coding of information involves the mathematics of point processes. Figure 1.7 shows the reduction of a measured signal into a point process. The information is encoded in the time interval between occurrences or in the interaction between different channels (neurons).

### 1.2.3 Basic Signal Forms

There are different general types of forms for signals. One concerns periodicity. A signal $x(t)$ is *periodic* if it exists for all time, $t$, and

$$x(t) = x(t + P) \tag{1.1}$$

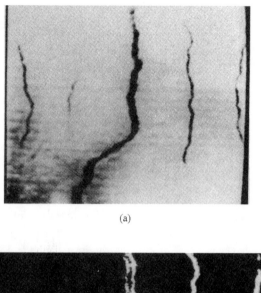

(a)

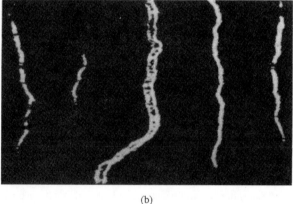

(b)

**FIGURE 1.5**   (a) Test image of a steel slab with imperfections. (b) Processed image with defects located. [Adapted from Suresh et al., figs. 12 and 13, with permission.]

where $P$ is the duration of the period. These signals can be constituted as a summation of periodic waveforms that are harmonically related. The triangular waveform in Figure 1.8 is periodic. Some signals can be constituted as a summation of periodic waveforms that are not harmonically related. The signal itself is not periodic and is called *quasiperiodic*. Most signals are neither periodic nor quasiperiodic and are called *aperiodic*. Aperiodic signals can have very different waveforms, as shown in the next two figures. In Figure 1.9 is shown a biomedical signal, an electroencephalogram, with some spike features indicated by dots. In Figure 1.10 is shown the output voltage of an electrical generator.

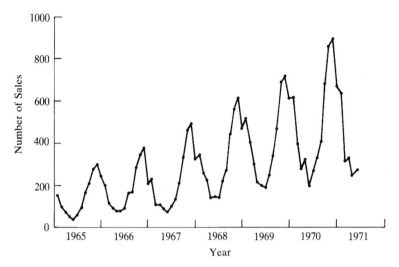

**FIGURE 1.6**   Monthly sales from an engineering company. [Adapted from Chatfield, fig. 1.3, with permission.]

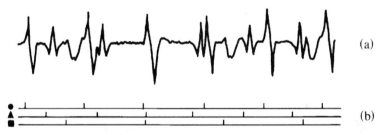

**FIGURE 1.7**   (a) An EMG signal containing three different waveform complexes. (b) Three impulse trains showing the times of occurrence of these complexes. [Adapted from Guiheneuc et al., fig. 1, with permission.]

The time span over which a signal is defined is also important. If a signal has zero value or is nonexistent during negative time, $t < 0$, then the signal is called *causal*. The unit step function is a causal waveform. It is defined as

$$U(t) = \begin{cases} 1, & t \geq 0 \\ 0, & t < 0 \end{cases} \tag{1.2}$$

Any signal can be made causal by multiplying it by $U(t)$.

If a signal's magnitude approaches zero after a relatively short time, it is *transient*. An example of a transient waveform is a decaying exponential function defined during positive time; that is,

$$x(t) = e^{-at}U(t), \quad a > 0 \tag{1.3}$$

The wind gust velocity measurement shown in Figure 1.11 is a transient signal.

Periodic Triangular Waveform

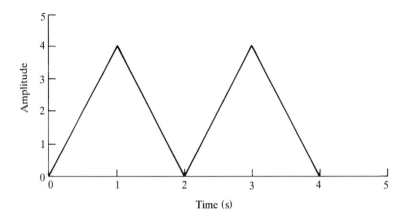

**FIGURE 1.8**   Two periods of a periodic triangular waveform.

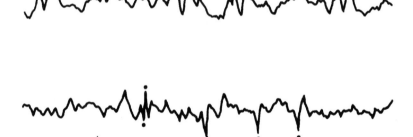

**FIGURE 1.9**   Electroencephalographic signal with sharp transients marked with dots. [Adapted from Glover et al., fig. 1, with permission.]

### 1.2.4   The Transformed Domain, The Frequency Domain

Other domains for studying signals involve mathematical transformations of the signal. A very important domain over which the information in signals is considered is the *frequency domain*. Knowledge of the distribution of signal strength or power over different frequency components is an essential part of many engineering endeavors. At this time it is best understood in the form of the Fourier series. Recall that any periodic function, with a period of $P$ units, as

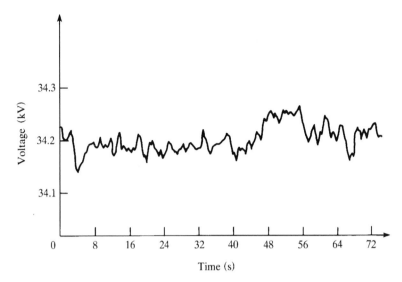

**FIGURE 1.10**   Output voltage signal from an electrical generator used for process control. [Adapted from Jenkins and Watts, fig. 1.1, with permission.]

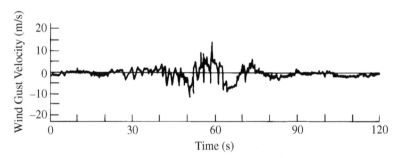

**FIGURE 1.11**   Record of a wind gust velocity measurement. [Adapted from Bendat and Piersol, fig. 1.3, with permission.]

plotted in Figure 1.8 can be mathematically modeled as an infinite sum of trigonometric functions. The frequency term in these functions are *harmonics*, integer multiples, of the *fundamental frequency*, $f_0$. The form is

$$x(t) = C_0 + \sum_{m=1}^{\infty} C_m \cos(2\pi m f_0 t + \theta_m) \qquad (1.4)$$

where $x(t)$ is the function, $f_0 = 1/P$, $C_m$ are the harmonic magnitudes, $f_m = mf_0$ are the harmonic frequencies, and $\theta_m$ are the *phase angles*. The signal can now be studied with the harmonic frequencies assuming the role of the independent variable. Information can be gleaned from the plots of $C_m$ vs $f_m$, called the

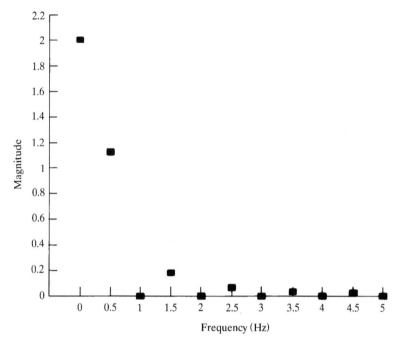

**FIGURE 1.12**   Magnitude spectrum of the periodic triangular waveform in Figure 1.8.

*magnitude spectrum*, and $\theta_m$ vs $f_m$, called the *phase spectrum*. The magnitude spectrum for the periodic waveform in Figure 1.8 is shown in Figure 1.12. Different signals have different magnitude and phase spectra. Signals that are aperiodic also have a frequency domain representation and are much more prevalent than periodic signals. This entire topic will be studied in great detail under the titles of frequency and spectral analysis.

### 1.2.5   General Amplitude Properties

There are two important general classes of signals that can be distinguished by waveform structure. These are deterministic and random. Many signals exist whose future values can be determined with certainty if their present value and some parameters are known. These are called *deterministic* and include all waveforms that can be represented with mathematical formulas, such as cosine functions, exponential functions, and square waves. *Random* or *stochastic* signals are those whose future values cannot be determined with absolute certainty based on known present and past values. Figure 1.13 is an example. Notice there is a decaying exponential trend that is the same as in the deterministic waveform; however, there are no methods to predict the exact amplitude values. Figures 1.9,

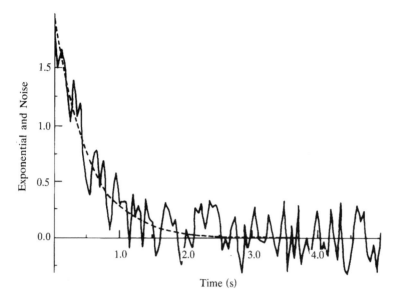

**FIGURE 1.13**  Random signal (–) with exponential trend (– – –).

1.10, and 1.11 are examples of actual random signals. The concepts of probability must be used to describe the properties of these signals.

A property associated with random signals is whether their characteristics change with time. Consider again the temperature signal in Figure 1.1. If one calculates an average value and maximum and minimum values over short periods of time and they do not change, then the signal is *stationary*. Contrast this with the trends in the sales quantities in Figure 1.6. Similar calculations show that these parameters change over time; this signal is *nonstationary*. In general, stationary signals have average properties and characteristics that do not change with time, whereas the average properties and characteristics of nonstationary signals do change with time. These concepts will be considered in detail in subsequent chapters.

## 1.3   ANALOG TO DIGITAL CONVERSION

As mentioned previously, most signals encountered in man-made or naturally occurring systems are continuous in time and amplitude. However, to perform computerized signal analysis, the computer must acquire the signal values. The input process is called *analog to digital* (A/D) *conversion*. This process is schematically diagrammed in Figure 1.14. The signal $g(t)$ is measured continuously by a sensor with a transducer. The transducer converts $g(t)$ into an electrical signal $f(t)$. Usually the transduction process produces a linear

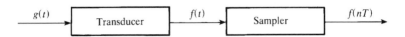

**FIGURE 1.14** Process to convert an analog signal to a discrete time signal.

relationship between these two signals. The sampler measures $f(t)$ every $T$ time units and converts it into a *discrete time sequence*, $f(nT)$. Typical A/D converters are capable of taking from 0 to 100,000 samples per second. The technology of telecommunications and radar utilizes A/D converters with sampling rates up to 100 MHz. Also inherent in the process is the *quantization* of the magnitude of the signal. Computer memory is composed of words with a finite bit length. Within the hardware of the A/D converter, each sampling, (measurement) of the analog signal's magnitude is converted into a digital word of a finite bit length. These integer words are then stored in memory. The word length varies from 4 to 32 bits. For many applications, a 12-bit quantization has sufficient accuracy to ignore the quantization error. For applications requiring extreme precision or when analyzing signals with a large *dynamic range*, defined as a range of magnitudes, converters with the longer word lengths are utilized.

For mathematical operations that are implemented in software, the set of numbers is transformed into a floating point representation that has proper units such as voltage, force, and degrees. For mathematical operations that are implemented in hardware, such as in mathematical coprocessors or in special purpose digital signal processing chips, the set of numbers remain in integer form but word lengths are increased up to 80 bits.

## 1.4 MEASURES OF SIGNAL PROPERTIES

There are many measures of signal properties that are used to extract information from or to study the characteristics of signals. A few of the useful simple ones will be defined in this section. Many others will be defined as the analytic techniques are studied throughout the text. Initially, both the continuous time and discrete time versions will be defined.

### 1.4.1 Time Domain

The predominant number of measures quantize some property of signal magnitude as it varies over time or space. The simplest measures are the maximum and minimum values. Another measure, which everyone uses intuitively, is the average value. The magnitude of the *time average* is defined as

$$x_{av} = \frac{1}{P} \int_0^P x(t)\, dt \tag{1.5}$$

in continuous time and

$$x_{av} = \frac{1}{N} \sum_{n=1}^{N} x(nT) \qquad (1.6)$$

in discrete time, where $P$ is the time duration, $N$ the number of data points, and $T$ is the sampling interval.

Signal energy and power are also important parameters. They provide a major classification of signals and sometimes determine the type of analyses that can be applied (Cooper and McGillem, 1999). Energy is defined as

$$E = \int_{-\infty}^{\infty} x^2(t)\, dt \qquad (1.7)$$

or, in discrete time,

$$E = T \sum_{n=-\infty}^{\infty} x^2(nT) \qquad (1.8)$$

An *energy signal* is one in which the energy is finite. Examples are pulse signals and transient signals, such as the wind gust velocity measurement in Figure 1.11. Sometimes signal energy is infinite as in periodic waveforms, such as the triangular waveform in Figure 1.8. However, for many of these signals the power can be finite. *Power* is energy averaged over time and is defined as

$$PW = \lim_{T \to \infty} \frac{1}{2T} \int_{-T}^{T} x^2(t)\, dt \qquad (1.9)$$

or, in discrete time,

$$PW = \lim_{N \to \infty} \frac{1}{2N+1} \sum_{n=-N}^{N} x^2(nT) \qquad (1.10)$$

Signals with nonzero and finite power are called *power signals*. The class of periodic functions always has finite power.

## 1.4.2 Frequency Domain

Power and energy as they are distributed over frequency are also important measures. Again, periodic signals will be used to exemplify these measures. From elementary calculus, the power in constant and sinusoidal waveforms are known. The power in the average component with a magnitude $C_0$ is $C_0^2$. For the sinusoidal components with amplitude $C_1$, the power is $C_1^2/2$. Thus for a periodic signal, the power, $PW_M$, within the first $M$ harmonics is

$$PW_M = C_0^2 + 0.5 \sum_{m=1}^{M} C_m^2 \qquad (1.11)$$

This is called the *integrated power*. A plot of $PW_M$ vs harmonic frequency is called the *integrated power spectrum*. More will be studied about frequency domain measures in subsequent chapters.

## REFERENCES

T. Anderson, *The Statistical Analysis of Time Series*, Wiley, New York, 1971.

J. Bendat and A. Piersol, *Engineering Applications of Correlation and Spectral Analysis*, Wiley, New York, 1980.

R. Blahut, *Fast Algorithms for Digital Signal Processing*, Addison-Wesley, Reading, MA, 1985.

J. Cadzow, *Foundations of Digital Signal Processing and Data Analysis*, Macmillan, New York, 1987.

C. Chatfield, *The Analysis of Time Series: Theory and Practice*, Wiley, New York, 1975.

C. Chen, *Signal Processing Handbook*, Dekker, New York, 1988.

A. Cohen, *Biomedical Signal Processing: Volume II—Compression and Automatic Recognition*, CRC Press, Boca Raton, FL, 1986.

G. Cooper and C. McGillem, *Probabilistic Methods of Signal and Systems Analysis*, Oxford University Press, New York, 1999.

W. Ebel and N. Younan, "Counting on Computers in DSP Education," *IEEE Signal Processing Magazine* **12**(6): 38–43 (1995).

K. Foster, "Math and Graphics," *IEEE Spectrum* November: 72–78 (1992).

J. Glover, P. Ktonas, N. Raghavan, J. Urunuela, S. Velamuri, and E. Reilly, "A Multichannel Signal Processor for the Detection of Epileptogenic Sharp Transients in the EEG," *IEEE Trans. Biomed. Eng.* **33**: 1121–1128 (1986).

C. Granger, "Acronyms in Time Series Analysis (ATSA)," *J. Time Series Analysis* **3**: 103–107, (1982).

P. Guiheneuc, J. Calamel, C. Doncarli, D. Gitton, and C. Michel, "Automatic Detection and Pattern Recognition of Single Motor Unit Potentials in Needle EMG," in *Computer-Aided Electromyography*, J. Desmedt, Ed., Karger, Basel, 1983.

G. Inbar, *Signal Analysis an Pattern Recognition in Biomedical Engineering*, Wiley, New York, 1975.

G. Jenkins and D. Watts, *Spectral Analysis and its Applications*, Holden-Day, San Francisco, 1968.

A. Oppenheim, *Applications of Digital Signal Processing*, Prentice-Hall, Englewood Cliffs, NJ, 1978.

W. Press, B. Flannery, S. Teukolsky, and W. Vetterling, *Numerical Recipes in C—The Art of Scientific Computing*, Cambridge University Press, New York, 1992.

W. Press, B. Flannery, S. Teukolsky, and W. Vetterling, *Numerical Recipes in FORTRAN—The Art of Scientific Computing*, Cambridge University Press, New York, 1993.

M. Schwartz and L. Shaw, *Signal Processing: Discrete Spectral Analysis, Detection, and Estimation*, McGraw-Hill, New York, 1975.

S. Stearns and R. David, *Signal Processing Algorithms*, Prentice-Hall, Englewood Cliffs, NJ, 1988.

J. Strohbehn and E. Douple, "Hyperthermia and Cancer Therapy: A Review of Biomedical Engineering Contributions and Challenges," *IEEE Trans. Biomed. Eng.* **31**: 779–787 (1984).

B. Suresh, R. Fundakowski, T. Levitt, and J. Overland, "A Real-Time Automated Visual Inspection System for Hot Steel Slabs," *IEEE Trans. Pattern Analy. Mach. Intell.* **5**: 563–572 (1983).

# II

# EMPIRICAL MODELING
# AND APPROXIMATION

## 2.1 INTRODUCTION

In many situations it is necessary to discover or develop a relationship between two measured variables. Such situations occur in the study of physics, biology, economics, engineering, etc. Often, however, neither *a priori* nor theoretical knowledge is available regarding these variables, or their relationship is very complicated. Examples include the relationship between the heights of parents and children, cigarette smoking and cancer, operating temperature and rotational speed in an electric motor, and resistivity and deformation in a strain gauge transducer. Thus one must resort to *empirical modeling* of the relationship. Techniques for empirical modeling have been available for quite some time and are alternatively called *curve fitting* in engineering, *regression analysis* in statistics, or *time series forecasting* in economics. The fundamental principles for modeling are the same in all of these areas.

Examine now a few applications that are interesting to physiologists, engineers, or educators. In Figure 2.1 are plotted two sets of measurements from walking studies relating step length to body height in men and women. The plot of the measured data points is called a *scatter diagram*. There is clearly an increasing trend in the data even though there may be several values of step length for a certain height. Straight lines, also drawn in the figure, should be good

16

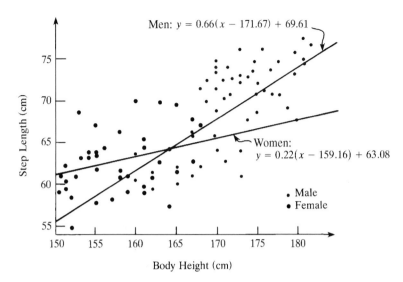

**FIGURE 2.1**   Scatter diagrams and linear models relating walking step length in men and women and their body height. [Adapted from Hirokawa and Matsumara, fig. 11, with permission.]

approximations for these sets of data. Notice that for a certain height, no step length values either for men or women are predicted exactly. There is some error. How does this occur? There are two possibilities: (1) There is some error in the measurements, and (2) more independent variables are necessary for a more accurate approximation. Perhaps if step length were modeled as a function of both height and age, the modeling could be more accurate, but the model would becomes more complex. We will therefore restrict ourselves to bivariate relationships. For example, Figure 2.2 shows the data relating the electrical activity in a muscle and the muscular force generated when the muscle's nerve is stimulated. Notice that this data has a curvilinear trend, so a polynomial model would be more suitable than a linear one. A suitable quadratic model is also shown in the figure. Another example is shown in Figure 2.3, which shows a third-order relationship from protein data.

Models provide the ability to estimate values of the dependent variable for desired values of the independent variable when the actual measurements are not available. For instance, in Figure 2.3, one can estimate that a protein concentration of 8 g/dl will produce an oncotic pressure of 30 Torr. This is a common situation when using tables—for example, when looking up values of trigonometric functions. In this situation the available information or data is correct and accurate but one still needs to accurately determine unavailable values of a dependent variable. Techniques for accomplishing this task are called *interpolation*. Finally, Figure 2.4 shows the daily viscosity values of a produced

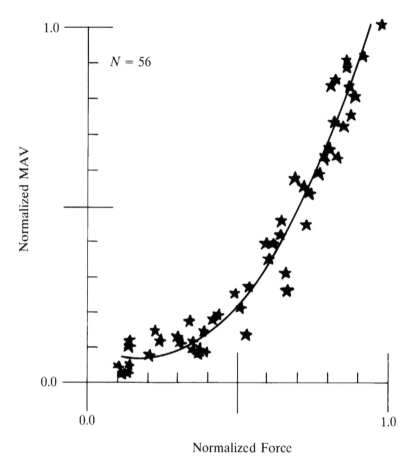

**FIGURE 2.2** A scatter diagram and quadratic model relating normalized electrical activity in a muscle, MAV, and the normalized force, $F$, generated by the muscle when it's nerve is stimulated. The model is $MAV = 0.12 - 0.60F + 1.57F^2$. [Adapted from Solomonow et al., fig. 7, with permission.]

chemical product. Notice that the viscosity values fluctuate cyclically over time. Handling this type of data requires different techniques, which will be considered in subsequent chapters. In this chapter we will be focusing on developing linear and curvilinear relationships between two measured variables for the purposes of modeling or interpolation.

## 2.2   MODEL DEVELOPMENT

Whatever variables are being examined, they exist in pairs of values of the independent and dependent variables and a qualitative relationship can be

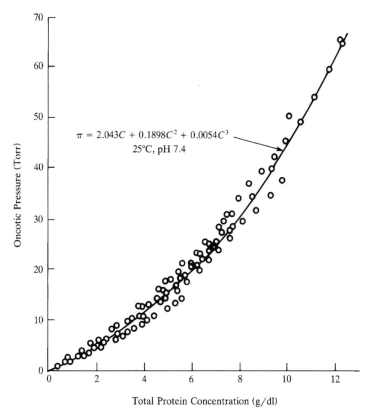

**FIGURE 2.3**   The scatter diagram and cubic model relating protein osmotic, oncotic, pressure, and protein concentration. [Adapted from Roselli et al., with permission.]

appreciated by examining the Cartesian plot of the point pairs. Figure 2.5 shows a scatter diagram and two models for some stress–strain data (Dorn and McCracken, 1972). The points seem to have a straight-line trend, and a line that seems to be an appropriate model has been drawn. Notice that there is an error between the ordinate values and the model. The error for one of the point pairs is indicated as $e_i$. Some of the errors are smaller than others. If the line's slope is changed slightly, the errors change, but the model still appears appropriate. The obvious desire is of course for the line to pass *as close as possible* to all the data points. Thus some definition of model accuracy must be made. An empirical relationship can be determined quantitatively through the use of curve-fitting techniques. An essential notion with these techniques is that the measure of accuracy or error is part of the model development. There are many error measures, but the most general and easily applicable are those that involve the minimization of the total or average squared error between the observed data

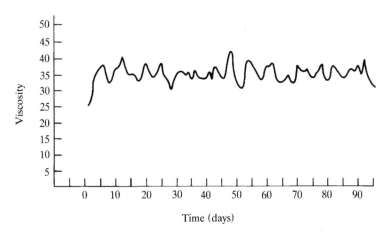

**FIGURE 2.4.**  Daily reading of viscosity of chemical product. [From Bowerman and O'Connell, fig. 2.11, with permission.]

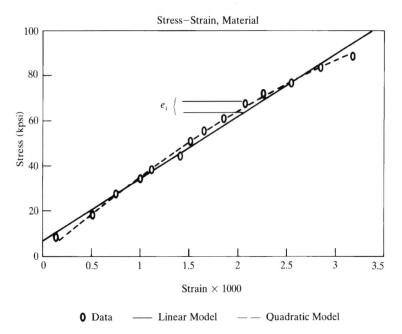

**O** Data        ——— Linear Model      — — Quadratic Model

**FIGURE 2.5**  The scatter plot and two models of stress–strain data for a certain material.

and the proposed model. The resultant model then satisfies the least squares error criterion.

Consider now the modeling of a data set with polynomial function $y = f(x) = a + bx + cx^2 + \cdots$. The straight line is the simplest function and will be used to demonstrate the general concept of model development. Higher-order functions will then be used to improve the modeling, but the general concept and procedure is the same. Begin by considering the data in Figure 2.5. The straight line with parameters $a$ and $b$ is

$$y = a + bx \tag{2.1}$$

The value, $\hat{y}_i$, estimated by equation 2.1 for point pair $(y_i, x_i)$ is

$$\hat{y}_i = a + bx_i \tag{2.2}$$

The error of estimation, $e_i$, is then

$$e_i = y_i - \hat{y}_i = y_i - a - bx_i \tag{2.3}$$

The total squared error, $E_1$, for this model is

$$E_1 = \sum_{i=1}^{N} e_i^2 = \sum_{i=1}^{N} (y_i - a - bx_i)^2 \tag{2.4}$$

where $N$ is the number of data points and the subscript of $E_1$ indicates the order of the model. The quantity $E_1$ is a measure of how well the line fits the entire set of points and is sometimes called *the sum of squared residuals*. $E_1$ will be zero if and only if each of the points are on the line, and the farther the points are, on the average, from the line, the larger the errors will become. The least squares error criterion *selects the parameters a and b so that the total squared error is as small as possible.* (Note: It is assumed that the surface produced by $E_1$ has a well-defined minimum.)

To accomplish this, the usual procedure for minimizing a function of several variables is performed. The two first partial derivatives of equation 2.4 with respect to the unknown parameters, $a$ and $b$, are equated to zero. The first equation is derived in detail. The derivation and verification of the second one is left as an exercise for the reader. Remember that the operations of summation and differentiation are commutable (interchangeable).

Taking the first partial derivative yields

$$\frac{\partial E_1}{\partial a} = \frac{\partial}{\partial a} \sum_{i=1}^{N} (y_i - a - bx_i)^2 = 0$$

$$= -2 \sum_{i=1}^{N} (y_i - a - bx_i) \tag{2.5}$$

After rearranging by collecting terms with the unknown terms on the right side of the equation, the result is

$$\sum_{i=1}^{N} y_i = aN + b \sum_{i=1}^{N} x_i \tag{2.6}$$

Similarly, the solution of the second minimization yields the equation

$$\sum_{i=1}^{N} y_i x_i = a \sum_{i=1}^{N} x_i + b \sum_{i=1}^{N} x_i^2 \tag{2.7}$$

The solution of this set of simultaneous equations is a standard procedure.

---

### EXAMPLE 2.1

Fit the stress vs strain data of Figure 2.5 with a straight-line model by the method of least squares. The data are listed in Table 2.1.

The necessary sums are

$$\sum_{i=1}^{14} y_i = 728.87; \quad \sum_{i=1}^{14} x_i = 22.98; \quad \sum_{i=1}^{14} x_i^2 = 48.08; \quad \sum_{i=1}^{14} y_i x_i = 1480.76$$

As you study this example, verify the results with a calculator. Solving equations 2.6 and 2.7 produces $a = 6.99$ and $b = 27.46$. The resulting equation is

$$y = 6.99 + 27.46x \tag{2.8}$$

This equation is also plotted in Figure 2.5 with the measured data. Notice that although this is a good approximation, the data plot has some curvature. Perhaps a quadratic or higher-order model would be more appropriate. The squared error is 103.

**TABLE 2.1**    Stress, $y_i$ (psi $\times 10^3$), vs Strain, $x_i$ (strain $\times 10^{-3}$)

| $i$ | $y_i$ | $x_i$ |
|-----|-------|-------|
| 1 | 8.37 | 0.15 |
| 2 | 17.9 | 0.52 |
| 3 | 27.8 | 0.76 |
| 4 | 34.2 | 1.01 |
| 5 | 38.8 | 1.12 |
| 6 | 44.8 | 1.42 |
| 7 | 51.3 | 1.52 |
| 8 | 55.5 | 1.66 |
| 9 | 61.3 | 1.86 |
| 10 | 67.5 | 2.08 |
| 11 | 72.1 | 2.27 |
| 12 | 76.9 | 2.56 |
| 13 | 83.5 | 2.86 |
| 14 | 88.9 | 3.19 |

## 2.3  GENERALIZED LEAST SQUARES

Viewed in somewhat more general terms, the method of least squares is simply a process for finding the best possible values for a set of $M + 1$ unknown coefficients, $a, b, c, \ldots$, for an $M$th-order model. The general nonlinear model is

$$y = a + bx + cx^2 + dx^3 + \cdots \qquad (2.9)$$

The number of data points still exceeds the number of unknowns; that is, $N > M + 1$. The same general principal is utilized; the total squared error of the model is minimized to find the unknown coefficients. The individual error, $e_1$, is then

$$e_i = y_i - \hat{y}_i = y_i - a - bx_i - cx_i^2 - \cdots \qquad (2.10)$$

The total squared error, $E_M$, for this general model becomes

$$E_M = \sum_{i=1}^{N} e_i^2 = \sum_{i=1}^{N}(y_i - a - bx_i - cx_i^2 - \cdots)^2 \qquad (2.11)$$

To minimize $E_M$, obtain the first partial derivatives of equation 2.11 with respect to the $M + 1$ unknown parameters $(a, b, \ldots)$ and equate them to zero. For the coefficient $b$, the procedure is

$$\frac{\partial E_M}{\partial b} = \frac{\partial}{\partial b}\sum_{i=1}^{N}(y_i - a - bx_i - cx_i^2 - \cdots)^2 = 0$$

$$= -2\sum_{i=1}^{N}(y_i - a - bx_i - cx_i^2 - \cdots)x_i \qquad (2.12)$$

The results of the partial derivations will yield $M + 1$ equations similar in form to equation 2.12. The only difference will be in the power of $x_i$ on the extreme right end. Then one must solve the simultaneous equations for the parameters. For a quadratic model, $M = 2$, after rearranging terms and collecting unknowns on the right side of the equation, the equations are

$$\sum_{i=1}^{N} y_i = aN + b\sum_{i=1}^{N} x_i + c\sum_{i=1}^{N} x_i^2 \qquad (2.13)$$

$$\sum_{i=1}^{N} y_i x_i = a\sum_{i=1}^{N} x_i + b\sum_{i=1}^{N} x_i^2 + c\sum_{i=1}^{N} x_i^3 \qquad (2.14)$$

$$\sum_{i=1}^{N} y_i x_i^2 = a\sum_{i=1}^{N} x_i^2 + b\sum_{i=1}^{N} x_i^3 + c\sum_{i=1}^{N} x_i^4 \qquad (2.15)$$

Thus we have obtained a system of $M + 1$ simultaneous linear equations, called *normal* equations, in the $M + 1$ unknowns, $a, b, c, \ldots$, whose solution is now a routine matter. These equations can easily be structured into the matrix form

$$
\begin{bmatrix}
N & \sum\limits_{i=1}^{N} x_i & \sum\limits_{i=1}^{N} x_i^2 \\[2mm]
\sum\limits_{i=1}^{N} x_i & \sum\limits_{i=1}^{N} x_i^2 & \sum\limits_{i=1}^{N} x_i^3 \\[2mm]
\sum\limits_{i=1}^{N} x_i^2 & \sum\limits_{i=1}^{N} x_i^3 & \sum\limits_{i=1}^{N} x_i^4
\end{bmatrix}
\begin{bmatrix}
a \\ b \\ c
\end{bmatrix}
=
\begin{bmatrix}
\sum\limits_{i=1}^{N} y_i \\[2mm]
\sum\limits_{i=1}^{N} y_i x_i \\[2mm]
\sum\limits_{i=1}^{N} y_i x_i^2
\end{bmatrix}
\tag{2.16}
$$

Several numerical techniques and computer subroutine algorithms exist to solve this system of equations.

**EXAMPLE 2.2**

Since the stress–strain data seemed to have some nonlinear trend we will model it with the quadratic model from equations 2.13 to 2.16. The additional sums required are:

$$
\sum_{i=1}^{14} y_i x_i^2 = 3433.7; \quad \sum_{i=1}^{14} x_i^2 = 48.1; \quad \sum_{i=1}^{14} x_i^3 = 113.7; \quad \sum_{i=1}^{14} x_i^4 = 290.7
$$

The solution of these equations yields $a = 0.7$, $b = 37.5$, and $c = -2.98$ for the model

$$
y = 0.77 + 37.5x - 2.98x^2
$$

This is plotted also in Figure 2.5. The squared error, $E_2$, is 27.4. It is obvious that the quadratic model is a better model. The next question is how this can be quantitated. Recall that we stated that the squared error is also the criteria of "goodness of fit." Comparing the errors of the two equations shows that the quadratic model is much superior.

## 2.4   GENERALITIES

A problem in empirical curve approximation is the establishment of a criterion to decide the limit in model complexity. Reconsider the previous two examples. The stress–strain relationship is definitely nonlinear. However, is the quadratic model sufficient or is a cubic model needed? In general, the total squared error approaches zero as $M$ increases. A perfect fit for the data results when $M + 1$ equals $N$ and the resulting curve satisfies all the data points. However, the polynomial may fluctuate considerably between data points and may represent more artifact and inaccuracies than the general properties desired. In many situations a lower-order model is more appropriate. This will certainly be the case when measuring experimental data that theoretically should have a straight-line relationship but fails to show it because of errors in observation or measurement. To minimize the fluctuations, the degree of the polynomial should be much less than the number of data points.

Judgment and knowledge of the phenomena being modeled are important in the decision making. Important also is the trend in the $E_M$ versus the model order characteristic. Large decreases indicate that significant additional terms are being created, whereas small decreases reflect insignificant and unnecessary improvements. The error characteristic for modeling the stress–strain relationship is shown in Figure 2.6. Notice that the error decreases appreciably as $M$ increases from 1 to 2; thereafter the decrease has smaller decrements. This indicates that a model with $M = 2$ is probably sufficient. An approximate indicator is the magnitude of the additional coefficients. If they are small enough to make the effect of their respective term negligible, then the additional complexity is unnecessary. For instance, consider the fifth-order model in the equation

$$y = 3.43 + 32.51x - 8.83x^2 + 10.97x^3 - 4.75x^4 + 0.63x^5 \qquad (2.17)$$

Notice how the coefficients of the fourth- and fifth-order terms tend to become smaller.

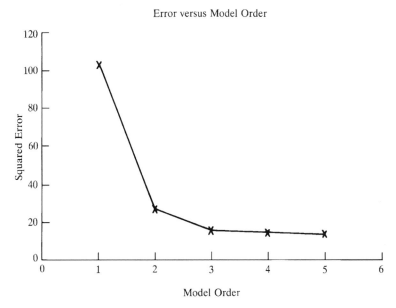

FIGURE 2.6 The total squared error model order characteristic for the stress–strain data.

Another error measure is the square root of the average squared error. This is called the *standard error of the estimate* and is defined by

$$\sigma_e = \sqrt{\frac{E_M}{N - (M + 1)}} \tag{2.18}$$

This parameter is also used to compare the accuracy of different models (Lapin, 1983).

Solution of these simultaneous linear equations is rather routine and many methods are available. Matrix methods using Cramer's rule or numerical methods such as Gauss elimination are solution approaches. While low-order models usually pose no difficulty, numerical difficulties can be encountered when inverting the matrices formed for a high-order model or when the matrices are ill-conditioned. There are many engineering-oriented numerical methods texts that explain these numerical techniques. For instance, refer to Hamming (1973), Pearson (1986), and Wylie (1995).

## 2.5  MODELS FROM LINEARIZATION

Many types of nonlinear relationships cannot be modeled adequately with polynomial functions. These include natural phenomena like bacterial or radioactive decay, which have exponentially behaving characteristics. Regression analysis can be applied to data of these types if the relationship is linearizable through some mathematical transformation (Bowerman and O'Connell, 1987; Chatterjee and Price, 1977). The exponential characteristic is one of the classical types. Consider the exponential model

$$y = \alpha e^{\beta x} \tag{2.19}$$

Applying the logarithm to both sides of equation 2.19 yields

$$\ln y = \ln \alpha + \beta x \tag{2.20}$$

Making the substitutions

$$w = \ln y, \qquad a = \ln \alpha, \qquad b = \beta$$

produces the linearized model

$$w = a + bx \tag{2.21}$$

where the data point pairs are $w_i = \ln y_i$ and $x_i$. Thus the logarithmic transformation linearized the exponential model. It will be seen in general that logarithmic transformations can linearize all multiplicative models.

### EXAMPLE 2.3
An interesting example is the data set concerning the survival of marine bacteria being subjected to X-ray radiation. The number of surviving bacteria and the

**TABLE 2.2** Surviving Bacteria, $y_i$ (number $\times 10^{-2}$), vs Time, $x_i$ (min)

| $i$ | $y_i$ | $x_i$ | $w_i$ |
|-----|-------|-------|-------|
| 1   | 355   | 1     | 5.87  |
| 2   | 211   | 2     | 5.35  |
| 3   | 197   | 3     | 5.28  |
| 4   | 166   | 4     | 5.11  |
| 5   | 142   | 5     | 4.96  |
| 6   | 106   | 6     | 4.66  |
| 7   | 104   | 7     | 4.64  |
| 8   | 60    | 8     | 4.09  |
| 9   | 56    | 9     | 4.02  |
| 10  | 38    | 10    | 3.64  |
| 11  | 36    | 11    | 3.58  |
| 12  | 32    | 12    | 3.47  |
| 13  | 21    | 13    | 3.04  |
| 14  | 19    | 14    | 2.94  |
| 15  | 15    | 15    | 2.71  |

duration of exposure time is listed in Table 2.2 and plotted in Figure 2.7 (Chatterjee and Price, 1977). The values of the linearized variable $w$ are also listed in the table. Applying the regression equations to the data points $w_i$ and $x_i$ yields $a = 5.973$ and $b = -0.218$. Thus, $\alpha = e^a = 392.68$ and $\beta = b = -0.218$. This model is also plotted in Figure 2.7 and shows good correspondence.

---

Another multiplicative model is the power law relationship

$$y = \alpha x^\beta \tag{2.22}$$

The application of this model is very similar to Example 2.3. Other types of transformations exist if they are single-valued. Sometimes which type can be suggested by the model itself. For instance, the model

$$y = a + b \ln x \tag{2.23}$$

is easily amenable to the transformation $z = \ln x$. Other situations are not so obvious. Consider the model

$$y = \frac{x}{ax - b} \tag{2.24}$$

The hyperbolic transformations $w = y^{-1}$ and $z = x^{-1}$ nicely produce the linear model

$$w = a + bz \tag{2.25}$$

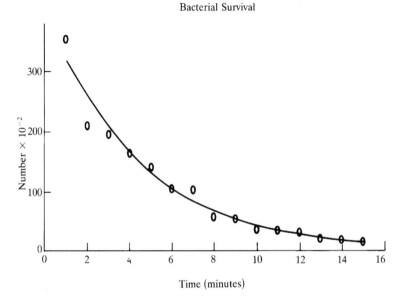

**FIGURE 2.7**   The number of surviving bacteria (in units of 100) plotted against X-ray exposure time (in minutes).

Additional applications of these types of transformations can be found in the exercises for this chapter.

### EXAMPLE 2.4

Another very useful multiplicative model is the product exponential form

$$y = \alpha x e^{\beta x}$$

The transformation necessary is simply to use the ratio of the two variables as the dependent variable. The model now becomes

$$w = \frac{y}{x} = \alpha e^{\beta x}$$

An appropriate application is the results of a compression test of a concrete cylinder. The stress–strain data is plotted in Figure 2.8 and the data are listed in Table 2.3 (James et al., 1985).

The resulting equation is

$$y = 4.41 x e^{-540.62x}$$

You will verify this solution in Exercise 2.7.

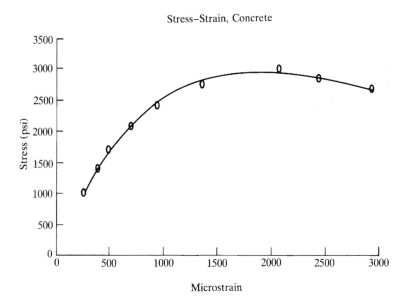

**FIGURE 2.8**   Stress–strain characteristic for a concrete block.

**TABLE 2.3**   Stress $y_i$ (psi), vs Microstrain, $x_i$

| $i$ | $y_i$ | $x_i$ | $w_i$ |
|---|---|---|---|
| 1 | 1025 | 265 | 3.86 |
| 2 | 1400 | 400 | 3.50 |
| 3 | 1710 | 500 | 3.42 |
| 4 | 2080 | 700 | 2.97 |
| 5 | 2425 | 950 | 2.55 |
| 6 | 2760 | 1360 | 2.03 |
| 7 | 3005 | 2080 | 1.44 |
| 8 | 2850 | 2450 | 1.16 |
| 9 | 2675 | 2940 | 0.91 |

## 2.6  ORTHOGONAL POLYNOMIALS

As has been shown in Section 2.4, a new matrix solution of increased dimension must be undertaken when it is necessary to increase the order of a model. There is a faster and more efficient method for determining the coefficients of the model when the interval or difference between values of the independent variable are equally spaced. This method requires the use of orthogonal polynomials. The results will be the same, but the complexity and labor involved in the least squares procedure that has just been described can be significantly reduced. In addition, the concept of orthogonality is important to learn in itself because it is an essential component in other areas. The mathematics of orthogonal functions defined when the independent variable is continuous is reviewed in Appendix 3.4.

First define a set of polynomial functions, $\{P_m(z_i)\}$, where $z_i$ is the independent variable that is a set of integers $\{0, 1, \ldots, N-1\}$, $m$ is the order of the polynomial, $N$ is the number of data points, and $0 \leq m \leq N - 1$. (As shall be seen, the independent variable can be any set of equally spaced values $\{x_i: 1 \leq i \leq N\}$.) The model or approximation for the data using polynomials up to order $M$ becomes

$$y = f(z) = A_0 P_0(z) + A_1 P_1(z) + \cdots + A_M P_M(z) \qquad (2.26)$$

The squared error is defined as

$$E_M = \sum_{i=1}^{N} (f(z_i) - y_i)^2$$

and substituting the model's values yields

$$E_M = \sum_{i=1}^{N} (A_0 P_0(z_i) + A_1 P_1(z_i) + \cdots + A_M P_M(z_i) - y_i)^2 \qquad (2.27)$$

Differentiating equation 2.27 with respect to each $A_j$ to minimize the error produces the set of equations

$$\frac{\partial E_M}{\partial A_j} = \sum_{i=1}^{N} (A_0 P_0(z_i) + A_1 P_1(z_i) + \cdots + A_M P_M(z_i) - y_i) P_j(z_i) = 0 \qquad (2.28)$$

where $0 \leq j \leq M$. Distributing the summations and collecting terms produces equations of the form

$$\sum_{i=1}^{N} y_i P_j(z_i) = A_0 \sum_{i=1}^{N} P_0(z_i) P_j(z_i) + A_1 \sum_{i=1}^{N} P_1(z_i) P_j(z_i) + \cdots + A_M \sum_{i=1}^{N} P_M(z_i) P_j(z_i)$$
$$(2.29)$$

There are in general $M + 1$ simultaneous equations, and the solution is quite formidable. By making the function set $\{P_m(z_i)\}$ satisfy orthogonality conditions, the solution for the coefficients, $A_j$, becomes straightforward. The orthogonality conditions are

$$\sum_{i=1}^{N} P_m(z_i) P_j(z_i) = \begin{cases} 0, & m \neq j \\ \lambda_j, & m = j \end{cases} \qquad (2.30)$$

The $\lambda_j$ equal the energy in the function $P_j(z_i)$. This reduces equation 2.29 to the form

$$\sum_{i=1}^{N} y_i P_j(z_i) = A_j \sum_{i=1}^{N} P_j^2(z_i) \qquad (2.31)$$

Thus the coefficients can now be directly solved by equations of the form

$$A_j = \frac{\sum\limits_{i=1}^{N} y_i P_j(z_i)}{\sum\limits_{i=1}^{N} P_j^2(z_i)}, \quad 0 \leq j \leq M \tag{2.32}$$

The polynomials for discrete data that satisfy this orthogonality condition are the *gram polynomials* and have the general form

$$P_j(z) = \sum_{i=0}^{j}(-1)^i \frac{(i+j)!(N-i-1)!z!}{i!i!(j-i)!(N-1)!(z-i)!} \tag{2.33}$$

when $z$ is an integer whose lower bound is zero. Some particular polynomials are

$$P_0(z) = 1$$

$$P_1(z) = 1 - \left(2\frac{z}{N-1}\right)$$

$$P_2(z) = 1 - \left(6\frac{z}{N-1}\right) + \left[6\frac{z(z-1)}{(N-1)(N-2)}\right]$$

In general, the energies for this set are

$$\lambda_j = \frac{(N+j)!(N-j-1)!}{(2j+1)(N-1)!(N-1)!} \tag{2.34}$$

For some particular values of $j$, they are

$$\lambda_0 = N$$

$$\lambda_1 = \frac{N(N+1)}{3(N-1)}$$

$$\lambda_2 = \frac{N(N+1)(N+2)}{5(N-1)(N-2)}$$

Because of the orthogonality conditions the total energy, $E_{\text{tot}}$, and the error, $E_M$, have simplified forms. These are

$$E_{\text{tot}} = \sum_{j=0}^{\infty} A_j^2 \lambda_j \tag{2.35}$$

and

$$E_M = \sum_{i=1}^{N} y(z_i)^2 - \sum_{j=0}^{M} \lambda_j A_j^2 \tag{2.36}$$

Equation 2.35 is one form of Parseval's Theorem, which will be encountered several times in this textbook. This set of polynomials can be applied to model any equally spaced data set by the simple linear transformation

$$z_i = \frac{(x_i - x_0)}{h} \tag{2.37}$$

where $x_0$ is the smallest value of the independent variable and $h$ is the spacing. An example will illustrate the methodology.

---

### EXAMPLE 2.5

When a muscle is active, it generates electrical activity that can be easily measured. This measurement is called an electromyogram (EMG). The relationship between a quantification of this electrical activity—the average absolute value (AAV)—and the load the muscle supports has been examined. To study this, a person was seated in a chair with the right foot parallel to but not touching the floor. The person kept the ankle joint at $90°$ while weights of different magnitudes were placed upon it. Table 2.4 lists the AAV, in millivolts, and the corresponding load in pounds (Shiavi, 1969). The data is plotted in Figure 2.9.

Because it was planned to use orthogonal polynomials, there was an equal increment of 1.25 pounds between successive loads. Polynomials of up to fourth order were to be used. For seven points, the polynomial equations are

$$P_0(z) = 1$$

$$P_1(z) = 1 - \frac{1}{3}z$$

$$P_2(z) = 1 - \frac{6}{5}z + \frac{1}{5}z^2$$

**TABLE 2.4**    Muscle Activity, $y_i$ (mv), vs Load, $x_i$ (lbs) and Parameter Values

| $i$ | $y_i$ | $x_i$ | $z_i$ | $P_1(z_i)$ | $P_2(z_i)$ | $y_i P_1$ | $y_i P_2$ |
|---|---|---|---|---|---|---|---|
| 1 | 10.27 | 0.00 | 0.00 | 1.00 | 1.00 | 10.27 | 10.27 |
| 2 | 14.07 | 1.25 | 1.00 | 0.67 | 0.00 | 9.38 | 0.00 |
| 3 | 20.94 | 2.50 | 2.00 | 0.33 | −0.60 | 6.98 | −12.56 |
| 4 | 24.14 | 3.75 | 3.00 | 0.00 | −0.80 | 0.00 | −19.31 |
| 5 | 22.5 | 5.00 | 4.00 | −0.33 | −0.60 | −7.50 | −13.50 |
| 6 | 24.91 | 6.25 | 5.00 | −0.67 | 0.00 | −16.61 | 0.00 |
| 7 | 27.06 | 7.50 | 6.00 | −1.00 | 1.00 | −27.06 | 27.06 |
| $\sum$ | 143.89 | | | | | −24.54 | −8.08 |

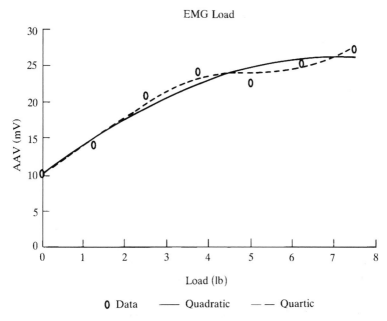

**FIGURE 2.9** The scatter plot and several models for data describing the relationship between AAV (in millivolts) and muscle load (in pounds) in a human muscle.

$$P_3(z) = 1 - \frac{10}{3}z + \frac{3}{2}z^2 - \frac{7}{6}z^3$$

$$P_4(z) = 1 - \frac{59}{6}z + \frac{311}{36}z^2 - \frac{7}{3}z^3 + \frac{7}{36}z^4$$

Let us now derive in particular the second-order model that is given in general by the equation

$$y(z) = A_0 P_0(z) + A_1 P_1(z) + A_2 P_2(z)$$

The important sums are given in Table 2.4 and the energies in Table 2.5. For the second-order term, equation 2.31 becomes

$$A_2 = \frac{\sum\limits_{i=1}^{7} y_i P_2(z_i)}{\sum\limits_{i=1}^{7} P_2^2(z_i)} = \frac{\sum\limits_{i=1}^{7} y_i P_2(z_i)}{\lambda_2} = \frac{-8.08}{3.36} = -2.40$$

where

$$\lambda_2 = \frac{N(N+1)(N+2)}{5(N-1)(N-2)} = \frac{7 \cdot 8 \cdot 9}{5 \cdot 6 \cdot 5} = 3.36$$

**TABLE 2.5** Model Parameters

| $j$ | $A_j$ | $\lambda_j$ | $E_j$ |
|---|---|---|---|
| 0 | 20.56 | 7.00 | — |
| 1 | −7.89 | 3.11 | 32.39 |
| 2 | −2.40 | 3.36 | 13.12 |
| 3 | −0.73 | 6.00 | 9.91 |
| 4 | 0.53 | 17.11 | 5.04 |

The parameters for the zeroth- and first-order polynomials are calculated similarly. The squared error is

$$E_2 = \sum_{i=1}^{7} y(x_i)^2 - \sum_{j=0}^{2} \lambda_j A_j^2 = 3184 - 7 \cdot 423 - 3.11 \cdot 62 - 3.36 \cdot 5.76 = 13.12$$

The associated energies, coefficients, and squared errors for models up to the fourth order are listed in Table 2.5.

The second-order model is given by the equation

$$\begin{aligned} y(z) &= A_0 P_0(z) + A_1 P_1(z) + A_2 P_2(z) \\ &= 20.56 P_0(z) - 7.89 P_1(z) - 2.40 P_2(z) \\ &= 10.28 + 5.5z - 0.48z^2 \end{aligned}$$

after substituting for $A_i$ and $P_i(z)$. The next step is to account for the change in scale of the independent variable with parameters $x_0 = 0$ and $h = 1.25$. The previous equation now becomes

$$y(x) = 10.28 + 4.4x - 0.307x^2$$

## 2.7 INTERPOLATION AND EXTRAPOLATION

After a suitable model of a data set has been created, an additional benefit has been gained: The value of the dependent variable can be estimated for any desired value of the independent variable. When the specific value of the independent variable lies within the range of magnitudes of the independent variable, the estimation is called *interpolation*. Consider again Figure 2.9. Although the EMG intensity for a weight of 5.5 lb was not measured, it can be estimated to be 24 mv. This is a very common procedure for instrument calibration, filling in missing data points, and estimation in general when using discrete data measurements. An application is the study of the dispersion of tracer dye in a nonhomogeneous fluid medium. Figure 2.10a shows the schematic of a column filled with glass beads in which a bolus of dye is injected at point A. The dye is dispersed in its travel (B)

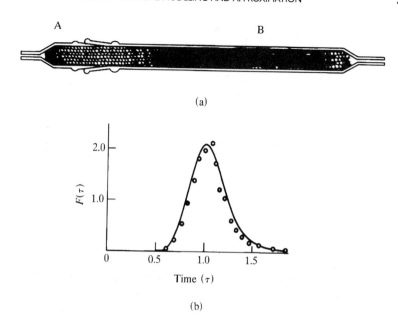

(a)

(b)

**FIGURE 2.10**   Schematic dispersion of a dye in a glass bead column (a) at the beginning (A) and at an intermediary position of its traversal (B): (b) dye concentration, $F(\tau)$, at the exit point vs time, $\tau$; measured data (o), fitted model (—). [Adapted from Sheppard, figs. 63 and 65, with permission.]

and its concentration is measured at the exit point of the column (Figure 2.10b). A model is fitted to the measured data and does not represent the data very well. Therefore, using interpolation to estimate nonmeasured values of concentration over a segment of the data would be more suitable.

The estimation or prediction of a value of the dependent variable when the magnitude of the independent variable is outside the range of the measured data is called *extrapolation*. This is used often in engineering and very often in economics and business management. For instance, in the management of factory production it is necessary to predict future production demands in order to maintain an inventory of sufficient raw materials and basic components and have sufficient trained personnel. In the business and economic fields, this prediction is also called *forecasting*.

Several polynomial-based methods exist for performing interpolation. There are several major differences between these methods of interpolation (to be described in this section) and the curve-fitting methods. First, it is assumed that the data is accurate and that the curve needs to have zero error at the acquired data points. This necessarily means that for $N$ points a polynomial of order $N - 1$ will be produced. For even a moderate value of $N$, a curve with large fluctuations can be produced, as illustrated in Figure 2.11. These fluctuations can be inordinately

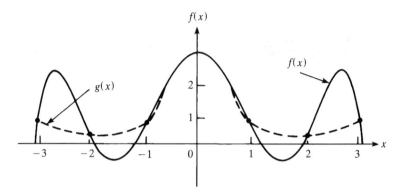

**FIGURE 2.11** Graphical illustration of the possible inaccuracies inherent in interpolation, $g(x)$ represents the ideal function and $f(x)$ represents the interpolator polynomial.

large and thus it is necessary to use a low-order polynomial to estimate the magnitude of the unknown function. The obvious implication here is that a characteristic curve will not be produced that will represent the entire range of magnitudes, but just a lesser range as defined by the series of $N$ points.

The second difference arises in the need to calculate the coefficients without matrix operations for data in which the independent variable is not equally spaced. This set of polynomials is called the *Lagrange polynomials*. Another difference is that certain conditions may be imposed on the polynomials. Sets in which this happens are called *spline functions*. Both of these basic methods will be studied in this section. Additional material can be found in the numerical methods literature (Chapra and Canale, 1988; Al-Khafaji and Tooley, 1986).

### 2.7.1 Lagrange Polynomials

Given $N = M + 1$ data points, $(x_i, y_i)$, and an unknown functional relationship, $y = g(x)$, the Lagrange interpolation function is defined as

$$\hat{y} = \hat{g}(x) = f(x) = L_0(x)f(x_0) + L_1(x)f(x_1) + \cdots + L_M(x)f(x_M)$$

or

$$f(x) = \sum_{i=0}^{M} L_i(x)f(x_i), \quad x_0 \le x \le x_M \tag{2.38}$$

Refer again to Figure 2.11. If $M = 3$, $x_0 = -1$, and $x_3 = 2$, then $f(x)$ is defined over the interval $-1 \le x \le 2$. The coefficient functions $L_i(x)$ have special properties, which are

$$L_i(x_j) = \begin{cases} 0, & i \ne j \\ 1, & i = j \end{cases} \tag{2.39}$$

and

$$\sum_{i=0}^{M} L_i(x) = 1 \tag{2.40}$$

The coefficient functions have a special factored form that directly produces these properties. The first condition of equation 2.39 is produced if

$$L_i(x) = C_i(x - x_0)(x - x_1) \cdots (x - x_M), \quad \text{excluding } (x - x_i) \tag{2.41}$$

where $C_i$ is a proportionality coefficient. For the second condition to be met, then

$$C_i = \frac{L_i(x_i)}{(x_i - x_0)(x_i - x_1) \cdots (x_i - x_M)} = \frac{1}{(x_i - x_0)(x_i - x_1) \cdots (x_i - x_M)} \tag{2.42}$$

Thus the coefficient functions have the form

$$L_i(x) = \frac{(x - x_0)(x - x_1) \cdots (x - x_M)}{(x_i - x_0)(x_i - x_1) \cdots (x_i - x_M)} = \prod_{\substack{j=0 \\ j \neq i}}^{M} \frac{x - x_j}{x_i - x_j}, \quad i = 0, 1, \ldots, M \tag{2.43}$$

Note that the number of factors in the numerator and denominator, and hence the order of each polynomial, is $M$. The explicit interpolation function is obtained by combining equation 2.38 with equation 2.43 and is

$$\hat{y} = f(x) = \sum_{i=0}^{M} \prod_{\substack{j=0 \\ j \neq i}}^{M} \frac{x - x_j}{x_i - x_j} y_i \tag{2.44}$$

This is a powerful function because it allows interpolation with unevenly spaced measurements of the independent variable without using matrix operations. (Refer to the least squares method in Section 2.3.)

---

### EXAMPLE 2.6

For the stress–strain measurements in Example 2.2, develop a second-order interpolation function for the data points $[(1.01, 34.2), (1.12, 38.8), (1.42, 44.8)]$. Compare this function with the regression equation. How does the estimate of the value of stress differ for $x = 1.30$?

Since $N = 3$, $M = 2$, and the coefficient functions are found using equation 2.43.

$$L_0(x) = \prod_{\substack{j=0 \\ j \neq i}}^{2} \frac{x - x_j}{x_i - x_j} = \frac{(x - x_1)(x - x_2)}{(x_0 - x_1)(x_0 - x_2)} = \frac{(x - 1.12)(x - 1.42)}{(1.01 - 1.12)(1.01 - 1.42)}$$

$$= \frac{x^2 - 2.54x + 1.59}{0.0451} = 22.173(x^2 - 2.54x + 1.59)$$

Similarly

$$L_1(x) = \frac{(x - x_0)(x - x_2)}{(x_1 - x_0)(x_1 - x_2)} = \frac{(x - 1.01)(x - 1.42)}{(1.12 - 1.01)(1.12 - 1.42)} = \frac{x^2 - 2.43x + 1.434}{-0.033}$$

$$L_2(x) = \frac{(x - x_0)(x - x_1)}{(x_2 - x_0)(x_2 - x_1)} = \frac{(x - 1.01)(x - 1.12)}{(1.42 - 1.01)(1.42 - 1.12)} = \frac{x^2 - 2.13x + 1.131}{0.123}$$

The polynomial is

$$\hat{y} = f(x) = L_0(x)y_0 + L_1(x)y_1 + L_2(x)y_2$$

Combining all the terms produces

$$\hat{y} = -53.125x^2 + 115.166x - 68.374, \quad 1.01 \le x \le 1.42$$

Compare this equation with the one developed in Example 2.2. Notice that the coefficients are different in magnitude. As expected for the three points used for the coefficient calculation, the ordinate values are exact values in the data. Compare now for $x = 1.3$: The Lagrange method estimates a value of 43.41, whereas the regression approach estimates a value of 44.48—a 2.5% difference.

---

By now you have probably realized that for interpolating between $M + 1$ points using the Lagrange method, an $M$th-order polynomial is produced. For a large data set, this would produce a high-order polynomial with the likelihood of containing large fluctuation errors. How is this situation handled? The approach is to segment the data set into regions and determine polynomials for each region.

Consider again the muscle activity vs load data in Example 2.5 and replotted in Figure 2.12. Assume that the measurements are accurate, and divide the range of the load variable into two regions. Each region contains four points, and two third-order Lagrange polynomials can be utilized to estimate the muscle activity. The resulting curve is plotted also in Figure 2.12. The points are interpolated with a load increment of 0.25 lb.

Many applications with times series involve thousands of signal points and hence hundreds of segments. Consider the need in human locomotion for producing an average of quasiperiodic signals when successive periods are not exactly equal. For example, Figures 2.13a, 2.13b, and 2.13c show the muscle signals from successive walking strides—the durations of each one are unequal (Shiavi and Green, 1983). The sampling rate is 250 samples per second and the duration of periods range between 0.95 s and 1.0 s. Before averaging, each signal must be converted into an equivalent signal with 256 points. This was done using cubic Lagrange interpolation polynomials.

The errors associated with Lagrange interpolation are due to many factors, including inaccurate data and *truncation*. A truncation error is incurred when the order of the interpolation polynomial is lower than that of the true data. In general, the errors are difficult to quantitate, and there are not any good

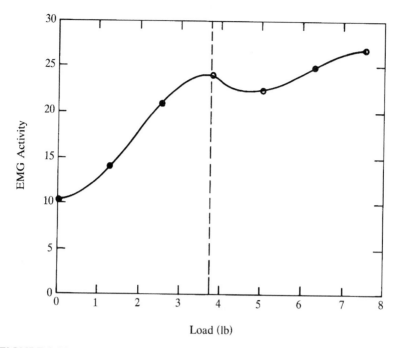

**FIGURE 2.12** The EMG vs load data is plotted with the load range divided into two regions. Each region has a separate Lagrange polynomial curve.

approximation schemes available. One can appreciate the effect of inaccurate data since the polynomials fit the acquired data exactly. For instance, if a particular $y_i$ contains a positive measurement error, the polynomial fitting the section of data containing it will be shifted upward. Since the Lagrange method is similar to the Newton divided difference method, a truncation error can be approximated using the error term from the Newton method (Al-Khafaji and Tooley, 1986). However, the error depends on the $M$th-order derivative of the unknown function and thus the error term is not helpful. One must use good judgment and ensure that the data is accurate and noise-free.

### 2.7.2 Spline Interpolation

The *spline functions* are another set of $M$th-order polynomials that are used as an interpolation method. These are also a succession of curves, $f_i(x)$, of the same order. However, a major difference between spline functions and the other interpolation methods is that an $f_i(x)$ exists only between data points $x_i$ and $x_{i+1}$, as schematically shown in Figure 2.14. Successive curves have common endpoints—the data points—called *knots*. Functional conditions are imposed between successive polynomial curves to develop enough equations to solve for

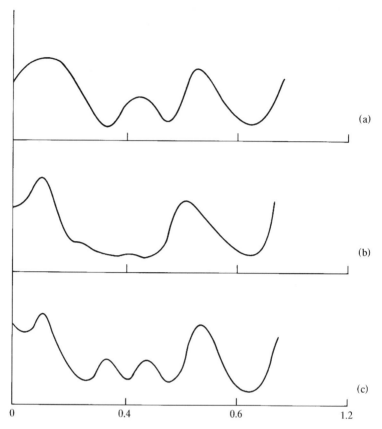

**FIGURE 2.13** Three segments of a random quasiperiodic signal, each with a different duration, are shown in (a), (b), and (c).

the unknown coefficients. The conditions are that successive curves must have the first $M - 1$ derivatives equal at the common knot. The result is that the range of magnitudes in the resulting functional approximation is less than that produced by the other interpolation methods. Figure 2.15 illustrates this.

### 2.7.2.1 Conceptual Development

The most used spline functions are the cubic splines. The development of their equations is quite laborious but not conceptually different from the other order splines. To give a better understanding of the procedure, we will develop the equations in detail for the quadratic spline and then give an example.

The general equation for a quadratic spline in interval $i$ is

$$f_i(x) = a_i + b_i x + c_i x^2, \quad 0 \le i \le N - 2, \quad x_i \le x \le x_{i+1} \tag{2.45}$$

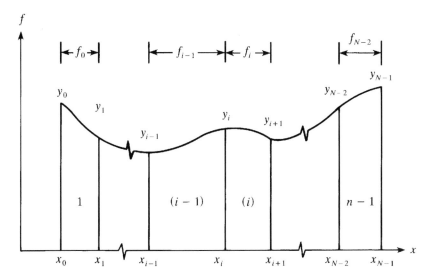

**FIGURE 2.14**   Schematic for spline interpolation for a set of $N$ data points.

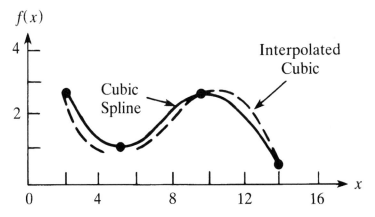

**FIGURE 2.15**   Schematic of a fit of four points with a cubic spline and an interpolating cubic polynomial.

For $N$ data points there are $N - 1$ equations and therefore $3(N - 1)$ unknown coefficients. The conditions for solving for these coefficients are produced as follows:

1. *The functional values must be equal at the interior knots.* That is,

$$f_i(x_i) = a_i + b_i x_i + c_i x_i^2$$

$$f_{i-1}(x_i) = a_{i-1} + b_{i-1} x_i + c_{i-1} x_i^2, \quad 1 \le i \le N - 2 \tag{2.46}$$

and produces $2(N - 2)$ conditions.

2. *The first and last functions must pass through the endpoints.*

$$f_0 = A_0 + b_0 x_0 + c_0 x_0^2$$

$$f_{N-2}(x_{N-1}) = a_{N-2} + b_{N-2} x_{N-1} + c_{N-2} x_{N-1}^2 \qquad (2.47)$$

Two additional conditions are produced.

3. *The first derivatives at the interior knots must be equal.* That is, the first derivatives of equations 2.46 are equal:

$$b_i + 2c_i x_i = b_{i-1} + 2c_{i-1} x_i, \quad 1 \le i \le N - 2 \qquad (2.48)$$

This produces $N - 2$ additional conditions. Now $3(N - 4)$ conditions exist, and one more must be sought.

4. The additional condition is that *the second derivative of $f_0(x_0)$ is zero*. This means that $c_0$ is zero.

## EXAMPLE 2.7

For the concrete cylinder data in Example 2.4, develop the quadratic splines to interpolate a value of stress for a microstrain of 2200. Use the last four points: (1360, 2760), (2080, 3005),(2450, 2850), and (2940, 2675).

For four points, three intervals are formed and the necessary equations are

$$f_0(x) = a_0 + b_0 x + c_0 x^2$$
$$f_1(x) = a_1 + b_1 x + c_1 x^2$$
$$f_2(x) = a_2 + b_2 x + c_2 x^2$$

There are nine unknown parameters that must be determined by the four sets of conditions. Condition 1 states that the equations must be equal at the interior knots and produces four equations:

$$3005 = a_0 + b_0 2080 + c_0 2080^2$$
$$3005 = a_1 + b_1 2080 + c_1 2080^2$$
$$2850 = a_1 + b_1 2450 + c_1 2450^2$$
$$2850 = a_2 + b_2 2450 + c_2 2450^2$$

Condition 2 states that the first and last functions must pass through the endpoints and

$$2760 = a_0 + b_0 1360 + c_0 1360^2$$
$$2675 = a_2 + b_2 2940 + c_2 2940^2$$

Two more equations are produced because the first derivatives must be equal at the interior knots:

$$b_0 + c_0 4160 = b_1 + c_1 4160$$
$$b_1 + c_1 4900 = b_2 + c_2 4900$$

The final condition stipulates that the second derivative of $f_0(x)$ is zero, or $c_0 = 0$. Thus now there are actually eight equations and eight unknowns. Solving them yields

$$(a_0, b_0, c_0) = (2.2972 \times 10^3, 3.4028 \times 10^{-1}, 0)$$
$$(a_1, b_1, c_1) = (-6.5800 \times 10^3, 8.8761, -2.0519 \times 10^{-3})$$
$$(a_2, b_2, c_2) = (-5.7714 \times 10^3, 6.7489, -1.3184 \times 10^{-3})$$

The strain value of 2200 is within the first interval, so by using the $f_1(x)$ spline we find that the interpolated stress value is 3016.29 psi.

### 2.7.2.2   Cubic Splines

For the cubic splines, the objective is to develop a set of third-order polynomials for the same intervals as illustrated in Figure 2.14. The equations now have the form

$$f_i(x) = a_i + b_i x + c_i x^2 + d_i x^3, \quad x_i \leq x \leq x_{i+1} \tag{2.49}$$

and there are $4(N - 1)$ unknown coefficients. The equations for solving for these coefficients are developed by expanding the list of conditions stated in the previous section. The additional equations are produced by imposing equality constraints on the second derivatives. The totality of conditions are as follows:

1. Functions of successive intervals must be equal at their common knot, $2(N - 2)$ conditions.
2. The first and last functions must pass through the endpoints, two conditions.
3. The first and second derivatives of functions must be equal at their common knots, $2(N - 2)$ conditions.
4. The second derivatives at the endpoints are zero, two conditions.

The specification of condition 4 leads to what are defined as *natural splines*. The $4(N - 1)$ conditions stated are sufficient for the production of the natural cubic splines for any data set. It is straightforward to solve this set of equations, which produces a $4(N - 1) \times 4(N - 1)$ matrix equation. However, there is an alternative approach that is more complex to derive but results in requiring the solution of only $N - 2$ equations.

The alternative method involves incorporating Lagrange polynomials into the formulation of the spline functions. This incorporation is contained in the first step, which is based on the fact that a cubic function has a linear second derivative. For any interval the second derivative, $f_i''(x)$, can be written as a linear interpolation of the second derivatives at the knots, or

$$f_{i-1}''(x) = \frac{x - x_i}{x_{i-1} - x_i} f''(x_{i-1}) + \frac{x - x_{i-1}}{x_i - x_{i-1}} f''(x_i) \tag{2.50}$$

This expression is integrated twice to produce an expression for $f_{i-1}(x)$. The integration will produce two constants of integration. Expressions for these two constants can be found by invoking the equality conditions at the two knots. If these operations are performed, the resulting cubic equation is

$$\begin{aligned} f_{i-1}(x) = {} & \frac{f''(x_{i-1})}{6(x_i - x_{i-1})} (x_i - x)^3 + \frac{f''(x_i)}{6(x_i - x_{i-1})} (x - x_{i-1})^3 \\ & + \left[ \frac{f(x_{i-1})}{x_i - x_{i-1}} - \frac{f''(x_{i-1})(x_i - x_{i-1})}{6} \right] (x_i - x) \\ & + \left[ \frac{f(x_i)}{x_i - x_{i-1}} - \frac{f''(x_i)(x_i - x_{i-1})}{6} \right] (x - x_{i-1}) \end{aligned} \tag{2.51}$$

This equation is much more complex than equation 2.49 but only contains two unknown quantities, $f''(x_{i-1})$ and $f''(x_i)$. These two quantities are resolved by invoking the continuity of derivatives at the knots. The derivative of equation 2.51 is derived for the intervals $i$ and $i-1$ and the derivatives are equated at $x = x_i$ since

$$f_{i-1}'(x_i) = f_i'(x_i) \tag{2.52}$$

This results in the equation

$$\begin{aligned} & (x_i - x_{i-1})f''(x_{i-1}) + 2(x_{i+1} - x_{i-1})f''(x_i) + (x_{i+1} - x_i)f''(x_{i+1}) \\ & = \frac{6}{x_{i+1} - x_i} [f(x_{i+1}) - f(x_i)] + \frac{6}{x_i - x_{i-1}} [f(x_{i-1}) - f(x_i)] \end{aligned} \tag{2.53}$$

Equation 2.53 is written for all interior knots and results in $N - 2$ equations. Since natural splines are used, $f''(x_0) = f''(x_{N-1}) = 0$, only $N - 2$ second derivatives are unknown and the system of equations can be solved. The results are inserted into the polynomial equations of equation 2.51 and the cubic spline functions are formed.

---

### EXAMPLE 2.8

The nine data points of the concrete cylinder data listed in Example 2.4 are used to find a set of eight natural cubic spline functions. The desired values of the independent variable, microstrain, are chosen in order to produce a set of values equally spaced from 600 to 2800 at increments of 200. The results are plotted in Figure 2.16. Compare this plot with the model presented in Figure 2.8. Notice how there is much more curvature incorporated in the set of points produced by the spline functions.

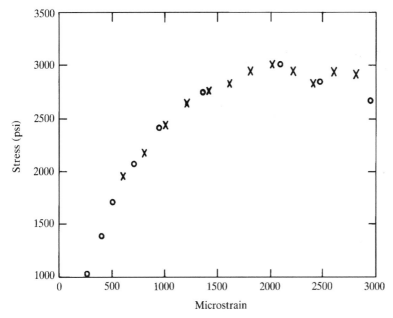

**FIGURE 2.16**  Stress vs microstrain data points (o) of a concrete cylinder and the points interpolated (x) using cubic splines.

## 2.8  OVERVIEW

There are many techniques for empirical modeling and approximation of values of unknown functions. All of them have their origin in the field of numerical methods and have been utilized extensively in engineering. One must be aware of the underlying assumptions and basic principles of development to implement them properly. One main distinction between curve fitting and interpolation is that in the former set of techniques, inaccurate or noisy measurements are used. Another important distinction is that interpolation techniques produce a set of estimated points whereas curve fitting actually produces a functional model of the measurements.

Recently the use of interpolation techniques has been extended to changing the number of points in a measured time series. When the number of resulting points has been increased, the procedure is called *interpolation*; when the number of resulting points has been decreased, the procedure is called *decimation*. Because this approach actually changes the sampling rate of the signal and because different techniques produce different values, the entire approach can be interpreted as a filtering procedure. This is a more advanced coverage reserved for advanced courses in digital signal processing. Consult textbooks such as DeFatta et al. (1988) for detailed explanations.

## REFERENCES

A. Al-Khafaji and J. Tooley, *Numerical Methods in Engineering Practice*, Holt, Rhinehart & Winston, New York, 1986.

B. Bowerman and R. O'Connell, *Time Series Forecasting*, Dunbury Press, Boston, 1987.

S. Chapra and R. Canale, *Numerical Methods for Engineers*, McGraw-Hill, New York, 1988.

S. Chatterjee and B. Price, *Regression Analysis by Example*, Wiley, New York, 1977.

D. DeFatta, J. Lucas, and W. Hodgkiss, *Digital Signal Processing: A System Design Approach*, Wiley, New York, 1988.

W. Dorn and D. McCracken, *Numerical Methods with FORTRAN IV Case Studies*, Wiley, New York, 1972.

R. Hamming, *Numerical Methods for Scientists and Engineers*, McGraw-Hill, New York, 1973.

S. Hirokawa and K. Matsumara, "Gait Analysis Using a Measuring Walkway for Temporal and Distance Factors," *Med. & Biol. Eng. & Comput.* **25**: 577–582 (1987).

M. James, G. Smith, and J. Wolford, *Applied Numerical Methods for Digital Computation*, Harper & Row, New York, 1985.

L. Lapin, *Probability and Statistics for Modern Engineering*. Brooks/Cole Engineering Division, Monterey, CA, 1983.

C. Pearson, *Numerical Methods in Engineering and Science*, Van Nostrand Reinhold, New York, 1986.

R. Roselli, R. Parker, K. Brigham, and T. Harris, "Relations between Oncotic Pressure and Protein Concentration for Sheep Plasma and Lung Lymph," *The Physiologist* **23**: 75 (1980).

C. Sheppard, *Basic Principles of the Tracer Method*, Wiley, New York, 1962.

R. Shiavi, *A Proposed Electromyographic Technique for Testing the Involvement of the Tibialis Anterior Muscle in a Neuromuscular Disorder*, Master's Thesis, Drexel Institute of Technology, 1969.

R. Shiavi and N. Green, "Ensemble Averaging of Locomotor Electromyographic Patterns Using Interpolation," *Med. & Biol. Eng. & Comput.* **21**: 573–578 (1983).

M. Solomonow, A. Guzzi, R. Baratta, H. Shoji, and D'Ambrosia, "EMG–Force Model of the Elbows Antagonistic Muscle Pair," *Am. J. Phys. Med.* **65**: 223–244 (1986).

C. Wylie, *Advanced Engineering Mathematics*, McGraw-Hill, New York, 1995.

## EXERCISES

**2.1** Derive the equation, summation formula, to estimate of the average value, $y_{ave}$, of a set of data points using the least squares criteria. [*Hint:* Start with $e_i = (y_i - y_{ave})$.]

**2.2** Derive equation 2.7, the second of the normal equations for solving the linear model.

**2.3** Data relating the volume rate of water discharging from a cooling system in a factory and the resultant height of a stationary water column gauge are listed in Table E2.3. What is the linear model for this relationship?

### TABLE E2.3

| $i$ | $y_i$ (m³/s) | $x_i$ (cm) |
|---|---|---|
| 1 | 15.55 | −23 |
| 2 | 15.46 | −22 |
| 3 | 20.07 | −16 |
| 4 | 21.99 | −16 |
| 5 | 36.11 | 14 |
| 6 | 59.82 | 33 |
| 7 | 86.58 | 46 |
| 8 | 110.96 | 69 |
| 9 | 136.52 | 88 |
| 10 | 204.40 | 120 |
| 11 | 232.87 | 136 |
| 12 | 492.50 | 220 |
| 13 | 1412.48 | 400 |

**2.4** An industrial engineer wishes to establish a relationship between the cost of producing a batch of laminated wafers and the size of the production run (independent variable). The data listed in Table E2.4 have been gathered from previous runs. [Adapted from Lapin (1983), p. 327, with permission.]

a. What is a suitable regression equation for this relationship?
b. What is the estimated cost of a run of 2000 units?
c. How is the data and model to be changed if one were interested in the relationship between cost per unit and size of the run?

**2.5** Estimating the length of time required to repair computers is important for a manufacturer. The data concerning the number of electronic components (units) repaired in a computer during a service call and the time duration to repair the computer are tabulated in Table E2.5.

a. Constuct the scatter plot and calculate the parameters of a second-order model to estimate the duration of a service call.

**TABLE E2.4**

| Size | Cost ($) | Size | Cost ($) |
|------|----------|------|----------|
| 1550 | 17,224 | 786 | 10,536 |
| 2175 | 24,095 | 1234 | 14,444 |
| 852 | 11,314 | 1505 | 15,888 |
| 1213 | 13,474 | 1616 | 18,949 |
| 2120 | 22,186 | 1264 | 13,055 |
| 3050 | 29,349 | 3089 | 31,237 |
| 1128 | 15,982 | 1963 | 22,215 |
| 1215 | 14,459 | 2033 | 21,384 |
| 1518 | 16,497 | 1414 | 17,510 |
| 2207 | 23,483 | 1467 | 18,012 |

**TABLE E2.5**

| Units (#) | Time (min) | Units (#) | Time (min) |
|-----------|------------|-----------|------------|
| 1 | 23 | 10 | 154 |
| 2 | 29 | 10 | 166 |
| 3 | 49 | 11 | 162 |
| 4 | 64 | 11 | 174 |
| 4 | 74 | 12 | 180 |
| 5 | 87 | 12 | 176 |
| 6 | 96 | 14 | 179 |
| 6 | 97 | 16 | 193 |
| 7 | 109 | 17 | 193 |
| 8 | 119 | 18 | 195 |
| 9 | 149 | 18 | 198 |
| 9 | 145 | 20 | 205 |

b. Plot the model overlaying the data. Now develop a third-order model. Is it much better than the quadratic model? Why or why not?

**2.6** Derive equation 2.13, the first of the normal equations for solving the quadratic model.

**2.7** Verify the fitting of the product exponential model in Example 2.4.

**2.8** Derive the normal equations analogous to those in Section 2.3 for the power law model $y = \alpha x^\beta$.

**2.9** The nonlinear equation

$$k = a\frac{x}{b+x}, \quad 0 \le x \le \infty$$

is often called the saturation–growth rate model. It adequately describes the relationship between population growth rate, $k$, of many species and the food supply, $x$. Sketch the curve and explain a reason for the name of this model. Linearize the model.

**2.10** Derive the normal equations for the function $y = a(\tan x)^b$.

**2.11** We want to determine the relationship between mass density, $\rho$ (slugs/ft$^3$), and altitude above sea level, $h$ (ft), using the exponential model

$$\rho = \alpha e^{\beta h}$$

The data are listed in Table E2.11. [Adapted from James et al., p. 357, with permission.]

a. Plot the data and find the model parameters (using a calculator) and the data points marked with an asterisk. Plot the model.
b. Find the model parameters using all the data and a computer program. Is this set of parameters better than those found in part (a)? Why or why not?
c. What is the estimated (interpolated) air density at 35,000 ft? What is the predicted (extrapolated) air density at 80,000 ft?

**TABLE E2.11**

| $h$ | $\rho \times 10^6$ | $h$ | $\rho \times 10^6$ |
|---|---|---|---|
| 0 | 2,377* | 15,000 | 1,497 |
| 1,000 | 2,308 | 20,000 | 1,267* |
| 2,000 | 2,241 | 30,000 | 891 |
| 4,000 | 2,117 | 40,000 | 587* |
| 6,000 | 1,987 | 50,000 | 364 |
| 10,000 | 1,755 | 60,000 | 224* |

**2.12** For equation 2.24, which requires that both variables have a hyperbolic transformation, plot the function with $a = 0.5$ and $b = -1.0$.

**2.13** Consider the following function with $a = 0.5$ and $b = 2.0$.

$$y = \frac{e^{a+bx}}{1 + e^{a+bx}}$$

a. Plot the function for $0 \le x \le 3$.
b. Find the linearizing transformation.

**2.14** In lung research, the relationship between the lymph-to-plasma protein ratio, LP, and pore size, R, is approximated by the equation

$$LP = a + \frac{b}{R} + \frac{c}{R^2}$$

a. What transformation will permit a polynomial regression?
b. Experimental data measured is listed in Table E2.14.

1. Find the model coefficients.
2. Plot the data and the model.

**TABLE E2.14**

| LP | R | LP | R |
|------|----|-------|----|
| 0.824 | 36 | 0.599 | 76 |
| 0.731 | 46 | 0.576 | 86 |
| 0.671 | 56 | 0.557 | 96 |
| 0.630 | 66 | | |

**2.15** Verify the solution found in Example 2.4.

**2.16** Prove Parseval's Theorem, equation 2.35, starting with equation 2.27. [*Hint:* Using equations 2.29 and 2.31 could be helpful.]

**2.17** Verify the final equations for $y(z)$ and $y(x)$ in Example 2.5.

**2.18** Fit a second-order polynomial to the data set in Table E2.18 using orthogonal polynomials.

**TABLE E2.18**

| $x_i$ | $f(x_i)$ |
|-----|-------|
| 0 | 1 |
| 0.5 | 3 |
| 1.0 | 6 |
| 1.5 | 10 |
| 2.0 | 18 |
| 2.5 | 24 |
| 3.0 | 35 |

a. What are the coefficients $A_i$?
b. What is the $f(z)$ equation?
c. What is the $f(x)$ equation?

**2.19** A metal pole for supporting a street light is installed erect with a height of 574 in. It has been struck by an automobile in an accident and badly misshapen. It is necessary to derive an equation for this shape to analyze the impact energies. To keep the model single-valued, the pole has been laid on its side. The $x$ coordinate represents the vertical coordinate of a spot on the pole and the $y$ direction the lateral coordinate of a spot. The base is at position $(0, 0)$. The data is shown in Table E2.19.

a. Plot a scatter diagram of the points.
b. What order polynomial seems sufficient from visual examination? Why?

c. Derive coefficients and squared errors necessary to create a fourth-order model using orthogonal polynomials.
d. What is the sufficient order for the model? Why?
e. Plot the model equation overlying the data.

**TABLE E2.19**

| $y_i$ | $x_i$ | $y_i$ | $x_i$ |
|------|------|------|------|
| 0 | 0 | −137 | 240 |
| −24 | 24 | −151 | 264 |
| −43 | 48 | −166 | 288 |
| −59 | 72 | −180 | 312 |
| −71 | 96 | −191 | 336 |
| −82 | 120 | −196 | 360 |
| −91 | 144 | −193 | 384 |
| −101 | 168 | −176 | 408 |
| −111 | 192 | −141 | 432 |
| −124 | 216 | −81 | 456 |

**2.20** Perform Exercise 2.5a with data of the even number of units repaired using orthogonal polynomials. Are the resulting models similar? (Use only the first point pair when a value of the number of units appears more than once.)

**2.21** Refer to the water discharge data in Exercise 2.3:

a. Find the Lagrange polynomial factors, $L_i(x)$, for the ranges $14 \le x \le 46$ and $-22 \le x \le 14$.
b. Find $f(x)$ for the range $14 \le x \le 46$.

**2.22** Derive a linear interpolation formula for interpolating between the point pairs $(x_i, y_i)$ and $(x_{i+1}, y_{i+1})$ with

$$y = Ay_i + By_{i+1}$$

where $A$ and $B$ are functions of $x$, $x_i$, and $x_{i+1}$. Show that this form is the same as that given by a first-order Lagrange formula.

**2.23** Use the EMG vs load data in Example 2.5.

a. Divide the load range into three regions.
b. In the first region, derive the Lagrange interpolating polynomial.
c. What EMG magnitudes does it produce for load values of 0.5, 1.0, and 1.5 lbs?

**2.24** Using the results in Exercise 2.23, predict the EMG magnitude for a load of 3.5 lbs. Notice that this load value is outside the design range of the polynomial. This is extrapolation. How does the extrapolated value compare with the measured value?

**2.25** In Table E2.25 are listed measurements on the force–displacement characteristics of an automobile spring.

   a. Plot the data.

   b. Divide the displacement range into two or three regions.

   c. Use a computer algorithm to interpolate for one value of displacement within each of the measured intervals.

   d. Plot the measured and interpolated values.

   e. Use a computer algorithm to interpolate over the range $0.1 \leq x \leq 0.45$, and produce a set of interpolated values that have displacement evenly spaced at intervals of $0.05$ m.

**TABLE E2.25**  Force, $N \times 10^4$, vs displacement, $m$

| $N$ | $m$ |
|---|---|
| 10 | 0.1 |
| 20 | 0.17 |
| 30 | 0.24 |
| 40 | 0.34 |
| 50 | 0.39 |
| 60 | 0.42 |
| 70 | 0.43 |

**2.26** For the natural cubic splines, write the general equations that satisfy the four conditions in Section 2.7.2.2.

**2.27** For the alternative cubic spline method, derive equation 2.53 from equation 2.51 and the general first derivative equality condition, equation 2.52.

**2.28** For the concrete cylinder example (Example 2.4), fit a set of quadratic splines to the last three data points and interpolate the stress for a microstrain of 2800. Show that the coefficients are

$$(a_0, b_0, c_0) = (3.8764 \times 10^3, -0.41891, 0)$$
$$(a_1, b_1, c_1) = (4.6331 \times 10^3, -1.0367, 1.2607 \times 10^{-4})$$

and that $f_1(2800) = 2718.8$.

**2.29** Fit quadratic spline functions to the first three points of the EMG vs load data in Example 2.5. Specifically, what are the equations satisfying conditions 1 through 4 in Section 2.7.2.1? Solve these equations.

**2.30** For the impact data listed in Exercise 2.19, you are asked to fit natural cubic splines.

   a. For $0 \leq i \leq 3$, write the equations for conditions 1, 2, and 3.

   b. For $i = 0$ and $N - 1$, write the equations for condition 4.

**2.31** A nonlinear resistor has the experimental current–voltage data listed in Table E2.31.

**TABLE E2.31**   Current, $i$, vs Voltage, $v$

| $i$ | 1.00 | 0.50 | 0.25 | −0.25 | −0.50 | −1.00 |
|---|---|---|---|---|---|---|
| $v$ | 200.0 | 40.0 | 12.5 | 12.5 | −40.0 | −200.0 |

a. Determine and plot the characteristic curve using a fifth-order polynomial.
b. Determine and plot the characteristic curve using a spline function (let $\Delta i = 0.2$).
c. Are there any differences?

**2.32** The kinematic viscosity of water is tabulated in Table E2.32.

**TABLE E2.32**

| $T$, °F | $v$, $10^{-5}$ ft$^2$/sec |
|---|---|
| 40 | 1.66 |
| 50 | 1.41 |
| 60 | 1.22 |
| 70 | 1.06 |
| 80 | 0.93 |

a. Plot this data and find $v$ for $T = 53$ using a linear interpolation.
b. Repeat part (a) using a second-order Lagrange interpolator.
c. Repeat part (a) using a second-order spline interpolator.
d. What are the percentage differences between each method?

**2.33** Values of complicated functions are often tabulated and interpolation is required to find the untabulated values. To demonstrate the utility of this procedure, use the saturation–growth rate characteristic for microbial kinetics that is given as

$$k = 1.23 \frac{x}{22.18 + x}, \quad x \geq 0$$

a. Using $N = 5$, $\Delta x = 4$, and the Lagrange polynomials, what is the interpolated value for $f(10)$? How does it compare to the value given by the model?
b. Repeat part (a) using spline functions.

# III

# FOURIER ANALYSIS

## 3.1 INTRODUCTION

Knowledge of the cyclic or oscillating activity in various physical and biological phenomena and in engineering systems has been recognized as essential information for many decades. In fact, interest in determining sinusoidal components in measured data through modeling began at least several centuries ago and was known as *harmonic decomposition*. The formal development of Fourier analysis dates back to the beginning of the eighteenth century and the creative work of Jean Fourier, who first developed the mathematical theory that enabled the determination of the frequency composition of mathematically expressible waveforms. This theory is called the *Fourier transform* and has become widely used in engineering and science. During the middle of the 19th century, numerical techniques were developed to determine the harmonic content of measured signals. Bloomfield (1976) summarizes a short history of these developments.

One of the oldest phenomenon to have been studied is the change in the intensity of light emitted from a variable star. A portion of such a signal is plotted in Figure 3.1. It is oscillatory with a period of approximately 25 days. Astronomers theorized that knowledge of the frequency content of this light variation could yield not only general astronomical knowledge but also information about the star's creation (Whittaker and Robinson, 1967). Another phenomenon that is usually recognized as oscillatory in nature is vibration.

Variable Star

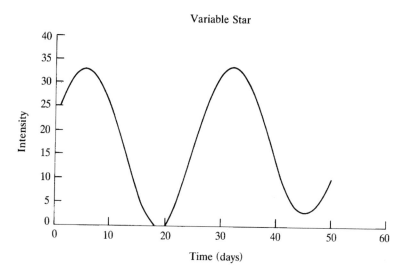

**FIGURE 3.1** Brightness changes of a variable star during 50 days.

Accelerometers, which must be tested to determine their accuracy, are used to measure the intensity and the frequencies of vibration. Figure 3.2 shows the electrical output of an accelerometer that has been perturbed sinusoidaly (Licht et al., 1987). Examination of the waveform reveals that the response is not a pure sinusoid. Fourier analysis will indicate what other frequency components, in addition to the perturbation frequency, are present. A biomedical application is

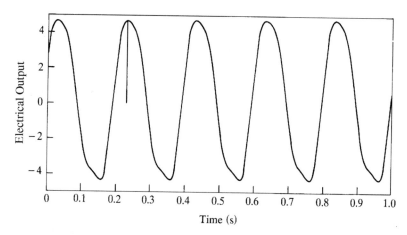

**FIGURE 3.2** Oscillatory response of an accelerometer to a sinusoidal perturbation of 4.954 Hz with amplitude of 250 microstrain. [Adapted from Licht et al., fig. 6, with permission.]

the analysis of electroencephalographic (EEG) waveforms to determine the state of mental alertness—i.e., drowsiness, deep sleeping, thinking—of an individual as determined by differences in frequency content (Bodenstein and Praetorius, 1977). Figure 3.3 shows a variety of EEG waveforms for a variety of conditions.

For different types of signals, there are different versions of Fourier analysis. Applications with deterministic periodic signals, such as in Figure 3.2, will require a Fourier series expansion; those with deterministic aperiodic signals, such as in Figure 3.1, will require a Fourier transform. The third and most complicated are those applications with random signals, such as in Figure 3.3. They require spectral analysis, which will be treated in subsequent chapters. For all these signal types there is the discrete time version of Fourier analysis that is utilized throughout this textbook.

The goal of this chapter is to provide a comprehensive review of the principles and techniques for implementing discrete time Fourier analysis. This is valuable because Fourier analysis is used to calculate the frequency content of deterministic signals and is fundamental to spectral analysis. Although aperiodic signals are much more prevalent than periodic ones, a review of Fourier series is warrented. This is because the concepts of frequency content and the treatment of complex numbers, which are used in all aspects of frequency analysis, are easier to understand in the series form. It is presumed that readers have had exposure to Fourier series and transforms through course work or experience. A review of the

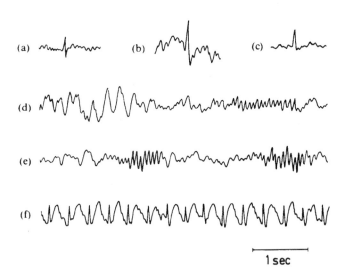

**FIGURE 3.3** Some examples of elementary EEG patterns occurring during various conditions. (a–c) Transients: (a) Biphasic spike. (b) Biphasic sharp wave. (c) Monophasic sharp wave. (d) and (e) Adult, light sleep. (f) Epilepsy. [Adapted from Bodenstein and Praetorius, fig. 2, with permission.]

definition of continuous time Fourier transformation is provided in the appendix of this chapter for the reader's convenience. If a comprehensive treatment is desired, please consult the reference listing. There are several books that treat Fourier transforms in continuous time extensively, like those of Papoulis (1962) and Bracewell (1978). However, there are many books on signals and systems that treat Fourier transforms in discrete time. Some of them are listed in the reference section. Although all of the referenced books treat many aspects of the Fourier transform, an attempt has been made to classify them according to the aspects they seem to emphasize.

## 3.2 REVIEW OF FOURIER SERIES

### 3.2.1 Definition

For periodic finite power signals, the trigonometric functions form a very powerful orthogonal function set. A review of the properites of orthogonal function sets can be found in Appendix 3.4. Most everyone with a technical background is familiar with the sine and cosine definition of the Fourier series. Another definition—one that is more useful for signal analysis—is composed of complex exponentials. The function set is

$$\Phi_m(t) = e^{jm\omega_0 t} \quad \text{for} \quad -\infty \leq m \leq \infty \quad \text{and} \quad \omega_0 = \frac{2\pi}{P} = 2\pi f_0$$

where $P$ is the time period and $f_0$ is the fundamental frequency in cycles per unit time (Hz). Figure 3.4 shows two periodic sawtooth waveforms with periods of 4 seconds. Thus $P = 4$ seconds, $f_0 = 0.25$ Hz. The orthogonality condition and the coefficient evaluation change slightly for complex function sets in that one term in the defining integral is a complex conjugate term (indicated by an asterisk). For this orthogonality condition, equation A3.2 becomes

$$\int_0^P \Phi_m(t)\Phi_n^*(t)\, dt = \begin{cases} \lambda_m, & m = n \\ 0, & m \neq n \end{cases} \tag{3.1}$$

It is easily verified that

$$\lambda_m = P \tag{3.2}$$

The coefficients are evaluated from

$$\begin{aligned} z_m &= \frac{1}{\lambda_m}\int_0^P f(t)\Phi_m^*(t) \\ &= \frac{1}{P}\int_0^P f(t)e^{-j2\pi m f_0 t}\, dt \quad \text{for} \quad -\infty \leq m \leq \infty \end{aligned} \tag{3.3}$$

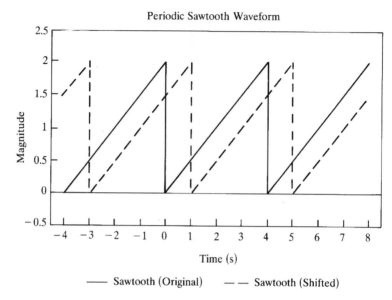

**FIGURE 3.4** Periodic sawtooth waveforms with period $= 4\,\text{s}$ and amplitude $= 2$ units.

The series expansion is now

$$f(t) = \sum_{m=-\infty}^{\infty} z_m \Phi_m(t) = \sum_{m=-\infty}^{\infty} z_m e^{j2\pi m f_0 t}$$

$$= \cdots + z_{-m}\Phi_{-m}(t) + \cdots + z_{-1}\Phi_{-1}(t) + z_0\Phi_0(t)$$
$$+ z_1\Phi_1(t) + \cdots + z_m\Phi_m(t) + \cdots$$
$$= \cdots + z_{-m}e^{-j2\pi m f_0 t} + \cdots + z_{-1}e^{-j\omega_0 t} + z_0 + z_1 e^{j\omega_0 t}$$
$$+ \cdots + z_m e^{j2\pi m f_0 t} + \cdots \tag{3.4}$$

The coefficients and the terms of this equation have some conjugate symmetries. If we expand equation 3.3 with Euler's formula for the complex exponential,

$$e^{-j2\pi m f_0 t} = \cos(2\pi m f_0 t) - j\sin(2\pi m f_0 t)$$

then

$$z_m = \frac{1}{P}\int_0^P f(t)(\cos(2\pi m f_0 t) - j\sin(2\pi m f_0 t))\,dt \tag{3.5}$$

Thus $z_m$ is a complex number and is a conjugate of $z_{-m}$ when $f(t)$ is a real signal.

## EXAMPLE 3.1

The coefficients of the complex Fourier series for the periodic sawtooth waveform indicated by the continuous line in Figure 3.4 are derived. For one cycle the equation is

$$f(t) = \tfrac{2}{4}t = \tfrac{1}{2}t, \quad 0 < t < 4; \quad P = 4$$

Now

$$z_m = \frac{1}{\lambda_m} \int_0^P f(t) e^{-j2\pi m f_0 t}\, dt = \frac{1}{4} \int_0^4 \frac{1}{2} t e^{-j0.5\pi m t}\, dt \quad \text{for } -\infty \le m \le \infty$$

$$= \frac{1}{8} \left( \frac{4 e^{-jm2\pi}}{-j0.5\pi m} - \frac{1}{(j0.5\pi m)^2} (e^{-jm2\pi} - 1) \right) \quad (3.6)$$

Since $e^{-j2\pi m} = 1$, then

$$z_m = \frac{1}{-j\pi m} = j \frac{1}{\pi m} = \frac{1}{\pi m} e^{j0.5\pi}$$

The zero frequency term has to be calculated separately, and it is

$$z_0 = \frac{1}{4} \int_0^4 \frac{1}{2} t\, dt = 1$$

The series is

$$f(t) = \frac{1}{\pi} \sum_{m=-\infty}^{-1} \frac{1}{m} e^{j0.5\pi(-mt+1)} + 1 + \frac{1}{\pi} \sum_{m=1}^{\infty} \frac{1}{m} e^{j0.5\pi(mt+1)} \quad (3.7)$$

Notice that the magnitude of the coefficients are inversely proportional to $m$, the harmonic number, and the series will converge.

---

The exponential form is directly related to the trigonometric form. The conversion is most directly implemented by expressing the complex coefficients in polar form and using the conjugate symmetry properties.

$$z_m = Z_m e^{j\theta_m} \quad \text{and} \quad Z_{-m} = |z_m|; \qquad \theta_{-m} = -\theta_m \quad (3.8)$$

Inserting the relations in equations 3.8 into equation 3.4, we have

$$f(t) = \sum_{m=-\infty}^{-1} Z_m e^{j\theta_m} e^{j2\pi m f_0 t} + z_0 + \sum_{m=1}^{\infty} Z_m e^{j\theta_m} e^{j2\pi m f_0 t} \quad (3.9)$$

Knowing to use Euler's formula for the cosine function and the symmetry properties of $z_{-m}$, equation 3.9 is rearranged to become

$$f(t) = z_0 + \sum_{m=1}^{\infty} Z_m(e^{j\theta_m}e^{j2\pi mf_0 t} + e^{-j\theta_m}e^{-j2\pi mf_0 t})$$

$$= z_0 + \sum_{m=1}^{\infty} 2Z_m \cos(2\pi mf_0 t + \theta_m)$$

$$= C_0 + \sum_{m=1}^{\infty} C_m \cos(2\pi mf_0 t + \theta_m) \tag{3.10}$$

Thus one can see that there are direct correspondences between the two forms. The average terms have the same magnitude, whereas the magnitudes in the complex coefficients have one-half the magnitude of the trigonometric form. The second equivalence is in the phase angle. The phase angle of the complex coefficients for positive $m$ are equal to the phase shift of the cosine terms. Remember that phase shifts translate into time shifts, $\tau_m$, of the cosine waves through the relationship

$$\theta_m = 2\pi mf_0 \tau_m \tag{3.11}$$

---

**EXAMPLE 3.1 (continued)**

The trigonometric series form for the sawtooth waveform can be easily formed. With

$$z_m = \frac{1}{\pi m}e^{j0.5\pi} \quad \text{for } m > 0, \qquad Z_m = \frac{1}{\pi m}, \quad \text{and } \theta_m = 0.5\pi$$

Thus

$$f(t) = 1 + \sum_{m=1}^{\infty} \frac{2}{\pi m}\cos(0.5m\pi t + 0.5\pi) \tag{3.12}$$

---

The information contained in the magnitudes of the coefficients and in the phase angles is very important in frequency analysis. Their values as a function of frequency are plotted and called the *magnitude and phase spectra*, respectively. Since these values exist only at harmonic frequencies, the spectra are *line spectra*. They are plotted in Figure 3.5 for the sawtooth waveform. In the magnitude spectrum, examine the trend in the magnitude: As frequency increases, the values are monotonically decreasing. This leads to a practical consideration—perhaps the waveform can be represented by a finite series instead of an infinite one as in equation 3.12. Suppose that the frequency component with the highest significant value is 3 Hz, or six harmonics. Then

$$\hat{f}(t) = 1 + \sum_{m=1}^{6} \frac{2}{\pi m}\cos(0.5m\pi t + 0.5\pi) \tag{3.13}$$

An interesting example of the utility of a finite series comes from a study of human walking, which is periodic by definition. Figure 3.6 shows the angle of

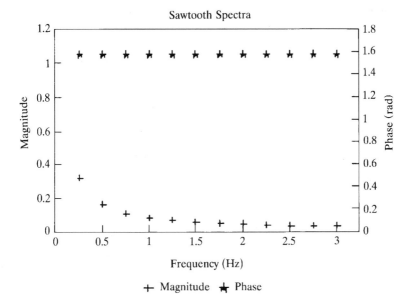

**FIGURE 3.5** Magnitude and phase line spectra for polar form of periodic sawtooth function indicated by the solid line in Figure 3.4.

flexion of the ankle joint over time. In the figure are also shown the approximations to the angle waveform using only 7 and 20 harmonics. Notice that they do provide a good representation of the actual waveform.

### 3.2.2 Convergence

The ability to represent waveforms with finite Fourier series is possible because of its *convergence property*. The Fourier series will converge to a periodically defined function in almost all conceivable practical situations. The function simply has to satisfy the Dirichlet Conditions:

1. $f(t)$ must have a finite number of discontinuities over the period.
2. $f(t)$ must have a finite number of maxima and minima over the period.
3. $f(t)$ must be bounded or absolutely integrable, that is

$$\int_0^P |f(t)|\, dt < \infty$$

In addition, the Fourier series exists as long as the range of integration is over any part of one full period. Equation 3.3 is generally written as

$$z_m = \frac{1}{P}\int_{t_i}^{t_1+P} f(t)e^{-j2\pi m f_0 t}\, dt \quad \text{for } -\infty \le t_1 \le \infty \tag{3.14}$$

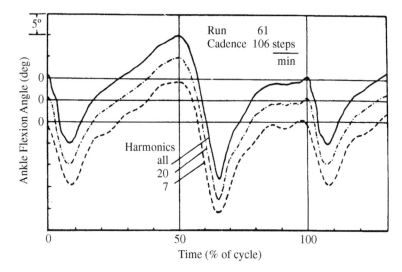

**FIGURE 3.6** Time histories of ankle flexion during human walking and the result of modeling the signal with finite Fourier series. [Adapted from Zarrugh and Radcliffe, fig. 11, with permission.]

At any points of discontinuity in $f(t)$, the series will converge to a value that is the average of the upper and lower values at the point of discontinuity. That is, if $f(t)$ is discontinuous at time $t_1$, the series value is

$$f(t_1) = \tfrac{1}{2}(f(t_1^-) + f(t_1^+))$$

The total energy in one period is simply expressed as

$$E_{\text{tot}} = \sum_{m=-\infty}^{\infty} A_m^2 \lambda_m = \sum_{m=-\infty}^{\infty} PZ_m^2$$

The average power is

$$P_{\text{av}} = \sum_{m=-\infty}^{\infty} Z_m^2$$

## 3.3 OVERVIEW OF FOURIER TRANSFORM RELATIONSHIPS

There are several forms of Fourier transforms when both continuous and discrete time and frequency domains are considered. Understanding their relationships is important when learning the procedures for implementing discrete time Fourier analysis. The relationships will be explained using the rectangular rule approximation for integration and the duality of formula relationships (Armbardar, 1995).

### 3.3.1 Continuous vs Discrete Time

The Fourier series defines the relationship between continuous time and discrete frquency domains. The continuous time and frequncy domains are related through the *continuous time Fourier transform (CTFT)*. The properties of the CTFT are summarized in Appendix 3.5. The transform and inverse transform pair are

$$X(f) = \int_{-\infty}^{\infty} x(t)e^{-j2\pi ft}\, dt \quad \text{and} \quad x(t) = \int_{-\infty}^{\infty} X(f)e^{j2\pi ft}\, df \qquad (3.15)$$

Now discretize the time domain by sampling the waveform every $T_1$ seconds; that is, consider $x(t)$ only at time intervals $t = nT_1$. The frequency domain remains continuous. The transform in equation 3.15 becomes

$$X_{\text{DT}}(f) = T_1 \sum_{n=-\infty}^{\infty} x(nT_1)e^{-j2\pi f_n T_1} \qquad (3.16)$$

where DT indicates discrete time. The right-hand side of this equation contains the weighted summation of a set of sampled values of $x(t)$ with a *sampling interval* of $T_1$ or, equivalently, with a *sampling frequency* of $f_s = 1/T_1$. The challenge now becomes to understand the properties of $X_{\text{DT}}(f)$. One way to accomplish this is to examine the Fourier series relationship since it is also a mixture of a discrete and continuous domains. Equation 3.4 is repeated below

$$\bar{f}(t) = \sum_{l=-\infty}^{\infty} f(t + lP) = \sum_{m=-\infty}^{\infty} z_m e^{j2\pi f_0 mt} \qquad (3.4)$$

The following variables play similar roles, *or duals*, in the last two equations; $t \to f, m \to n, z_m \to x(nT_1), f_0 \to T_1$. Thus $X_{\text{DT}}(f)$ has properties similar to $f(t)$ and is a periodic repetition of $X(f)$ or

$$\bar{X}_{\text{DT}}(f) = \sum_{l=-\infty}^{\infty} X(f + l/T_1) = \sum_{l=-\infty}^{\infty} X(f + lf_s) \qquad (3.17)$$

This transform is plotted schematically in Figure 3.7. Thus a sampled function has a Fourier transform that is periodic in frequency with the repetition occurring at integer multiples of the sampling frequency. This is called the *discrete time Fourier transform (DTFT)*.

The inverse DTFT is defined as

$$x(nT) = \int_{-f_N}^{f_N} \bar{X}_{DT}(f)e^{j2\pi f n T_1}\, df \qquad (3.18)$$

where $2f_N = f_s$. Examine Figure 3.7 again. For this integration to be unique, none of the periodic repetitions of $X(f)$ can overlap into other regions of the frequency

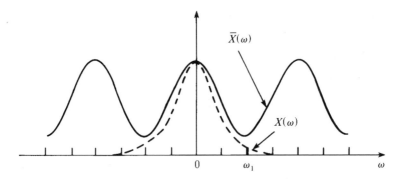

**FIGURE 3.7** Schematic of summation of periodically shifted Fourier transforms of a signal. [Adapted from Papoulis (1977), fig. 3-14, with permission.]

domain. Thus for the DTFT to accurately represent the CTFT, $x(t)$ must be *bandlimited*; that is,

$$X(f) = 0, \quad \text{for } |f| \geq f_N \tag{3.19}$$

This is consistent with the Shannon Sampling Theorem, which states that the sampling rate of a signal, $f_s = 1/T$, must be twice the highest frequency value, $f_N$, at which a signal has energy; otherwise an error called *aliasing* occurs (Cadzow, 1987). This minimum sampling frequency, $2f_N$, is called the *Nyquist rate*. Assuming that aliasing does not occur, the DTFT transform pair is written

$$X_{\text{DTFT}}(f) = T \sum_{n=-\infty}^{\infty} x(nT)e^{-j2\pi fnT} \quad \text{and} \quad x(nT) = \int_{-f_N}^{f_N} X_{\text{DTFT}}(f)e^{j2\pi fnT} \, df \tag{3.20}$$

### 3.3.2 Discrete Time and Frequency

For actual computer computation, both the time and frequency domains must be discretized. In addition, the number of sample points, $N$, must be a finite number. Define the duration of the signal as $P = NT$. The frequency domain is discretized at integer multiples of the inverse of the signal duration or

$$f = \frac{m}{P} = \frac{m}{NT} = mf_d \tag{3.21}$$

where $f_d$ is called the frequency spacing. Now the fourth version of the Fourier transform, the *discrete Fourier transform (DFT)* is derived by truncating the DTFT and discretizing the frequency domain and is defined as

$$X_{\text{DFT}}(mf_d) = T \sum_{n=0}^{N-1} x(nT)e^{-j2\pi fnT} = T \sum_{n=0}^{N-1} x(nT)e^{\frac{-j2\pi mnT}{NT}} \tag{3.22}$$

Note carefully that because the frequency domain has been discretized, the signal in the time domain has become periodic. Understanding these periodic repetitions is important when implementing the DFT; they will be studied in detail in the next section. The *inverse DFT (IDFT)* is defined as

$$x(nT) = \frac{1}{NT} \sum_{m=0}^{N-1} X(mf_d) e^{\frac{j2\pi mn}{N}} \tag{3.23}$$

Remembering the various versions of the Fourier transform and the effects of discretizing the time and/or frequency domains can be confusing initially. Figure 3.8 is provided as a schematical summary to help with this. Since the computation of the Fourier transform is the focus of this chapter, the DFT will be dealt with in

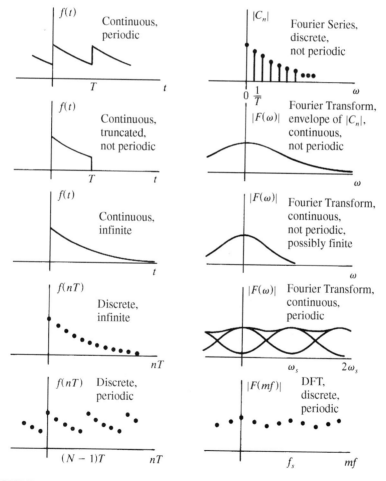

**FIGURE 3.8** Summary of the major characteristics of the various versions of the Fourier transform. [Adapted from Childers and Durling, fig. 2.14, with permission.]

the remaining sections. Other textbooks that treat these relationships in detail are Brigham (1988) and Marple (1987).

## 3.4  DISCRETE FOURIER TRANSFORM

### 3.4.1  Definition Continued

Since both domains of the DFT are discretized with sampling interval $T$ and frequency spacing $f_d$, these two parameters are dropped from the arguments in equation 3.23. A formal definition of the DFT–IDFT pair that is often used is

$$X_{\text{DFT}}(m) = T \sum_{n=0}^{N-1} x(n)e^{-\frac{j2\pi mn}{N}}, \qquad x(n) = \frac{1}{NT} \sum_{m=0}^{N-1} X(m)e^{\frac{j2\pi mn}{N}} \qquad (3.24)$$

Whether to use the radian or cyclic frequency scale is a matter of preference. From here on, the cyclic frequency scale will be used in this textbook.

In the previous section it was shown mathematically that when the frequency of the DTFT is discretized, $x(n)$ becomes periodic with period $P$. This is graphically summarized in Figure 3.9. Essentially the signal has become periodic inadvertently. In fact many texts derive a version of equation 3.24 from the definition of the Fourier series with $X_{\text{FS}}(m) = z_m$ and using the rectangular rule approximation for integration. Then

$$X_{\text{FS}}(m) = \frac{1}{P} \int_0^P x(t)e^{-jm\omega_0 t}\, dt \approx \frac{1}{NT} \sum_{n=0}^{N-1} x(nT)e^{-\frac{j2\pi mnT}{NT}} T$$

or

$$X_{\text{FS}}(m) = \frac{1}{N} \sum_{n=0}^{N-1} x(n)e^{-\frac{j2\pi mn}{N}} \qquad (3.25)$$

Notice that the only difference between the two definitions of a discrete time Fourier transform is the scaling factor. The former definition, equation 3.24, is used most often in the signal processing literature. One must be aware of the definition being implemented in an algorithm when interpreting the results of a computation. For instance, in the definition of equation 3.25, $X_{\text{FS}}(0)$ is the average of the samples, whereas the other definition will produce a weighted sum of the sampled values. The IDFT corresponding to equation 3.25, is

$$x(n) = \sum_{m=-N/2}^{N/2-1} X_{\text{FS}}(m)e^{\frac{j2\pi mn}{N}} = \sum_{m=0}^{N-1} X_{\text{FS}}(m)e^{\frac{j2\pi mn}{N}} \qquad (3.26)$$

when $N$ is even. In the first definition, equation 3.24, the IDFT contains the factor $1/NT$. Detailed derivations of the DFT–IDFT pair, their properties, important theorems, and transform pairs can be found in many texts. Refer to the

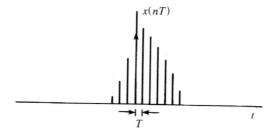

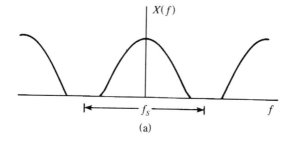

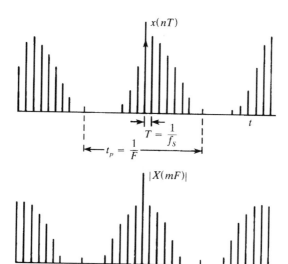

**FIGURE 3.9** Periodic repetitions: (a) Spectral repetition from sampling in the time domain. (b) Signal repetition from discretizing the frequency domain.

appropriate bibliographic section in this chapter for some references. The definition in equation 3.24 will be utilized in this text.

### 3.4.2 Partial Summary of DFT Properties and Theorems

Several theorems and properties of the DFT as pertain to real sequences will be briefly summarized because they are utilized in various sections of this textbook. The convolution relationships are often called cyclic convolutions because both the signal and its DFT have periodic repetitions.

a. *Linearity:* The DFT of a sum of signals equals the sum of the individual DFTs. For two time sequences $x(n)$ and $y(n)$,

$$X_{x+y}(m) = T \sum_{n=0}^{N-1} (ax(n) + by(n))e^{\frac{-j2\pi mn}{N}}$$

$$= T \sum_{n=0}^{N-1} ax(n)e^{\frac{-j2\pi mn}{N}} + T \sum_{m=0}^{N-1} by(n)e^{\frac{-j2\pi mn}{N}}$$

$$= aX(m) + bY(m) \tag{3.27}$$

b. *Periodicity:* $X(m) = X(m + kN)$, $k$ is an integer.

$$X(m + kN) = T \sum_{n=0}^{N-1} x(n)e^{\frac{-j2\pi n(m+kN)}{N}}$$

$$= T \sum_{n=0}^{N-1} x(n)e^{\frac{-j2\pi nm}{N}} \cdot e^{\frac{-j2\pi nkN}{N}}$$

$$= T \sum_{n=0}^{N-1} x(n)e^{\frac{-j2\pi nm}{N}} \cdot e^{-j2\pi nk}$$

Since $e^{-j2\pi nk} = 1$ for $k, n$ being integers,

$$X(m + kN) = T \sum_{n=0}^{N-1} x(n)e^{\frac{-j2\pi nm}{N}} \cdot 1 = X(m) \tag{3.28}$$

c. *Conjugate Symmetry:* $X(N - m) = X^*(m)$.

$$X(N - m) = T \sum_{n=0}^{N-1} x(n)e^{\frac{-j2\pi n(N-m)}{N}}$$

$$= T \sum_{n=0}^{N-1} x(n)e^{\frac{j2\pi nm}{N}} \cdot e^{\frac{-j2\pi nN}{N}}$$

$$= T \sum_{n=0}^{N-1} x(n)e^{\frac{j2\pi nm}{N}} \cdot e^{j2\pi n}$$

$$= T \sum_{n=0}^{N-1} x(n)e^{\frac{j2\pi nm}{N}} \cdot 1 = X^*(m). \tag{3.29}$$

d. *Time Shift Theorem:* $y(n) = x(n-k) \Leftrightarrow Y(m) = X(m) \cdot e^{\frac{-j2\pi mk}{N}}$.

$$Y(m) = T \sum_{n=0}^{N-1} x(n-k)e^{\frac{-j2\pi mn}{N}} = T \sum_{l=-k}^{N-1-k} x(l)e^{\frac{-j2\pi lm}{N}} \cdot e^{\frac{-j2\pi mk}{N}}$$

with the substitution $l = n - k$. Because of the periodicity occurring in the discretized domains, the complex exponentials are periodic and $x(l) = x(l + N)$.

$$= e^{\frac{-j2\pi mk}{N}} \cdot T \sum_{l=0}^{N-1} x(l)e^{\frac{-j2\pi lm}{N}} = e^{\frac{-j2\pi mk}{N}} \cdot X(m) \tag{3.30}$$

e. *Convolution in Frequency Theorem:* $x(n) \cdot y(n) \Rightarrow X(m)^{*}Y(m)$.

$$\text{DFT}[x(n) \cdot y(n)] = T \sum_{n=0}^{N-1} x(n) \cdot y(n) \cdot e^{\frac{-j2\pi nm}{N}}$$

Substituting for $y(n)$ its DFT with summing index $k$ and changing the order of summation gives

$$= T \sum_{n=0}^{N-1} x(n) \cdot \frac{1}{NT} \sum_{k=-N/2}^{N/2-1} Y(k)e^{\frac{j2\pi kn}{N}} \cdot e^{\frac{-j2\pi nm}{N}}$$

$$= \frac{1}{NT} \sum_{k=-N/2}^{N/2-1} Y(k) \cdot T \sum_{n=0}^{N-1} x(n) \cdot e^{\frac{-j2\pi n(m-k)}{N}}$$

$$= \frac{1}{NT} \sum_{k=-N/2}^{N/2-1} Y(k) \cdot X(m-k) = X(m)^{*}Y(m) \tag{3.31}$$

f. *Convolution in Time Theorem:* $X(m) \cdot Y(m) \Rightarrow x(n)^{*}y(n)$.

$$\text{DFT}[x(n)^{*}y(n)] = T \sum_{n=0}^{N-1} \left( T \sum_{k=0}^{N-1} x(k)y(n-k) \right) e^{\frac{-j2\pi nm}{N}}$$

Rearranging the summations and using the Time Shift Theorem yields

$$= T \sum_{k=0}^{N-1} x(k) \left( T \sum_{n=0}^{N-1} y(n-k)e^{\frac{-j2\pi nm}{N}} \right)$$

$$= T \sum_{k=0}^{N-1} x(k)Y(m)e^{\frac{-j2\pi km}{N}}$$

$$= Y(m) \cdot T \sum_{k=0}^{N-1} x(k)e^{\frac{-j2\pi km}{N}} = Y(m)X(m). \tag{3.32}$$

## 3.5 FOURIER ANALYSIS

The DFT is used computationally to perform Fourier analysis on signals. There are, in general, two types of algorithms. One is direct implementation of either equation 3.24 or equation 3.25 using either complex arithmetic or cosine-sine equivalent. These are rather slow computationally and clever algorithms using lookup tables were used to increase speed of calculation of trigonometric functions. Fortunately much faster algorithms were developed and have been routinely implemented during the last 20 years. The fast Fourier transform (FFT) algorithm as introduced by Cooley and Tukey has revolutionized the utility of the DFT. Interestingly, the earliest development and usage of fast DFT algorithms occurred around 1805 and is attributable to Gauss. Heideman et al. (1984) have written a history of the development of these algorithms. The radix 2 FFT, the most often used algorithm, is described in almost every signals and systems textbook, and its software implementation is available in most signal processing libraries. There are also mixed radix and other specialized fast algorithms for calculating the DFT (Blahut, 1985; Elliot and Rao, 1982). The usage of the DFT is so prevalent that special hardware signal processors that implement some of these algorithms are also available. The theory and description of these algorithms and their software is widespread and can easily be found in the IEEE special publications and textbooks on algorithms listed in the reference section.

Whatever algorithms are used, they provide the same information and the same care must be taken to implement them. The next several sections describe the details necessary for implementation. They are necessary because of the properties of the DFT and the possible errors that may occur because signals of finite duration are being analyzed. The book by Brigham (1988) explains comprehensively all the details of the properties and errors involved with the DFT. The next six subsections explain these properties through examples. Several of the many current applications are described in Section 3.7.

### 3.5.1   Frequency Range and Scaling

The DFT is uniquely defined for signals bandlimited by the folding frequency range, that is, $-f_N \leq f \leq f_N$, where $f_N = \frac{1}{2}f_s = 1/2T$. Refer again to Figure 3.9. In addition, because of conjugate symmetry (refer to 3.4.2c), the spectrum in the negative frequency region is just the complex conjugate of that in the positive region. Therefore, the frequency range with unique spectral values is only $0 \leq f \leq f_N$. Since the frequency spacing is $fd = 1/NT$, the range of frequency numbers is $0 \leq m \leq N/2$ with the specific frequency values being $f = mf_d$. These details will be exemplified on a small data set.

### EXAMPLE 3.2

In Figure 3.23 on page 86 is shown the electrogastrogram measured from an active muscle and its DFT. For this application $N = 256$ and $f_s = 1$ Hz. The scale above the horizontal axis is used. Thus the highest frequency representable is $f_N = 0.5$ Hz and the frequency spacing is $f_d = 1/NT = \frac{1}{256}$ Hz. Each point of the DFT is calculated at the actual frequencies, which are multiples of $\frac{1}{256}$ Hz. Notice that the tic marks are conveniently drawn at multiples of 0.1 Hz for ease of comprehension but do not correspond to points of the DFT. For 0.1 Hz the closest point in the DFT would have a frequency index of approximately $f/f_d = 0.1 \times 256 = 25.6$. That is, it lies in between $m = 25$ and $m = 26$. The first major frequency component occurs at a frequency of 0.054 Hz. It has a frequency index of $f/f_d = 0.54 \times 256 = 14$. If the sampling frequency had units of cycles/minute, the scale below the horizontal axis would be used.

---

### EXAMPLE 3.3

The signal intensity measurements starting at midnight of ionospheric reflections from the E-layer are listed in Table 3.1.

The signal is plotted in Figure 3.10. The DFT was calculated directly using equation 3.24. The real and imaginary parts of $X(m)$ are plotted in Figure 3.11 for frequency numbers in the range $0 \le m < N - 1$, $N = 12$ and listed in Appendix 3.1. The symmetry line for the conjugate symmetry is also shown in the figure. The magnitude and phase spectra were calculated with the polar coordinate transformation

$$|X(m)| = \sqrt{\text{Re}(X(m))^2 + \text{Im}(X(m))^2} \quad \text{and} \quad \theta = \arctan \frac{\text{Im}(X(m))}{\text{Re}(X(m))}$$

and are plotted in Figure 3.12 for the positive frequency range. The sampling interval used is $T = \frac{1}{24}$ day. Thus the frequencies on the frequency axis are $f = m(1/NT) = 2m$ cycles/day.

**TABLE 3.1** Ionospheric Reflections

| Time (hrs) | Intensity |
|:---:|:---:|
| 0 | 6 |
| 1 | 20 |
| 2 | 28 |
| 3 | 8 |
| 4 | 1 |
| 5 | −7 |
| 6 | 20 |
| 7 | 6 |
| 8 | 7 |
| 9 | −14 |
| 10 | −19 |
| 11 | −12 |

## 3.5.2 The Effect of Discretizing Frequency

One of the main concerns with the DFT is that it provides values of $X(f)$ only at a specific set of frequencies. What if an important value of the continuous spectrum exists at a frequency $f \ne mf_d$. This value would not be appreciated. This error is

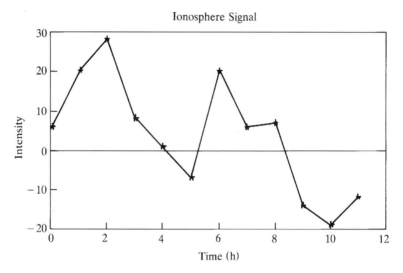

**FIGURE 3.10** The intensity of ionospheric reflections for a twelve-hour period.

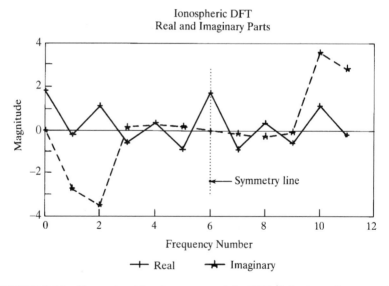

**FIGURE 3.11** The real and imaginary parts of the DFT of the ionospheric signal are plotted against frequency number. The dotted vertical line indicates the point of symmetry about the Nyquist frequency number.

called the *picket fence effect* in the older literature. The analogy is that when one looks at a scene through a picket fence, one only sees equally spaced parts of the total scene. Naturally, an obvious solution is to increase the number of signal points. This would not only increase $N$, and hence decrease the frequency

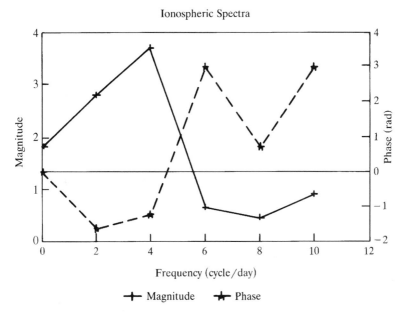

**FIGURE 3.12** The magnitude and phase spectra of the ionospheric signal.

spacing, but also the accuracy of the DFT. What if this were not possible? Fortunately, the frequency spacing can be reduced by *zero padding*. This is adding a string of 0-valued data points to the end of a signal. If $M$ zeros were added to increase the signal length to $LT$, the new frequency spacing is $f'_d = 1/LT = 1/(N + M)T$ and the DFT is calculated for frequency values $f = kf'_d = k(1/LT)$. The effect of the zero padding on the magnitude of the DFT can be appreciated directly by using the defining equation 3.24. The DFT for $L$ points becomes

$$X_L(k) = T \sum_{n=0}^{L-1} x(n)e^{\frac{-j2\pi kn}{L}} = T \sum_{n=0}^{N-1} x(n)e^{\frac{-j2\pi kn}{L}}; \quad 0 \le k \le \frac{L}{2} \qquad (3.33)$$

For the same frequency in the unpadded and zero padded calculations, $f = m(1/NT) = k(1/LT)$, and the exponents have the same value. Thus $X_L(k) = X(m)$ and there is no change in accuracy. Essentially this is an *interpolation* procedure. The effect of zero padding on the calculations is best shown by an example.

## EXAMPLE 3.4

Calculate the spectra again of the ionospheric signal with the number of signal points increased to 24 by padding with 12 zeros. The resulting spectra are shown in Figure 3.13. The frequency range has not changed since the sampling interval has not changed. Since the number of signal points has doubled, the number of

Ionospheric Padded Spectra

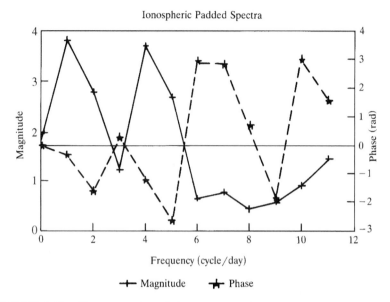

**FIGURE 3.13**   The magnitude and phase spectra of the ionospheric signal after zero padding.

spectral points has doubled. Notice that now the spectrum has a large magnitude at 1 cycle per day. This was not evident with $f_d = 2$. However, the values of the magnitude and phase spectra at the frequencies used in the previous example have not changed.

An important additional use for zero padding has developed because of the implementation of fast algorithms, which often restrict the number of signal points to specific sets of values. For instance, the radix 2 FFT algorithm must have $N = 2^l$, where $l$ is an integer. If the required number of signal points is not available, simply padding with zeros will suffice.

### 3.5.3   The Effect of Truncation

Another major source of error in calculating the DFT is a consequence of truncation. This is implicit because a finite sample is being made from a theoretically infinite duration signal. The truncation is represented mathematically by having the measured signal, $y(n)$, be equal to the actual signal, $x(n)$, multiplied by function called a *rectangular data window*, $d_R(n)$. That is,

$$y(n) = x(n) \cdot d_R(n)$$

where

$$d_R(n) = \begin{cases} 1, & \text{for } 0 \leq n \leq N - 1. \\ 0, & \text{for other } n \end{cases} \tag{3.34}$$

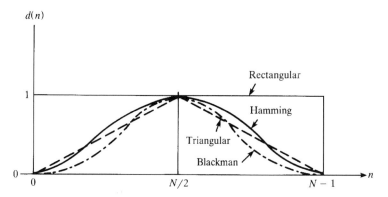

**FIGURE 3.14** Plots of several data windows.

There are several types of *data windows*, $d(n)$, that will be discussed, and in general

$$y(n) = x(n) \cdot d(n) \qquad (3.35)$$

Several data windows are plotted in Figure 3.14. It is known from the Convolution Theorem that the Fourier transforms of the three time sequences in equation 3.35 have a special relationship. However, it must be remembered that this still occurs while the frequency domain is continuous. The frequency domain convolution is rewritten for this situation and is

$$Y(f) = X(f) * D(f). \qquad (3.36)$$

where $D(f)$ is called a *spectral window*. The DTFT of the rectangular data window is well documented and is

$$D_R(f) = T \frac{\sin(\pi f NT)}{\sin(\pi f T)} e^{j\pi f(N-1)T}. \qquad (3.37)$$

Thus in general the DFT that is calculated, $Y(m)$, is not equal to $X(m)$ because of the convolution in equation 3.36. The error or discrepancy caused by the truncation effect is called *leakage error*. The convolution causes signal energy that exists at a given frequency in $X(m)$ to cause nonzero values of $Y(m)$ at frequencies in which no energy of $x(n)$ exists. The characteristics of the leakage error depend on the spectral window. An illustration of this is presented in the next example.

## EXAMPLE 3.5

Let $x(n)$ be a cosine sequence. Its transform is a sum of two delta functions and the transform pair is

$$\cos(2\pi f_0 nT) \Leftrightarrow 0.5 \, \delta(f - f_0) + 0.5 \, \delta(f + f_0). \qquad (3.38)$$

Figure 3.15 shows the signal and DFT for $y(n)$ and its transform when $P$ is an integral number of periods of $1/f_0$. As expected for $f_0 = \frac{1}{8}$, $N = 32$, and $T = 1$, the unit impulse exists at frequency number 4, $m = 4$. Notice what happens when $f_0 \neq m f_d$. When $f_0 = 1/9.143$, $Y(m)$ becomes much different and is shown in Figure 3.16. There are two discrepancies between this DFT and the theoretical one. First of all, there are nonzero magnitudes of $Y(m)$ at frequencies at which there is no signal component. This is the effect of the leakage error of the rectangular window. The second discrepancy is that the magnitude of $Y(m)$ surrounding the actual frequency is less than 0.5. This happens because the DFT

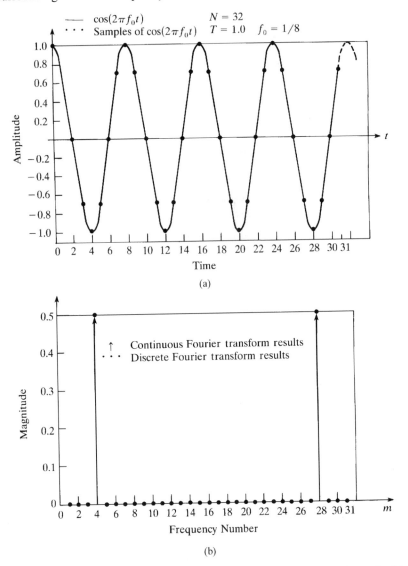

**FIGURE 3.15** (a) Cosine sequence, $N = 32$, with integral multiples of periods. (b) Its DFT vs frequency numbers. [Adapted from Brigham, fig. 9-6, with permission.]

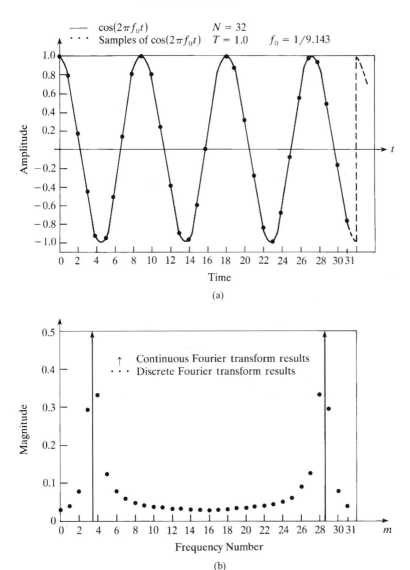

**FIGURE 3.16** (a) Cosine sequence, $N = 32$, with nonintegral multiples of periods. (b) Its DFT vs frequency numbers. [Adapted from Brigham, fig. 9-7, with permission.]

is not calculated at the exact frequency; that is, the spectral window is not evaluated at its peak. This discrepancy will always occur and is called *scalloping loss*. This is not an important factor in situations when a continuum of frequencies exist; that is, when $X(f)$ is continuous. One benefit is that without some leakage the occurrence of this frequency component would not be evident in the DFT.

Leakage error is unacceptable and must be minimized. The procedure for minimizing leakage can be understood by considering the shape of the magnitude of $D_R(f)$. It is an even function and only the positive frequency portion is plotted in Figure 3.17. There are many crossings of the frequency axis and the portions of $D_R(f)$ between zero crossings are called *lobes*. The lobe centered around the zero frequency is called the *main lobe*, and the other lobes are called *side lobes*. The convolution operation spreads the weighting effect of side lobes over the entire frequency range. One way to minimize their effect is to increase $N$. Figure 3.18 shows the rectangular spectral window for various values of $N$. As $N$ increases,

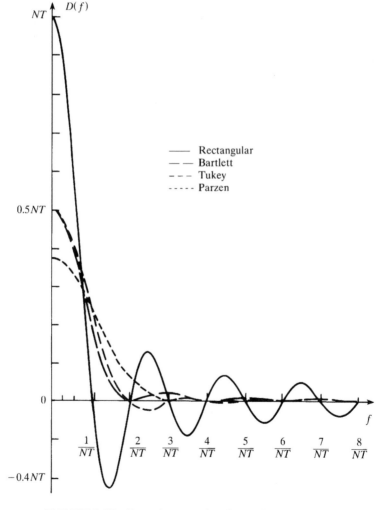

**FIGURE 3.17** Plots of magnitudes of several spectral windows.

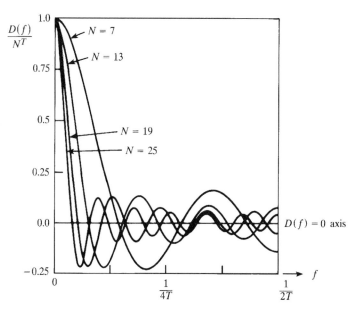

**FIGURE 3.18** The magnitude of the rectangular spectral window for various numbers of signal points. [Adapted from Roberts and Mullis, fig. 6.3.3, with permission.]

the width of the main lobe decreases and the magnitude of the side lobes decreases at any frequency value.

### 3.5.4 Windowing

Since reducing the spread of the leakage error by increasing the number of signal points is not usually possible, another method is necessary. An effective and much used method is to change the shape of the data window. The purpose is to produce spectral windows whose side lobes contain much less energy than the main lobe. Several spectral windows are also plotted in Figure 3.17, and their Fourier transforms are listed in Appendix 3.6. It is evident from studying the figure that the magnitudes of the side lobes of these other spectral windows fit this specification. The quantification of this property is the *side lobe level*, that is, the ratio of the magnitudes of the largest side lobe to the main lobe. These are also listed in the table. The use of the other windows should reduce the leakage error considerably.

### EXAMPLE 3.6

One of the commonly used windows is the Hanning or Tukey window; it is sometimes called the cosine bell window for an obvious reason. Its formula is

$$d_T(n) = 0.5(1 - \cos(2\pi n/(N-1))), \quad 0 \le n \le N-1. \tag{3.39}$$

Its side lobe level is $-31$ db, 2.8%, a factor of 10 less than the level of the rectangular window. The Hanning window has been applied to the cosine sequence of Example 3.5, and the resulting $y(n)$ sequence and its DFT are plotted in Figure 3.19. Notice that the windowing has removed the discontinuity between the periodic repetitions of the time waveform and has made $Y(m)$ much more closely resemble its theoretical counterpart. The magnitude spectrum is zero for many of its components where it should be, and the leakage is spread to only adjacent frequency components. Consider the peak magnitude of the spectrum and realize that it is less than that in Figure 3.16. This additional reduction in amplitude occurs because nonrectangular windowing, called *process loss*, removes signal energy.

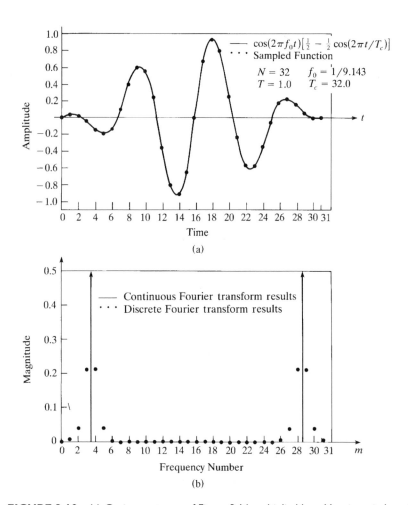

**FIGURE 3.19**   (a) Cosine sequence of Figure 3.16 multiplied by a Hanning window. (b) Its DFT with frequency numbers. [Adapted from Brigham, fig. 9-9, with permission.]

### 3.5.5  Resolution

Examine again the plots of the spectral windows in Figure 3.17. Particularly notice the lowest frequencies at which zero crossings of the frequency axis occur for each window. They are different. The *width of the main lobe* is defined as the frequency range of these zero crossings and are listed in Appendix 3.6. In the previous example, a compromise has occurred. The width of the main lobe of the Hanning window is twice as large as that of the rectangular window. The window width affects the ability to detect the existence of two sinusoids with closely spaced frequencies in a magnitude spectrum. A simple example can illustrate this point. Consider a sequence of two sinusoids with equal amplitudes and frequencies $f_1$ and $f_1 + \Delta f$,

$$x(n) = \cos(2\pi f_1 nT) + \cos(2\pi (f_1 + \Delta f)nT). \tag{3.40}$$

The DFT of a finite sample with $\Delta f$ much larger than the main lobe width is plotted in Figure 3.20a. The presence of the two sinusoids is evident. The shape of the DFT as $\Delta f$ becomes smaller is sketched Figure 3.20b. As one can envision, the spectral peaks are not as easily resolved as $\Delta f$ approaches 50% of the main lobe width. In fact, for $\Delta f < 1/NT$, the peaks are not distinguishable (Figure 3.20c). For sinusoids with unequal amplitudes, the frequency difference at which they are resolvable is more complex to define. The rule of thumb is that the *resolution* of a data window is one-half of its main lobe width.

---

#### EXAMPLE 3.7
Compute the DFT magnitude spectrum for the function

$$x(t) = e^{-0.063t} \sin(2\pi t) + e^{-0.126t} \sin(2.2\pi t)$$

with $t = nT$ when $T = 0.175$ and $N = 50$, 100, 200, and 400. The results are plotted in Figure 3.21. The frequency difference, $\Delta f$, is 0.1 Hz. For $N = 50$, the main lobe width is 0.228 Hz and the frequency resolution is approximately 0.114 Hz. Notice that the peaks of the two sinusoids are not present in the spectrum. If the number of signal points is increased, the main lobe width will be reduced and the frequency components resolvable. This can be seen for the DFT calculated when $N \geq 100$.

---

In general, there is a compromise when applying data windows. Study Appendix 3.6. It shows that when the side lobe level is reduced, the width of the main lobe is increased. Thus when reducing the leakage error, one also reduces the spectral resolution. This is usually not a problem when the spectra are continuous and do not have narrow peaks. Other windows have been designed for special applications such as resolving the frequencies of two simultaneously occurring sinusoids with close frequencies. The most comprehensive source of information is the review article by Harris (1978). It reviews the concepts and

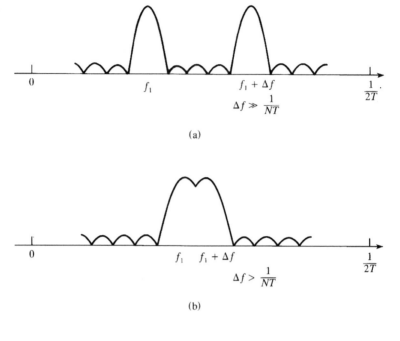

(a)

(b)

(c)

**FIGURE 3.20** Plots of the DTFT of the sequence of two cosine functions whose frequencies differ by $\Delta f$ Hz: (a) $\Delta f \gg 1/NT$, (b) $\Delta f > 1/NT$, (c) $\Delta f \leq 1/NT$.

purposes of windowing and summarizes the characteristics of over 20 window functions that have been developed.

### 3.5.6 Detrending

Another aspect similar to windowing is the removal of polynomial time trends and average value. Both of these time functions have energies around zero frequency that are spread into the low frequency ranges of $Y(m)$. This error is easily removed by fitting the time series with a low-order polynomial and then removing the values estimated by it. This process is called *detrending*. A good

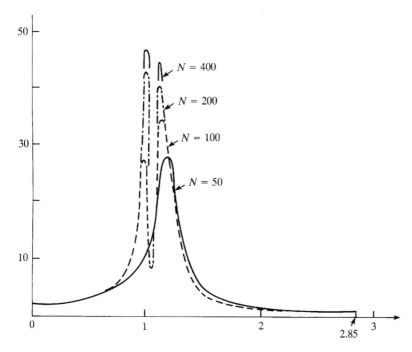

**FIGURE 3.21**    The magnitude spectrum of the DFT of the sampled function in Example 3.7 with $T = 0.175$ and $N = 50$, 100, 200, 400. [Adapted from Chen, fig. 4.18, with permission.]

example of a signal with a linear trend is the record of airline passenger data, which is listed in Appendix 3.3 and plotted in Figure 3.22a. A linear regression curve fitted for this data has the equation

$$f(t) = 198.3 + 3.01t. \tag{3.41}$$

where $f(t)$ is the number of thousands of passengers using air travel at an airport and $t$ is the number of months relative to January, 1953. The month-by-month subtraction of $f(t)$ from the measured data produces the detrended data, which is plotted in Figure 3.22b and has an average value of zero. Most situations in engineering require only removing the average value.

## 3.6  PROCEDURAL SUMMARY

In summary, the procedure for calculating the DFT and minimizing the potential sources of error is as follows.

1. Detrend the signal.
2. Multiply the signal with a suitable data window.

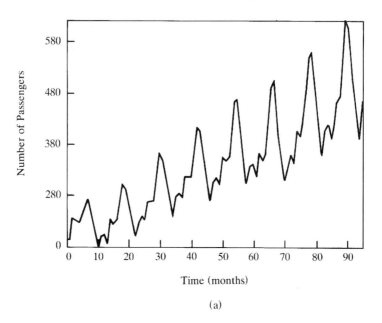

(a)

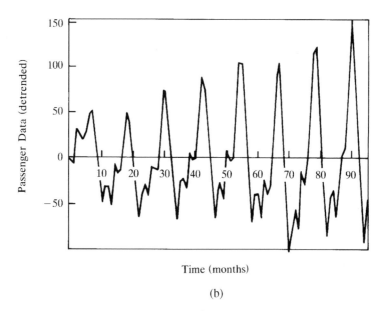

Time (months)

(b)

**FIGURE 3.22** Plot of number of airline passengers, in thousands, for years 1953 through 1960: (a) Measured data. (b) Detrended data.

3. Zero pad as necessary to reduce frequency spacing or to implement a fast algorithm.
4. Perform the DFT operation on the resulting signal.

Exercise of these concepts on actual signals will help the learning process. The sampled values of several signals are tabulated in the appendix of this chapter. Exercises requiring computer application are suggested in the exercise section. They have been chosen because the approximate periodicities can be appreciated through visual inspection of plots of the signals.

## 3.7 SELECTED APPLICATIONS

Several applications of the DFT are briefly summarized below to show its utility over a broad range of engineering fields.

- Food and waste matter are moved through the digestive system to nourish living systems. Muscles encircle the tube or tract of the digestive system and contract to move the food along the tract. The electrical signals from the muscles can be measured to study their contractile characteristics. In general, these signals are called electromyograms and in particular for the digestive system, they are called electrogastrograms. Figure 3.23a shows a 4.24-minute recording of the electrogastrogram from the ascending colon of a person. Visual examination of the electrogastrogram reveals that there are periods of greater and lessor activity, which indicates that the muscle contracts and relaxes approximately three times per minute. The signal was sampled once per second for 4.24 minutes and the DFT was calculated using a radix 2 FFT algorithm. The magnitude spectrum of the DFT in Figure 3.23b shows several well-defined peaks. The largest magnitude is 0.1 millivolts and occurs at a frequency of 3.1 cycles per minute. Other major cyclic components oscillate at 6.4 and 12.8 cycles per minute.
- The wear and incipient faults in gears of rotating machinery can be monitored with frequency analysis. The method is to measure the vibration in the gearbox and study its magnitude spectrum. Figure 3.24a shows a vibration signal and it's spectrum for a machine exhibiting uniform gear wear. The tooth-meshing frequency (TM) is the number of teeth meshing per unit time and is 60 per second. The harmonics of TM are always present. As the wear becomes nonuniform, the energy in the harmonics increase with respect to the fundamental frequency. Figure 3.24b shows such a situation. This indicates that a fault will occur soon and some corrective steps must be taken.
- In the production of high-quality audio systems, the characteristics of loudspeakers is very important. It is desired that they have a flat frequency

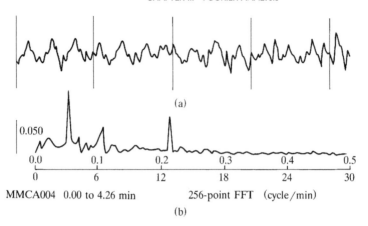

MMCA004  0.00 to 4.26 min          256-point FFT  (cycle/min)

(b)

**FIGURE 3.23**    (a) The electrogastrogram from the muscle of the ascending colon. Its duration is 4.24 minutes and the vertical lines demarcate one-second segments. (b) The 256-point FFT of the electrogastrogram. The signal was sampled at I Hz and the magnitude bar indicating a 0.05 millivolt amplitude is plotted on the left side. [Adapted from Smallwood et al., fig. 4, with permission.]

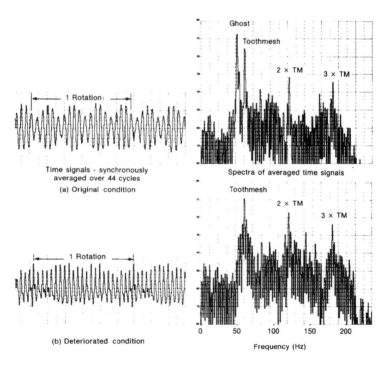

**FIGURE 3.24**    (a) The vibration signal and its magnitude spectrum for a rotating gearbox exhibiting normal wear. (b) The vibration signal and its magnitude spectrum for a rotating gearbox exhibiting an incipient fault. [Adapted from Angelo, fig. 8, with permission.]

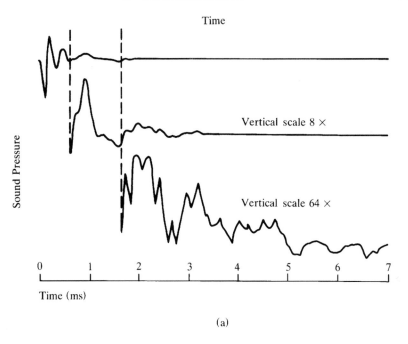

(a)

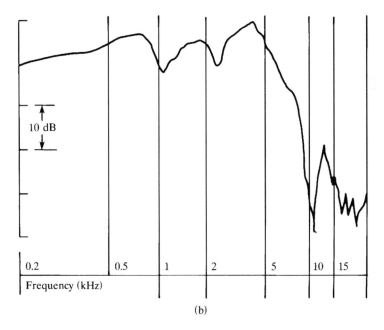

(b)

**FIGURE 3.25**    (a) Impulse response of a loudspeaker being tested. (b) The magnitude spectrum of its DFT. [Adapted from Blesser and Kates, figs. 2-25 & 2-26, with permission].

response so that all frequencies of sound are faithfully reproduced. If the response is not flat, then electronic equalizers can be designed to make the sound energy produced have a flat spectrum. The testing is done by exciting the loudspeaker with an impulse response and calculating the DFT of the resulting sound. Figures 3.25a and 3.25b show the impulse response and magnitude spectrum, respectively. Notice that the spectrum has dips at frequencies of 1.1 and 2.2 kHz that will need to be compensated.

## REFERENCES

### Discrete Fourier Transforms and Fast Algorithms

R. Blahut, *Fast Algorithms for Digital Signal Processing*, Addison-Wesley, Reading, MA, 1985.

E. Brigham, *The Fast Fourier Transform and Its Applications*, Prentice-Hall, Englewood Cliffs, NJ, 1988.

C. Burrus and T. Parks, *DFT/FFT and Convolution Algorithms*, Wiley, New York, 1985.

J. Cadzow, *Foundations of Digital Signal Processing and Data Analysis*, Macmillan, New York, 1987.

D. Childers and A. Durling, *Digital Filtering and Signal Processing*, West, St. Paul, 1975.

D. Elliot and K. Rao, *Fast Transforms—Algorithms, Analyses, Applications*, Academic Press, New York, 1982.

M. Heideman, D. Johnson, and C. Burrus, "Gauss and the History of the Fast Fourier Transform,", *IEEE Acous., Speech, and Sig. Proc. Mag.* **1**(4): 14–21 (1984).

*Programs for Digital Signal Processing*, IEEE Acoustics, Speech, and Signal Processing Society Special Committee, Ed., IEEE Press, New York, 1979.

L. Rabiner and B. Gold, *Theory and Application of Digital Signal Processing*, Prentice-Hall, Englewood Cliffs, NJ, 1975.

### Discrete Fourier Transforms

A. Ambardar, *Analog and Digital Signal Processing*, PWS, Boston, 1995.

C. Chen, *One-Dimensional Signal Processing*, Dekker, New York, 1979.

F. Harris, "On the Use of Windows for Harmonic Analysis with the Discrete Fourier Transform," *Proc. IEEE* **66**: 51–83 (1978).

R. Kuc, *Introduction to Digital Signal Processing*, McGraw-Hill, New York, 1988.

S. Marple, *Digital Spectral Analysis with Applications*, Prentice-Hall, Englewood Cliffs, NJ, 1987.

R. Roberts and C. Mullis, *Digital Signal Processing*, Addison-Wesley, Reading MA, 1987.

W. Stanley, G. Dougherty, and R. Dougherty, *Digital Signal Processing*, Reston, Reston, VA, 1984.

C. Williams, *Designing Digital Filters*, Prentice-Hall, Englewood Cliffs, NJ, 1986.

## Fourier Series and Transforms

N. Ahmed and K. Rao, *Orthogonal Transforms for Digital Signal Processing*, Springer-Verlag, New York, 1975.

P. Bloomfield, *Fourier Analysis of Time Series—An Introduction*, Wiley, New York, 1976.

R. Bracewell, *The Fourier Transform and its Applications*, McGraw-Hill, New York, 1978.

M. Cartwright, *Fourier Methods for Scientists and Engineers*, Ellis Horwood, London, 1990.

F. de Coulon, *Signal Theory and Processing*, Artech House, Dedham, MA, 1986.

H. Hsu, *Fourier Analysis*, Simon and Schuster, New York, 1970.

A. Papoulis, *The Fourier Integral and Its Applications*, McGraw-Hill, New York, 1962.

A. Papoulis, *Signal Analysis*, McGraw-Hill, New York, 1977.

R. Ziemer, W. Tranter, and D. Fannin, *Signals and Systems—Continuous and Discrete*, Macmillan, New York, 1983.

## Applications

M. Angelo, "Vibration Monitoring of Machines," *Bruel & Kjaer Techn. Rev.* No. 1: 1–36 (1987).

G. Bodenstein and H. Praetorius, "Feature Extraction from the Electroencephalogram by Adaptive Segmentation," *Proc. IEEE* **65**: 642–652 (1977).

B. Blesser and J. Kates, "Digital Processing in Audio Signals," in *Applications of Digital Signal Processing*, A. Oppenheim, Ed., Prentice-Hall, Englewood Cliffs, NJ, 1978.

T. Licht, H. Andersen, and H. Jensen, "Recent Developments in Accelerometer Design," *Bruel & Kjaer Techn. Rev.* No. 2: 1–22 (1987).

R. Smallwood, D. Linkins, H. Kwak, and C. Stoddard, "Use of Autoregressive—Modeling Techniques for the Analysis of Colonic Myoelectric Activity," *Med. & Biol. Eng. & Comput.* **18**: 591–600 (1980).

E. Whittaker and G. Robinson, *The Calculus of Observations*, Dover, New York, 1967.

M. Zarrugh and C. Radcliffe, "Computer Generation of Human Gait Kinematics," *J. Biomech.* **12**: 99–111 (1979).

## EXERCISES

**3.1** Prove Parseval's Theorem, such as equation A3.6, for orthogonal function sets.

**3.2** Show that the squared error in a truncated Fourier series using $N + 1$ terms is $2 \sum\limits_{m=N+1}^{\infty} \lambda_m z_m^2$.

**3.3** Verify equation 3.2, $\lambda_m = P$.

**3.4** Find the period for the following functions:
a. $\cos(kt)$    b. $\cos(2\pi t)$    c. $\sin(2\pi t/k)$    d. $\sin(t) + \sin(t/3) + \sin(t/5)$
e. $|\sin(\omega_0 t)|$

**3.5** Consider the periodic function defined by

$$f(t) \begin{cases} 1, & -3 < t < 0 \\ 0, & 0 < t < 3 \end{cases}$$

a. Find the trigonometric and complex Fourier series.
b. Plot the magnitude and phase spectra for the first 10 harmonics.

**3.6** Consider the periodic function defined by

$$f(t) = t^2, \quad -1.5 < t < 1.5$$

a. Find the trigonometric and complex Fourier series.
b. Plot the magnitude and phase spectra for the first 10 harmonics.

**3.7** Consider the periodic function defined by

$$f(t) = e^t, \quad 0 < t < 2$$

a. Find the trigonometric and complex Fourier series.
b. Plot the magnitude and phase spectra for the first 10 harmonics.

**3.8** Reexamine the time-shifted sawtooth waveform in Figure 3.4.

a. Form the complex Fourier series.

b. How do the coefficient magnitudes and phase angle compare to the ones from the unshifted version?

c. Plot the line spectra for $-10 < m < 10$

d. Calculate the time shifts, $\tau_m$, and plot a time shift line spectrum.

**3.9**  Show that the average power of the Fourier series can be expressed in terms of the trigonometric series coefficients as

$$C_0^2 + \sum_{m=1}^{\infty} \frac{1}{2} C_m^2$$

**3.10** It is obvious that changing the magnitude of the harmonic coefficients will change the shape of a periodic function. Changes in only the phase angle can do that also without changing the average power. Build a function $g(t)$ with two cosine waveforms of amplitude 1.0, phase angles equal to 0, and fundamental frequency of 1.0 Hz; that is,

$$g(t) = C_1 \cos(\omega_0 t) + C_2 \cos(2\omega_0 t)$$

a.  Plot $g(t)$.
b.  Change $C_1$ to 2.0 and replot it as $g_1(t)$.
c.  Take $g(t)$ and make $\theta_2 = 90°$. Replot it as $g_2(t)$. Notice it is different from $g(t)$ also.
d.  Take $g(t)$ and make $\tau_1 = \tau_2 = 0.25$. Notice how the shape has remained the same and $g(t)$ is shifted in time.

**3.11** The function $\exp(5t)$ is defined over the epoch $-0.1 < t < 0.1$ second and is periodic.
a.  Sketch three cycles of this waveform.
b.  Does this waveform satisfy the convergence conditions listed in Section 3.2.2?
c.  Determine the average value, $z_0$, and general term, $z_m$, for the complex Fourier series.
d.  Write explicitly the approximate series for $n$ ranging from $-2$ through $+2$.
e.  Sketch the magnitude and phase spectra for $m = 0$, 1, and 2.

**3.12** In Figure 3.23 the DFT has peaks at frequencies of approximately 6.7 and 13.1 cycles/minute. What frequency numbers correspond to these frequencies?

**3.13** Draw and label only the frequency axes over the valid frequency range for the following conditions. Show the first five or ten major tic marks and the one at the folding frequency.
a.  $T = 1$ s;  $N = 10, 20, 40$
b.  $T = 0.1$ s;  $N = 10, 20, 40$
c.  $T = 1$ ms;  $N = 10, 20, 40$.

**3.14** Using the sequence $x(n) = [1, 1, 0, 0, 0, 0, \ldots]$ of length $N$ with $T = 1$.
a.  Derive the DTFT.
b.  Calculate the DFT for $N = 4$ and $N = 8$.

c. Plot the three magnitude spectra found in parts (a) and (b) and compare them. What is the effect of the number of data points?

**3.15** Let $f(t) = e^{-t} \cdot U(t)$.

   a. Find or derive its Fourier transform, $X(f)$.
   b. Compute the DFT of $f(t)$ with $T = 0.25$ for $N = 4$ and 8.
   c. How do the two $X(m)$ compare with $X(f)$?
   d. Use zero padding to create $f_p(t)$, $N = 8$, from $f(t)$, $N = 4$, and compute the DFT.
   e. How do the results of part (d) compare with the other estimates of $X(f)$?

**3.16** Verification of Example 3.3.

   a. In Figure 3.11, examine the conjugate symmetry of the real and imaginary parts of the DFT.
   b. What would be the frequency axis in Figure 3.12 if the sampling interval were 1 hour?
   c. In Figure 3.12, verify the values of the magnitude and phase spectra for $f = 0$, 2, and 6 cycles/day.

**3.17** Using the spectra in Example 3.3, perform the inverse transform using equation 3.24 for $0 \leq n \leq 24$. Is the signal periodically repeated? (Remember that $N = 12$.)

**3.18** Repeat Exercise 3.15b with $N = 64$. Plot the spectra. How do they compare with $X(f)$?

**3.19** Consider the signals in Exercise 3.15.

   a. Detrend the sampled signals described in part (b) and plot them.
   b. Repeat part (b) on the detrended signals. What values have changed?

**3.20** Derive the spectral window, equation 3.37, of the rectangular data window, equation 3.34.

**3.21** Consider the cosine sequence in Example 3.5.

   a. Prove that the DTFT of the truncated sequence is

$$Y(f) = 0.5D_R(f - f_0) + 0.5D_R(f + f_0)$$

   where

$$D_R(f) = T\frac{\sin(\pi f NT)}{\sin(\pi f T)}e^{-j\pi f(N-1)T}$$

   b. Verify the magnitude values at frequency numbers 3, 4, and 10 when $f_0 = 1/9.143$.

**3.22** Repeat Exercise 3.18 after applying a Hanning window. Have the calculated spectra become more like the theoretical spectra?

**3.23** Calculate the DFT of the ionospheric signal (Examples 3.3 and 3.4) after detrending, windowing, and padding with 12 zeros. How does it compare with the spectrum in Figure 3.13?

## Computer Exercises

**3.24** There is rectangular waveform $x = [1\ 1\ 1\ 1\ 0\ 0\ 0\ 0]$. Calculate its DFT manually. Initially assume $T = 1$.
   a. Now calculate its DFT with some computerized method. How do the values compare to those calculated manually?
   b. Which values of the DFT are equal, complex conjugates? Use the frequency number to identify the spectral values.
   c. Calculate the magnitude and phase spectra of this DFT. For each spectrum, which values are equal? Where are the points of symmetry? Which parts are redundant?
   d. Plot the magnitude and phase spectra and label both axes.
   e. Now let $T = 0.5$ and redo part (d).
   f. The inverse discrete Fourier transform, IDFT, can reproduce a waveform from its DFT. Apply the IDFT to the DFT of part (b). Does it reproduce $x(n)$?

**3.25** Verification of Examples 3.3 and 3.4.
   a. Reproduce Figures 3.11, 3.12, and 3.13 with the correct scaling.
   b. Now calculate the magnitude spectrum with $T = \frac{1}{12}$ day (use every other point, $N = 6$). How does it compare with Figure 3.12? What are the major differences?

**3.26** Explore the effect of the sampling interval and the repetition of the spectra.
   a. Find the CTFT of $x(t) = \exp(-2t)U(t)$ and plot the magnitude spectrum from 0 to 4 Hz.
   b. Let $T = 0.125$ s and $N = 100$. Create $x(n)$ and plot it; label the axes.
   c. Calculate the DFT up to the frequency number $N - 1$. Plot the magnitude spectrum. Where is the point of symmetry?
   d. Repeat part (c) for $T = 0.5$ s. Note that the sampling frequency has been changed. What has changed (notice the frequency axis)?
   e. For the frequency corresponding to $m = N/2$, how accurate is $X(m)$? [*Hint:* Compare to $X(f)$.] This is the effect of aliasing or undersampling. What would happen if $T = 1.0$ s?

The following series of exercises illustrates the leakage error that can occur in calculating the DFT. Now let us examine the necessities of computing the DFT accurately. All axes must be labeled correctly.

**3.27** a. Derive the CTFT of the function defined as

$$x(t) = \begin{cases} 6t, & 0 \le t \le 1\,\text{s} \\ 0, & \text{elsewhere} \end{cases}$$

It is $X(f) = 6/(2\pi f)^2[(1 + j2\pi f)\exp(-j2\pi f) - 1]$. (Note: Refer to a book of tables.)

   b. Plot the magnitude spectrum. Let $f$ range from 0 to 20 Hz and make the frequency increment be at least 0.5 Hz. (Be careful, $X(0)$ is infinite.)

**3.28** a. Create the signal

$$x(t) = \begin{cases} 6t, & 0 \le t \le 1\,\text{s} \\ 0, & 1 \le t \le 2\,\text{s} \end{cases}$$

in discrete time. Let $T = 0.01$ s or less. Plot it.

   b. Calculate the DFT $X(m)$, of the signal formed in part (a) and plot its magnitude spectrum. How does it compare to $X(f)$ from Exercise 3.27?

**3.29** a. Create the signal

$$x(t) = \begin{cases} 6t, & 0 \le t \le 1\,\text{s} \\ 0, & 1 \le t \le 2\,\text{s} \\ 6(t-2), & 2 \le t \le 2.5\,\text{s} \end{cases}$$

in discrete time. Let $T = 0.01$ s or less. Plot it.

   b. Calculate the DFT of the signal formed in part (a) and plot its magnitude spectrum. How does it compare to $X(m)$ from Exercise 3.28? You should see the effect of leakage.

**3.30** a. Consider the window functions *hanning* and *hamming*, which are used to create data windows. Use the signal created in Exercise 3.29a and apply a Hanning or Hamming window to it. Plot the window that is chosen.

   b. Calculate the DFT of the windowed signal formed in part (a) and plot its magnitude spectrum. How does it compare to the $X(m)$ from Exercises 3.28 and 3.29? (It should have lessened the leakage considerably. Focus on the frequency range 0 to 20 Hz.)

**3.31** Analyze the star light intensity signal in Appendix 3.2.
   a. Plot at least 25% of the signal.
   b. What are the approximate periodicities present?
   c. Detrend and window the entire signal; plot the result.
   d. Calculate the DFT using an algorithm of your choice.
   e. Plot the magnitude and phase spectra.
   f. What major frequency components or ranges are present? Do they agree with your initial estimation?
   g. Choose another data window and repeat steps (c) to (f). Are there any differences in the two spectra?

**3.32** Analyze the passenger travel signal in Appendix 3.3.
   a. Plot the entire signal.
   b. What are the approximate harmonics present? At what time of the year do peaks occur?
   c. Detrend and window the entire signal; plot the result. How does the plot compare to Figure 3.22b?
   d. Calculate the DFT using an algorithm of your choice.
   e. Plot the magnitude and phase spectra.
   f. What major frequency components are present? What phase angles and time lags do these components have? Are these results consistent with what is expected from part (b)?

**3.33** Explore losses using the passenger travel signal in Appendix 3.3.
   a. Detrend the entire signal and pad with zeros to quadruple the signal duration.
   b. Calculate and plot the magnitude spectrum of the zero padded signal in part (a) and assume it is the "ideal" one.
   c. Calculate and plot the magnitude spectrum of the original detrended signal. Compare it with that of part (b). Is there any scalloping loss? Why or why not?
   d. Take the first 90 points of the original signal and calculate and plot the magnitude spectrum. Compare it with that of part (b). Is there any scalloping loss? Why or why not?
   e. Detrend and window the original signal. Calculate and plot the magnitude spectrum and compare it to that of part (b). Is there any process loss? Why or why not?

## APPENDIX 3.1   DFT OF IONOSPHERE DATA

Real and Imaginary Parts of Ionosphere Reflections

| Frequency Cycles/day | Re($X(f)$) | Im($X(f)$) |
|---|---|---|
| 0.0 | 1.83 | 0.00 |
| 2.0 | −0.24 | −2.79 |
| 4.0 | 1.23 | −3.54 |
| 6.0 | −0.63 | 0.13 |
| 8.0 | 0.33 | 0.29 |
| 10.0 | −0.89 | 0.17 |
| 12.0 | 1.75 | 0.00 |
| 14.0 | −0.89 | −0.17 |
| 16.0 | 0.33 | −0.29 |
| 18.0 | −0.63 | −0.13 |
| 20.0 | 1.23 | 3.54 |
| 22.0 | −0.24 | 2.79 |

## APPENDIX 3.2   DATA FOR THE BRIGHTNESS OF A VARIABLE STAR AT MIDNIGHT

Brightness of a variable star at midnight on 600 successive days, time across rows

| | | | | | | | | | | | | | | | | | | | |
|---|---|---|---|---|---|---|---|---|---|---|---|---|---|---|---|---|---|---|---|
| 25 | 28 | 31 | 32 | 33 | 33 | 32 | 31 | 28 | 25 | 22 | 18 | 14 | 10 | 7 | 4 | 2 | 0 | 0 | 0 |
| 2 | 4 | 8 | 11 | 15 | 19 | 23 | 26 | 29 | 32 | 33 | 34 | 33 | 32 | 30 | 27 | 24 | 20 | 17 | 13 |
| 10 | 7 | 5 | 3 | 3 | 3 | 4 | 5 | 7 | 10 | 13 | 16 | 19 | 22 | 24 | 26 | 27 | 28 | 29 | 28 |
| 27 | 25 | 24 | 21 | 19 | 17 | 15 | 13 | 12 | 11 | 11 | 10 | 10 | 11 | 12 | 12 | 13 | 14 | 15 | 16 |
| 17 | 18 | 19 | 19 | 19 | 19 | 20 | 20 | 20 | 20 | 20 | 20 | 20 | 20 | 21 | 20 | 20 | 20 | 20 | 19 |
| 18 | 17 | 16 | 15 | 13 | 12 | 11 | 10 | 9 | 9 | 10 | 10 | 11 | 12 | 14 | 16 | 19 | 21 | 24 | 25 |
| 27 | 28 | 29 | 29 | 28 | 27 | 25 | 23 | 20 | 17 | 14 | 11 | 8 | 5 | 4 | 2 | 2 | 2 | 4 | 6 |
| 9 | 12 | 16 | 19 | 23 | 27 | 30 | 32 | 33 | 34 | 33 | 32 | 30 | 27 | 24 | 20 | 16 | 12 | 9 | 5 |
| 3 | 1 | 0 | 0 | 1 | 3 | 6 | 9 | 13 | 17 | 21 | 24 | 27 | 30 | 32 | 33 | 33 | 32 | 31 | 28 |
| 25 | 22 | 19 | 15 | 12 | 9 | 7 | 5 | 4 | 4 | 5 | 5 | 7 | 9 | 12 | 14 | 17 | 20 | 22 | 24 |
| 25 | 26 | 27 | 27 | 26 | 25 | 24 | 22 | 20 | 18 | 17 | 15 | 14 | 13 | 13 | 12 | 12 | 12 | 13 | 13 |
| 13 | 14 | 14 | 15 | 15 | 16 | 17 | 17 | 17 | 17 | 18 | 18 | 19 | 19 | 20 | 20 | 21 | 21 | 22 | 22 |
| 22 | 22 | 22 | 21 | 20 | 19 | 17 | 16 | 14 | 12 | 11 | 9 | 8 | 7 | 8 | 8 | 9 | 10 | 12 | 14 |
| 17 | 20 | 23 | 25 | 27 | 29 | 30 | 30 | 30 | 29 | 27 | 25 | 22 | 19 | 16 | 12 | 9 | 6 | 4 | 2 |
| 1 | 1 | 2 | 4 | 7 | 10 | 14 | 17 | 21 | 25 | 29 | 31 | 33 | 34 | 34 | 33 | 31 | 29 | 26 | 22 |
| 19 | 14 | 11 | 7 | 4 | 2 | 1 | 0 | 1 | 2 | 5 | 7 | 11 | 15 | 19 | 22 | 25 | 28 | 30 | 32 |
| 32 | 32 | 31 | 29 | 26 | 23 | 21 | 17 | 14 | 11 | 9 | 7 | 6 | 5 | 6 | 6 | 7 | 9 | 11 | 13 |
| 15 | 18 | 20 | 22 | 23 | 24 | 25 | 25 | 25 | 24 | 24 | 22 | 21 | 19 | 18 | 17 | 16 | 15 | 15 | 14 |
| 14 | 14 | 14 | 14 | 14 | 14 | 14 | 14 | 14 | 15 | 15 | 15 | 15 | 16 | 16 | 17 | 18 | 19 | 20 | |
| 21 | 22 | 23 | 23 | 24 | 24 | 24 | 23 | 22 | 21 | 19 | 17 | 15 | 13 | 11 | 9 | 7 | 6 | 6 | 6 |
| 7 | 8 | 10 | 12 | 15 | 18 | 22 | 24 | 27 | 29 | 31 | 31 | 31 | 31 | 29 | 27 | 24 | 21 | 18 | 14 |
| 10 | 7 | 5 | 2 | 1 | 0 | 1 | 2 | 5 | 8 | 12 | 15 | 19 | 23 | 27 | 30 | 32 | 34 | 34 | 34 |
| 32 | 30 | 28 | 24 | 20 | 16 | 13 | 9 | 6 | 3 | 2 | 1 | 1 | 2 | 4 | 6 | 9 | 13 | 17 | 20 |
| 23 | 26 | 28 | 30 | 31 | 31 | 31 | 29 | 27 | 24 | 22 | 19 | 16 | 13 | 11 | 9 | 8 | 7 | 7 | 7 |
| 8 | 9 | 11 | 12 | 14 | 16 | 18 | 20 | 21 | 22 | 23 | 23 | 23 | 23 | 23 | 22 | 21 | 20 | 19 | 18 |
| 18 | 17 | 17 | 16 | 16 | 16 | 16 | 15 | 15 | 15 | 14 | 14 | 13 | 13 | 13 | 13 | 13 | 13 | 14 | 14 |
| 15 | 16 | 18 | 19 | 21 | 22 | 24 | 24 | 25 | 26 | 26 | 25 | 24 | 23 | 21 | 19 | 16 | 14 | 12 | 9 |
| 7 | 5 | 5 | 4 | 5 | 6 | 8 | 10 | 13 | 16 | 20 | 23 | 26 | 29 | 31 | 32 | 32 | 32 | 31 | 29 |
| 26 | 23 | 20 | 16 | 12 | 8 | 6 | 3 | 1 | 0 | 0 | 1 | 3 | 6 | 10 | 13 | 17 | 21 | 25 | 28 |
| 31 | 33 | 34 | 34 | 33 | 31 | 29 | 26 | 22 | 18 | 15 | 11 | 8 | 5 | 3 | 2 | 2 | 2 | 4 | 5 |

## APPENDIX 3.3   INTERNATIONAL AIRLINE PASSENGER DATA

Monthly total (in 1000s) of international airline passengers

|       | 1953 | 1954 | 1955 | 1956 | 1957 | 1958 | 1959 | 1960 |
|-------|------|------|------|------|------|------|------|------|
| Jan.  | 196  | 204  | 242  | 284  | 315  | 340  | 360  | 417  |
| Feb.  | 196  | 188  | 233  | 277  | 301  | 318  | 342  | 391  |
| Mar.  | 236  | 235  | 267  | 317  | 356  | 362  | 406  | 419  |
| Apr.  | 235  | 227  | 269  | 313  | 348  | 348  | 396  | 461  |
| May   | 229  | 234  | 270  | 318  | 355  | 363  | 420  | 472  |
| June  | 243  | 264  | 315  | 374  | 422  | 435  | 472  | 535  |
| July  | 264  | 302  | 364  | 413  | 465  | 491  | 548  | 622  |
| Aug.  | 272  | 293  | 347  | 405  | 467  | 505  | 559  | 606  |
| Sept. | 237  | 259  | 312  | 355  | 404  | 404  | 463  | 508  |
| Oct.  | 211  | 229  | 274  | 306  | 347  | 359  | 407  | 461  |
| Nov.  | 180  | 203  | 237  | 271  | 305  | 310  | 362  | 390  |
| Dec.  | 201  | 229  | 278  | 306  | 336  | 337  | 405  | 432  |

## APPENDIX 3.4   REVIEW OF PROPERTIES OF ORTHOGONAL FUNCTIONS

There are several function sets that can be used to describe deterministic signals in continuous time over a finite time range with duration $P$. They can be described with a series of weighted functions such as

$$f(t) = A_0\Phi_0(t) + A_1\Phi_1(t) + \cdots + A_m\Phi_m(t) + \cdots$$

$$= \sum_{m=0}^{\infty} A_m\Phi_m(t) \tag{A3.1}$$

The members of these sets have an orthogonality property similar to the orthogonal polynomials. The Fourier series is composed of one of these sets. The function set $\{\Phi_m(t)\}$ is orthogonal if

$$\int_0^P \Phi_m(t)\Phi_n(t)\,dt = \begin{cases} \lambda_m, & m = n \\ 0, & m \neq n \end{cases} \tag{A3.2}$$

The usual range for $m$ is $0 \leq m \leq \infty$; sometimes the range is $-\infty \leq m \leq \infty$ and equation A3.1 is defined over this extended range. The coefficients, $A_m$, are found by using the least squares error principle.

Define an approximation of the function $f(t)$ using a finite number of terms, $N + 1$, as

$$\hat{f}(t) = A_0\Phi_0(t) + A_1\Phi_1(t) + \cdots + A_m\Phi_M(t)$$

$$= \sum_{m=0}^{M} A_m\Phi_m(t) \tag{A3.3}$$

The squared error for the approximation is

$$E_M = \int_0^P (f(t) - \hat{f}(t))^2 \, dt = \int_0^P \left( f(t) - \sum_{m=0}^{M} A_m \Phi_m(t) \right)^2 dt \tag{A3.4}$$

Performing the indicated operations gives

$$E_M = \int_0^P \left( f(t)^2 - 2f(t) \sum_{m=0}^{M} A_m \Phi_m(t) + \left( \sum_{m=0}^{M} A_m \Phi_m(t) \right)^2 \right) dt$$

and taking the partial derivative with respect to a particular $A_m$ yields

$$\frac{\partial E_M}{\partial A_m} = -2 \int_0^P f(t) \Phi_m(t) \, dt + 2 \int_0^P \left( \sum_{n=0}^{M} A_n \Phi_n(t) \right) \Phi_m(t) \, dt = 0$$

Rearranging the above equation and using the orthogonality conditions produces

$$\int_0^P f(t) \Phi_m(t) \, dt = \sum_{n=0}^{M} \int_0^P A_n \Phi_n(t) \Phi_m(t) \, dt = A_m \lambda_m$$

or

$$A_m = \frac{1}{\lambda_m} \int_0^P f(t) \Phi_m(t) \, dt, \quad \text{for } 0 \le m \le \infty \tag{A3.5}$$

Thus as with the orthogonal polynomials there is a simple method to calculate the coefficients in a series. The range of $A_m$ would be $-\infty \le m \le \infty$ when the function set is also defined over that range. Parseval's Theorem states that the total energy for a waveform is

$$E_{\text{tot}} = \sum_{m=-\infty}^{\infty} A_m^2 \lambda_m \tag{A3.6}$$

There are many orthogonal function sets that can be used to approximate finite energy signals. The continuous sets include the Legendre, Laguerre, and Hermite functions (de Coulon, 1986). More recently, discontinuous sets composed of square and rectangular waveforms—the Walsh and Radamaker functions—have become utilized (Ahmed and Rao, 1975).

## APPENDIX 3.5   THE FOURIER TRANSFORM

The Fourier transform is suitable for finding the frequency content of aperiodic signals. Conceptually and mathematically, the Fourier transform can be derived from the Fourier series relationships by a limiting operation. Consider the waveform from one cycle of a periodic process, let it stay unchanged and let the period approach an infinite duration. Figure A3.1 shows the sawtooth waveform after such a manipulation. Several other properties change concurrently. First, the signal becomes an energy signal. Second, the spectra become continuous. This is

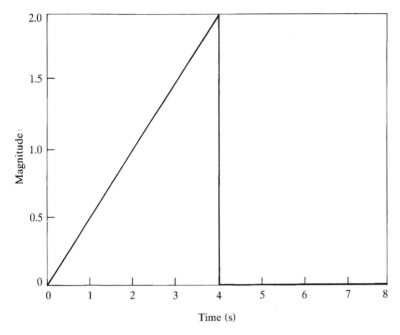

**FIGURE A3.1**   The sawtooth waveform as an aperiodic waveform.

understood by considering the frequency difference between harmonics of the line spectra during the limiting operation. This frequency difference is the fundamental frequency. So

$$\Delta f = f_0 = \lim_{P \to \infty} \frac{1}{P} = 0 \qquad (A3.7)$$

and the line spectra approach continuous spectra, and the summation operation of equation A3.4 becomes an integration operation. The resulting relationships for the time function become

$$x(t) = \int_{-\infty}^{\infty} X(f)e^{j2\pi ft}\,df = \frac{1}{2\pi}\int_{-\infty}^{\infty} X(\omega)e^{j\omega t}\,d\omega \qquad (A3.8)$$

For the frequency transformation in cycles or radians per unit time, the interval of integration becomes infinite or equation A3.14 becomes

$$X(f) = \int_{-\infty}^{\infty} x(t)e^{-j2\pi ft}\,dt \quad \text{or} \quad X(\omega) = \int_{-\infty}^{\infty} x(t)e^{-j\omega t}\,dt \qquad (A3.9)$$

The derivation of these Fourier transform pairs is presented in great detail in all of the cited references. The properties of the Fourier transform are the same as those of the complex coefficients of the Fourier series. Writing equation A3.9 by expanding the complex exponential it becomes

$$X(f) = \int_{-\infty}^{\infty} x(t)(\cos(2\pi ft) - j\sin(2\pi ft))\,dt \qquad \text{(A3.10)}$$

Thus $X(f)$ is a complex function. The fact that its real part is an even function and its imaginary part is an odd function can be easily proved for real signals by inserting $-f$ for every $f$ in the integrand. Thus

$$\text{Re}(X(\mathrm{f})) = \Re[X(f)] = \Re[X(-f)] \quad \text{and} \quad \text{Im}(X(\mathrm{f})) = \Im[X(f)] = -\Im[X(-f)] \qquad \text{(A3.11)}$$

In polar form this becomes

$$X(f) = |X(f)|\exp(\theta(f))$$

where

$$|X(f)| = \sqrt{(\text{Re}^2(X(f)) + \text{Im}^2(X(f)))} \quad \text{and} \quad \theta(f) = \arctan\frac{\text{Im}(X(f))}{\text{Re}(X(f))} \qquad \text{(A3.12)}$$

Because of the even and odd properties of $\text{Re}(X(\mathrm{f}))$ and $\text{Im}(X(\mathrm{f}))$ it is easily seen that $|X(f)|$ is an even function of frequency and that $\theta(f)$ is an odd function of frequency. The transforms for many waveforms can be found in the references. The spectra for the sawtooth waveform in Figure A3.1 are plotted in Figure A3.2.

In aperiodic signals energy is defined as

$$E_{\text{tot}} = \int_{-\infty}^{\infty} x^2(t)\,dt = \int_{-\infty}^{\infty} x(t)\left(\int_{-\infty}^{\infty} X(f)e^{j2\pi ft}\,df\right)dt$$

with one expression of the signal described in its transform definition. Now reorder the order of integrations and the energy expression becomes

$$E_{\text{tot}} = \int_{-\infty}^{\infty} X(f)\left(\int_{-\infty}^{\infty} x(t)e^{j2\pi ft}\,dt\right)df \qquad \text{(A3.13)}$$

The expression within the brackets is the Fourier transform of $x(t)$ with the substitution of $-f$ for $f$ or it is equivalent to $X(-f)$. Thus

$$E_{\text{tot}} = \int_{-\infty}^{\infty} X(f)X(-f)\,df = \int_{-\infty}^{\infty} X(f)X^*(f)\,df = \int_{-\infty}^{\infty} |X(f)|^2\,df \qquad \text{(A3.14)}$$

In other words, the energy equals the area under the squared magnitude spectrum. The function $|X(f)|^2$ is the *energy per unit frequency* or the *energy density spectrum*.

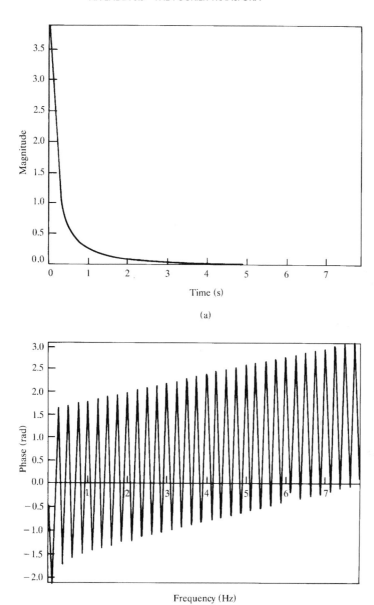

(a)

Frequency (Hz)

**FIGURE A3.2**   The magnitude (a) and phase (b) spectra for the sawtooth waveform in Figure A3.1.

# APPENDIX 3.6   DATA AND SPECTRAL WINDOWS

Data and Spectral Windows—Characteristics and Performance*

| Data Window | Spectral Window | Main Lobe Width | Side Lobe Level |
|---|---|---|---|
| | *RECTANGULAR* | | |
| $d_R(n) = 1, \quad 0 \leq n \leq N-1$ | $D_R(f) = T \dfrac{\sin(\pi f NT)}{\sin(\pi f T)} e^{-j\pi f T(N-1)}$ | $\dfrac{2}{NT}$ | 22.4% |
| | *TRIANGULAR–BARTLETT* | | |
| $d_B(n) = \begin{cases} \dfrac{2n}{N-1} & 0 \leq n \leq \dfrac{N-1}{2} \\ 2 - \dfrac{2n}{N-1} & \dfrac{N-1}{2} \leq n \leq N-1 \end{cases}$ | $D_B(f) = \dfrac{2T}{N} \left( \dfrac{\sin(\pi f NT/2)}{\sin(\pi f T)} \right)^2 e^{-j\pi f T(N-1)}$ | $\dfrac{4}{NT}$ | 4.5% |
| | *HANNING–TUKEY* | | |
| $d_T(n) = \frac{1}{2}(1 - \cos(2\pi n/(N-1)))$ | $D_T(f) = 0.5 D_R(f) + 0.25 D_R\left(f + \dfrac{1}{NT}\right)$ $+ 0.25 D_R\left(f - \dfrac{1}{NT}\right)$ | $\dfrac{4}{NT}$ | 2.8% |

*HAMMING*

$d_H(n) = 0.54 - 0.46\cos(2\pi n/(N-1))$

$D_H(f) = 0.54 D_R(f) + 0.23 D_R\left(f + \dfrac{1}{NT}\right)$
$\qquad\qquad + 0.23 D_R\left(f - \dfrac{1}{NT}\right)$

$\dfrac{4}{NT}$          0.9%

*BLACKMAN*

$d_{BL}(n) = 0.42 - 0.5\cos(2\pi n/(N-1))$
$\qquad\qquad + 0.08\cos(4\pi n/(N-1))$

$D_{BL}(f) = 0.42 D_R(f) + 0.25 D_R\left(f + \dfrac{1}{NT}\right)$
$\qquad\qquad + 0.25 D_R\left(f - \dfrac{1}{NT}\right)$
$\qquad\qquad + 0.04 D_R\left(f + \dfrac{2}{NT}\right)$
$\qquad\qquad + 0.04 D_R\left(f - \dfrac{2}{NT}\right)$

$\dfrac{6}{NT}$          0.1%

*PARZEN*

$d_P(n) = \begin{cases} 1 - 6\left(1 - 2\dfrac{n}{N}\right)^2 + 6\left(\left|1 - 2\dfrac{n}{N}\right|\right)^3, \\ \qquad\qquad\qquad N/4 \le n \le 3N/4 \\[2mm] 2\left(1 - \left|1 - 2\dfrac{n}{N}\right|\right)^3, \\ \qquad 0 \le n < N/4,\ 3N/4 < n \le N \end{cases}$

$D_P(f) = \dfrac{64T}{N^3}\left(\dfrac{3}{2}\dfrac{\sin^4(\pi f NT/4)}{\sin^4 \pi f T} - \dfrac{\sin^4(\pi f NT/4)}{\sin^2(\pi f T)}\right)$
$\qquad\qquad \cdot e^{-j\pi f T(N-1)}$

$\dfrac{6}{NT}$          0.22%

* All data windows have $d(n) = 0$ for $n < 0$ and $n \ge N$.

# IV

# PROBABILITY CONCEPTS AND SIGNAL CHARACTERISTICS

## 4.1  INTRODUCTION

**M**any situations occur that involve nondeterministic or random phenomena. Some common ones are the effects of wind gusts on the position of a television antenna, air turbulence on the bending of an airplane's wing, and rough terrain on the motion of a wheeled vehicle. Similarly, there are many single and time series measurements with random characteristics. Their behavior is not predictable with certainty because either it is too complex to model, knowledge is incomplete, or, as with some noise processes, it is essentially indeterminate. To analyze and understand these phenomena, a probabilistic approach must be used. The concepts and theory of probability and estimation provide a fundamental mathematical framework for the techniques of analyzing random signals.

It is assumed that the reader has had an introduction to probability and statistics. Hence this chapter will provide a brief summary of the relevant concepts of probability and random variables before introducing some concepts of estimation essential for signal analysis. If one desires a comprehensive treatment of probability and random variables from an engineering and scientific viewpoint, the books by Papoulis (1984) and Ochi (1990) are excellent; a less comprehensive but also a good source is the book by Stark and Woods (1986). If one desires a review, any introductory textbook is suitable. A few are listed in the reference section (Maisel, 1971; O'Flynn, 1982; Peebles, 1987). If one is interested in a comprehensive treatment of probability and statistics from an engineering and scientific viewpoint, refer to the books written by Milton and Arnold (1990) and Vardeman (1994).

## 4.2   INTRODUCTION TO RANDOM VARIABLES

Signals and data that possess random characteristics arise in a broad array of technical, scientific, and economic fields. Consider the record of voltage deviations in a turboalternator, shown in Figure 4.1. The average value over time seems constant and there are many undulations. Notice that these undulations do not have a consistent period. The frequency analysis methodologies treated in Chapter 3 would not be applicable for analyzing the frequency content of this signal. Noise in electrical circuits and measurements is another common random signal. In the field of reliability and quality control, the concern is with the characteristics of the output of a production line or a device. Figure 4.2 represents an example showing the size of paint droplets produced by a new design of a spray painting nozzle. In this case the concern is the distribution of droplet sizes and a histogram description is appropriate.

These examples concern *continuous random variables*, but there are many applications that concern *discrete random variables*. These are mostly counting processes. For instance, Figure 4.3 shows the daily census over time of the number of patients in a hospital. Other counting processes include numbers of

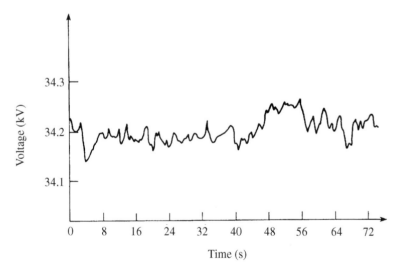

**FIGURE 4.1**   Voltage deviations from the stator terminals of a 50 Megawatt turboalternator. [Adapted from Jenkins and Watts, fig. 1.1, with permission.]

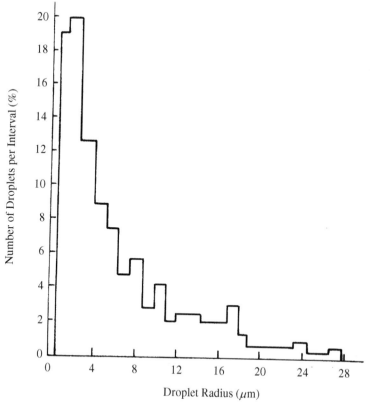

**FIGURE 4.2** Paint droplets produced by a spray gun, percentage of droplets with different radii in microns.

particles produced during radioactive decay, numbers of calls in a telephone switching and routing unit, and the frequency of heartbeats.

## 4.2.1 Probability Descriptors

### 4.2.1.1 Sample Space and Axioms of Probability

In the theory of probability, one is concerned with the *sample space of events*, that is, the set of all possible outcomes of experiments. These events are mapped numerically to values on the real line; these values are the *random variable*. In engineering, almost all measurements or procedures produce some numerical result such that a numerical value or sequence is inherently assigned to the outcome. For example, the production of a paint droplet is an outcome of an experiment and its size a random variable. Some outcomes are *nominal*, like the colors produced by a flame or gray levels in an image from an X ray. These nominal variables can be mapped to a numeric field so that in general the values of random variables will be defined over some real number field.

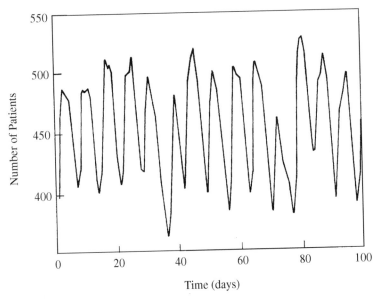

**FIGURE 4.3**   Daily inpatient census in a hospital. [Adapted from Pandit and Wu, fig. 9.12, with permission.]

Continuous and discrete random variables obey the same general laws and have the same general descriptors. The range of all possible values of a random variable comprises its *sample space*, S. For the paint droplets of Figure 4.2, this is the range of radii continuous from 0 to 28 microns, and for a die, it is the integers from 1 to 6 inclusive. A collection of values is called a *set* and is usually labeled with a capital letter.

All the values in a sample space are associated with probabilities of occurrence and obey the *axioms of probability*. Each value in a discrete sample space and any set, A, of values in a continuous or discrete sample space are associated with a probability, P[A]. The axioms of probability are

1. $0 \leq P[A] \leq 1$.
2. $P[S] = 1$.
3. If sets A and B are mutually exclusive—that is, they have no common values—$P[A \cup B] = P[A] + P[B]$.

An empty set contains none of the values in the sample space, is called a *null set*, and has a probability of zero.

The easiest interpretation of probability is the concept of *relative frequency*, developed from the *Law of Large Numbers* (Papoulis, 1984). Consider the situation of removing a black marble, B, from a box of marbles with an assortment of colors. If there are N marbles and $N_B$ black marbles, the probability,

$P[B]$, of choosing a black marble is the relative frequency of black marbles, $N_B/N$.

### 4.2.1.2 Probability Density and Cumulative Distribution Functions

The frequency of occurrence of values in a sample space is described by a pair of complementary functions called probability density and cumulative distribution functions. Their basic properties satisfy the axioms of probability. The *cumulative distribution function (cdf)*, sometimes also called the *cumulative probability function*, is defined as follows. For either a continuous or a discrete random variable, $x$, the probability that it has a magnitude equal to or less than a specific value, $\alpha$, is

$$P[x \le \alpha] = F_x(\alpha). \tag{4.1}$$

The properties of a cdf $F_x(\alpha)$ are

1. $0 \le F_x(\alpha) \le 1, \quad -\infty \le \alpha \le \infty$     (4.2)
2. $F_x(-\infty) = 0, \quad F_x(\infty) = 1.$
3. $F_x(\alpha)$ is nondecreasing with $\alpha$.
4. $P[\alpha_1 \le x \le \alpha_2] = F_x(\alpha_2) - F_x(\alpha_1).$

For a continuous random variable, the *probability density function (pdf)*, $f_x(\alpha)$, is essentially a derivative of $F_x(\alpha)$. Its properties are

1. $f_x(\alpha) \ge 0, \quad -\infty \le \alpha \le \infty.$     (4.3)
2. $\int_{-\infty}^{\infty} f_x(u)\, du = 1.$
3. $F_x(\alpha) = \int_{-\infty}^{\alpha} f_x(u)\, du.$
4. $P[\alpha_1 \le x \le \alpha_2] = \int_{\alpha_1}^{\alpha_2} f_x(u)\, du.$

For discrete random variables, the cdf is discontinuous and $f_x(\alpha) = P[x = \alpha]$. Equations 4.3 are also correct if the delta function is used in the definition of the probability density function. However, discrete random variables are not a concern; for more information refer to O'Flynn (1982) or Peebles (1987). Probabilities are most often calculated using equation 4.3.4.

Probability density functions often can be represented with formulas as well as graphs. The *uniform* pdf describes values that are equally likely to occur over a finite range.

$$f_x(\alpha) = \begin{cases} \dfrac{1}{b-a}, & a \le \alpha \le b \\ 0, & \text{for } \alpha \text{ elsewhere} \end{cases} \tag{4.4}$$

The *exponential* random variable has a semi-infinite range with the pdf

$$f_x(\alpha) = \frac{1}{b} \begin{cases} e^{-(\alpha-a)/b}, & a \le \alpha \le \infty \\ 0, & \alpha < a \end{cases} \tag{4.5}$$

This is commonly used to describe failure times in equipment and times between calls entering a telephone exchange.

At this point it is important to state that there is a diversity in notation for cdfs and pdfs. When there is no confusion concerning the random variable and its specific values, a simpler notation is *often* utilized (Papoulis, 1984). Specifically, the alternative notation is

$$f(x) = f_x(\alpha) \quad \text{and} \quad F(x) = F_x(\alpha) \tag{4.6}$$

This simpler notation will be used when the context of the discussion is clear. Another pdf is the *Gaussian or normal* pdf. It has the formula

$$f(x) = \frac{1}{\sqrt{2\pi}\sigma} \exp\left(-\frac{(x-m)^2}{2\sigma^2}\right), \qquad -\infty \leq x \leq \infty \tag{4.7}$$

The parameters $m$ and $\sigma$ have direct meanings that are explained in the next section. This pdf is useful for describing many phenomena such as random noise and biological variations.

The density and cumulative distribution functions for uniform, exponential, and normal random variables are plotted in Figures 4.4 to 4.6, respectively. There are many other pdfs; several are presented in Appendix 4.1. A comprehensive treatment of the many types of probability functions can be found in Larson and Shubert (1979).

Probabilities often can be calculated from distribution functions within computer applications and programs. The derivation of the cdf for some models is easy and have closed-form solutions. The cdfs without closed solutions can be calculated with infinite or finite asymptotic series. These are complex, and numerical precision is important. The handbook by Abramowitz and Stegun (1965) presents several solutions for important distribution functions.

---

**EXAMPLE** 4.1

The failure time for a certain kind of lightbulb has an exponential pdf with $b = 3$ and $a = 0$ months. What is the probability that one will fail between 2 and 5 months?

$$P[2 \leq x \leq 5] = \int_2^5 \frac{1}{3} e^{-x/3} \, dx = -(e^{-5/3} - e^{-2/3}) = 0.324$$

---

## 4.2.2  Moments of Random Variables

Not only are the forms of the pdfs very important but so are some of their moments. The moments not only quantify useful properties but in many situations only the moments are available. The general moment of a function $g(x)$

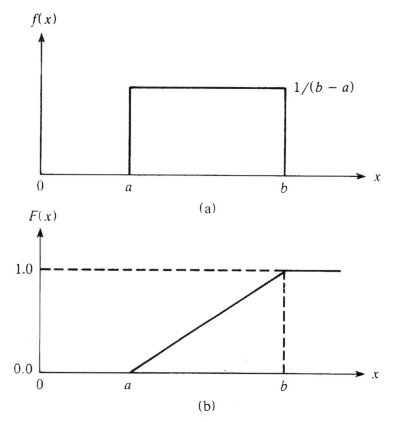

**FIGURE 4.4** The probability density (a) and distribution (b) functions for the uniform random variable. [Adapted from Peebles, fig. 2.5-2, with permission.]

of the random variable $x$ is symbolized by the *expectation operator*, $E[g(x)]$, and is

$$E[g(x)] = \int_{-\infty}^{\infty} g(x)f(x)\, dx \qquad (4.8)$$

The function $g(x)$ can take any form but is usually a polynomial, most often of the first or second order. In addition to using the expectation operator, some *moments* are given special symbols. The *mean* or average value is

$$E[x] = m_1 = m = \int_{-\infty}^{\infty} xf(x)\, dx \qquad (4.9)$$

The mean squared value is similarly defined as

$$E[x^2] = m_2 = \int_{-\infty}^{\infty} x^2 f(x)\, dx \qquad (4.10)$$

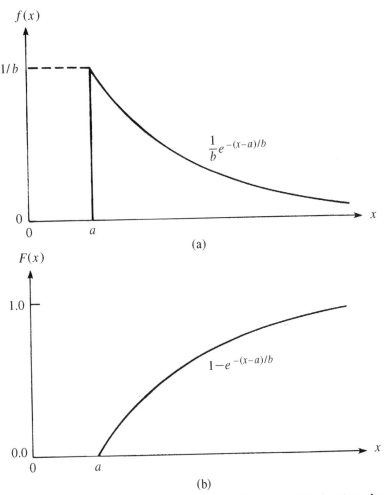

**FIGURE 4.5** The probability density (a) and distribution (b) functions for the exponential random variable. [Adapted from Peebles, fig. 2.5-3, with permission.]

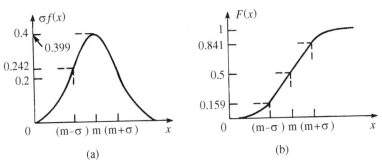

**FIGURE 4.6** The probability density (a) and distribution (b) functions for the normal variable. [Adapted from Papoulis, fig. 4.7, with permission.]

The higher-order moments of a random variable are used in advanced statistics. Similarly, *central moments*, moments about the mean, are defined for $g(x) = (x - m)^k$ and symbolized with the notation $\mu_k$. The most used central moment is for $k = 2$ and is the *variance* with the symbol $\sigma^2$. The definition is

$$E[(x - m)^2] = \sigma^2 = \int_{-\infty}^{\infty} (x - m)^2 f(x)\, dx \qquad (4.11)$$

The mean, the mean square, and the variance are interrelated by the equation

$$\sigma^2 = m_2 - m^2 \qquad (4.12)$$

The variance is an important parameter because it indicates the spread of magnitudes of a random variable. Its square root, $\sigma$—the *standard deviation*, is used synonymously. This indication can be easily seen in Figure 4.7, which shows two uniform density functions with the same mean and different ranges of values. The variance for the density function with the broader spread is greater.

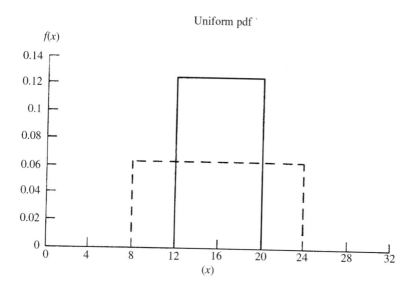

Uniform pdf

FIGURE 4.7   The plots of two uniform density functions; for $(-)m = 16$ and $\sigma^2 = 5.33$, for $(- -)m = 16$ and $\sigma^2 = 21.33$.

**EXAMPLE 4.2**

Find the mean and variance of the uniform density function in Figure 4.4 with $a = 2$ and $b = 7$.

$$m = \int_{-\infty}^{\infty} xf(x)\, dx = \int_{2}^{7} x \cdot 0.2\, dx = 0.1(49 - 4) = 4.5$$

$$\sigma^2 = \int_{-\infty}^{\infty} (x - m)^2 f(x)\, dx = \int_{2}^{7} (x - 4.5)^2 \cdot 0.2\, dx = 2.0833$$

**EXAMPLE 4.3**

Find the mean of the Rayleigh density function as shown in Appendix 4.1.

$$m = \int_{-\infty}^{\infty} xf(x)\, dx = \int_{0}^{-\infty} x \cdot \frac{x}{\alpha^2} e^{-x^2/2\alpha^2}\, dx = \frac{\alpha}{\sqrt{2}} \Gamma(0.5) = \frac{\alpha\sqrt{\pi}}{\sqrt{2}}$$

where $\Gamma(y)$ is the gamma function.

### 4.2.3   Gaussian Random Variable

The Gaussian or normal density function is extremely important because of its many applications and its tractability. The *Central Limit Theorem* makes it important because under certain conditions many processes formed by summing together several random variables with finite variances can be described in the limit by a normal pdf. Thus it is a rational representation for many noise and biological processes that naturally consist of a summation of many random events. An efficient aspect of the normal pdf is that its mean, $m$, and variance, $\sigma^2$, are written directly as parameters of the function. Reexamine equation 4.7. It is often abbreviated as $N(m, \sigma^2)$. Derivations of these moments can be found in Fante (1988). Calculating probabilities for this pdf is not simple because there is no closed-form solution for integrating over a finite range. To assist in this problem, a variable transformation called *standardization* is performed. The transformation is linear and is

$$t = \frac{x - m}{\sigma} \tag{4.13}$$

Performing this transformation on equation 4.7 produces

$$f(t) = \frac{1}{\sqrt{2\pi}} \exp\left(-\frac{t^2}{2}\right), \qquad -\infty \le t \le \infty \tag{4.14}$$

This is a Gaussian pdf with a mean of zero and a variance of one and is called the *standard normal* pdf, $N(0, 1)$. Probabilities over a range of values of $t$ are calculated mathematically with the standard normal cdf, $\Phi(z)$:

$$\Phi(z) = P[-\infty \le t \le z] = \int_{-\infty}^{z} f(t)\, dt \tag{4.15}$$

The solution of this integral is an infinite series; thus its values are usually tabulated as in Appendix 4.2. Similar tables can be found in most basic statistics textbooks and in books of tables. Rational polynomial approximations for calculating values of $N(0, 1)$ can be found in Abramowitz and Stegun (1965). Another function closely related to $\Phi(z)$ and sometimes used to calculate its values is the *error function (erf)*. It is defined as

$$\text{erf}(Y) = \frac{2}{\sqrt{\pi}} \int_0^Y \exp(-t^2)\, dt, \qquad -\infty \le t \le \infty \qquad (4.16)$$

These two functions are related through the equation

$$\Phi(z) = 0.5 + 0.5\,\text{erf}(z/\sqrt{2}) \qquad (4.17)$$

## EXAMPLE 4.4

For $f(x) = N(3, 4)$, find $P[x \le 5.5]$. In detail this is

$$P[x \le 5.5] = \int_{-\infty}^{5.5} -\infty\ \frac{1}{\sqrt{2\pi} \cdot 2} \exp\left(-\frac{(x-3)^2}{2.4}\right)$$

Standardizing using equation 4.13, the transformation is $y = \dfrac{(x-3)}{2}$ and the probabilities become

$$P[x \le 5.5] = P[y \le 1.25]$$

$$= \int_{-\infty}^{1.25} N(0, 1)\, dy = \Phi(1.25) = 0.89434$$

## EXAMPLE 4.5

A manipulation more commonly encountered for finding probabilities occurs when both bounds are finite. For $f(x) = N(2, 4)$, find $P[1 \le x \le 4]$. The standardization is $y = \dfrac{(x-2)}{2}$. Therefore,

$$P[1 \le x \le 4] = P[-0.5 \le y \le 1]$$

$$= \int_{-\infty}^{1} N(0, 1)\, dy - \int_{-\infty}^{-0.5} N(0, 1)\, dy = \Phi(1) - \Phi(-0.5)$$

$$= \Phi(1) - (1 - \Phi(0.5)) = 0.84134 - (1 - 0.69146) = 0.5328$$

## 4.3  JOINT PROBABILITY

### 4.3.1  Bivariate Distributions

The concept of *joint probability* is a very important one in signal analysis. Very often the values of two variables from the same or different sets of measurements are being compared or studied and a *two-dimensional sample space* exists. The joint or bivariate probability density function and its moments are the basis for describing any interrelationships or dependencies between the two variables. A simple example is the selection of a resistor from a box of resistors. The random variables are the resistance, $r$, and wattage, $w$. If a resistor is selected, it is desired to know the probabilities associated with ranges of resistance and wattage values; that is,

$$P[(r \leq R) \quad \text{and} \quad (w \leq W)] = P[r \leq R, w \leq W] \tag{4.18}$$

where $R$ and $W$ are particular values of resistance and wattage, respectively. For signals, this concept is extended to describe the relationship between values of a process $x(t)$ at two different times, $t_1$ and $t_2$, and between values of two continuous processes, $x(t)$ and $y(t)$, at different times. These probabilities are written

$$P[x(t_1) \leq \alpha_1, x(t_2) \leq \alpha_2] \quad \text{and} \quad P[x(t_1) \leq \alpha_1, y(t_2) \leq \beta_2] \tag{4.19}$$

respectively. The joint probabilities are functionally described by *bivariate cumulative distribution and density functions*. The bivariate cdf is

$$F_{xy}(\alpha, \beta) = P[x \leq \alpha, y \leq \beta] \tag{4.20}$$

It is related to the bivariate pdf through the double integration

$$F_{xy}(\alpha, \beta) = \int_{-\infty}^{\beta} \int_{-\infty}^{\alpha} f_{xy}(u, v) \, du \, dv \tag{4.21}$$

These functions have the following important properties.

1. $F_{xy}(\alpha, \infty) = F_x(\alpha)$, the marginal cdf for variable $x$. $\tag{4.22}$
2. $F_{xy}(\infty, \infty) = 1$.
3. $f_{xy}(\alpha, \beta) \geq 0$.
4. $f_x(\alpha) = \int_{-\infty}^{\infty} f_{xy}(\alpha, v) \, dv$, the marginal pdf for variable $x$.

Again, when there is no confusion concerning the random variable and its specific values, a simpler notation is *often* utilized. The alternative notation is

$$f(x, y) = f_{xy}(\alpha, \beta) \quad \text{and} \quad F(x, y) = F_{xy}(\alpha, \beta) \tag{4.23}$$

Related to this is the notion of *conditional probability*, that is, given the knowledge of one variable, what are the probability characteristics of the other variable? A conditional pdf for resistance, knowing that a resistor with a particular value of wattage has been selected, is written $f(r|w)$. The conditional, *marginal*, and *joint* pdf are related through *Bayes' rule* by

$$f(r|w) = \frac{f(w|r)f(r)}{f(w)} = \frac{f(r, w)}{f(w)} \tag{4.24}$$

Conditional density functions have the same properties as marginal density functions. Bayes' rule and conditional probability provide a foundation for the concept of *independence* of random variables. If knowledge of wattage does not contain any information about resistance, then

$$f(r|w) = f(r) \tag{4.25}$$

That is, the conditional density function equals the marginal density function. Using equation 4.25 with equation 4.24 gives

$$f(r, w) = f(r)f(w) \tag{4.26}$$

That is, the bivariate density function is the product of the marginal density functions. Two random variables $r$ and $w$ are independent if either equation 4.25 or equation 4.26 is true. Comprehensive treatments of joint probability relationships can be found in books such as Ochi (1990) and O'Flynn (1982).

### 4.3.2 Moments of Bivariate Distributions

Moments of bivariate distributions are defined with a function $g(x, y)$. The expectation is

$$E[g(x, y)] = \int_{-\infty}^{\infty} \int_{-\infty}^{\infty} g(x, y) f(x, y) \, dx \, dy \tag{4.27}$$

Two moments are extremely important: the *mean product*, $E[xy]$, with $g(x, y) = x \cdot y$ and the *first-order central moment*, $\sigma_{xy}$, with $g(x, y) = (x - m_x)(y - m_y)$. The moment $\sigma_{xy}$ is also called the *covariance*. They are all related by

$$E[xy] = \sigma_{xy} + m_x m_y \tag{4.28}$$

The covariance is an indicator of the strength of linear relationship between two variables and defines the state of *correlation*. If $\sigma_{xy} = 0$, then $x$ and $y$ are *uncorrelated* and $E[xy] = m_x m_y$. If variables $x$ and $y$ are independent, then also $E[xy] = m_x m_y$, and the covariance is zero. However, the converse is not true. In other words, independent random variables are uncorrelated, but uncorrelated random variables can still be dependent. This linear relationship will be explained in the section on estimation.

When $\sigma_{xy} \neq 0$, the random variables are correlated. However, the strength of the relationship is not quantified because the magnitude of the covariance depends on the units of the variables. For instance, the usage of watts or milliwatts as units for $w$ will make a factor of $10^3$ difference in the magnitude of the covariance $\sigma_{rw}$. This situation is solved by using a unitless measure of *linear dependence* called the *correlation coefficient*, $\rho$, where

$$\rho = \frac{\sigma_{rw}}{\sigma_r \sigma_w} \tag{4.29}$$

and

$$-1 \leq \rho \leq 1 \qquad (4.30)$$

The measure is linear because if $r = k_1 w + k_2$, where $k_1$ and $k_2$ are constants, then $\rho = \pm 1$. For uncorrelated random variables, $\rho = 0$.

**EXAMPLE 4.6**

An example of a two-dimensional pdf is

$$f(x, y) = \begin{cases} abe^{-(ax+by)}, & x \geq 0, \ y \geq 0 \\ 0, & \text{elsewhere} \end{cases}$$

It can be seen that the pdf is separable into two functions

$$f(x, y) = ae^{-(ax)} \cdot be^{-(by)} = f(x) \cdot f(y).$$

Thus the variables $x$ and $y$ are independent and also $\rho = 0$.

**EXAMPLE 4.7**

Another bivariate pdf is

$$f(x, y) = \begin{cases} xe^{-x(y+1)}, & x \geq 0, \ y \geq 0 \\ 0, & \text{elsewhere} \end{cases}$$

These variables are dependent since the pdf $f(x, y)$ cannot be separated into two marginal pdfs. The marginal density function for $x$ is

$$f(x) = \int_0^\infty f(x, y) \, dy = xe^{-x} \int_0^\infty e^{-xy} \, dy = e^{-x}, \qquad x \geq 0$$

The conditional density function for $y$ is

$$f(y|x) = \frac{f(x, y)}{f(x)} = \frac{xe^{-x(y+1)}}{e^{-x}} = \begin{cases} xe^{-xy}, & x \geq 0, \ y \geq 0 \\ 0, & \text{elsewhere} \end{cases}$$

**EXAMPLE 4.8**

An extremely important two-dimensional pdf is the bivariate normal distribution. It has the form

$$f(x, y) = \frac{1}{2\pi\sigma_x\sigma_y\sqrt{1 - \rho^2}} e^{-a/2}$$

with

$$a = \frac{1}{1 - \rho^2} \left( \frac{(x - m_x)^2}{\sigma_x^2} - 2\rho \frac{(x - m_x)(y - m_y)}{\sigma_x \sigma_y} + \frac{(y - m_y)^2}{\sigma_y^2} \right)$$

The means and variances for the random variables $x$ and $y$ and their correlation coefficient are explicitly part of the pdf. Sketches of the surface for two different values of $\rho$ are shown in Figure 4.8. If $\rho = 0$, then the middle term in the

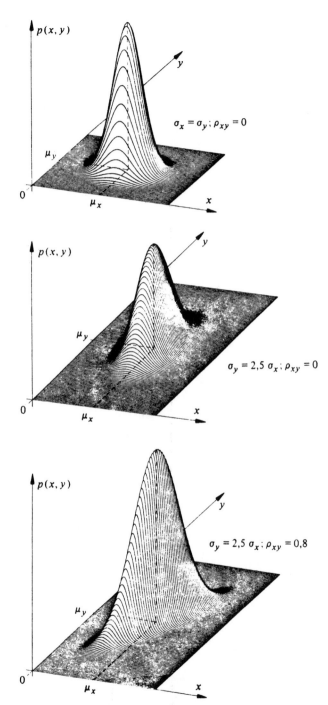

**FIGURE 4.8**  Sketches of surfaces for two-dimensional normal probability density functions with two different values of $\rho$. [Adapted from Coulon, fig. 5.26, with permission.]

exponent is also zero and $f(x, y) = f(x) \cdot f(y)$. Thus uncorrelated normal random variables are also independent. This is not true in general for other two-dimensional pdfs.

---

## 4.4   CONCEPT OF SAMPLING AND ESTIMATION

The study or investigation of a random phenomenon usually requires the knowledge of its statistical properties, that is, its probabilistic description. Most often this description is not available and must be discovered from experimental measurements. The measurements produce a *sample* of $N$ data values $\{x_1, x_2, \ldots, x_N\}$. Mathematical operations are performed on the sample to determine the statistical properties. These operations are called *estimators*, and this entire process is called *estimation*. An important aspect of estimation is its accuracy, which can be quite difficult to determine. This section will introduce the notion of *sample moments* and some general characteristics of estimation. All of this will be exemplified by focusing on estimating the mean and variance of a data sample.

### 4.4.1   Sample Moments

The procedure for estimating many properties is often obtained by directly translating the mathematical definition of its theoretical counterpart. Consider approximating the general expectation operator of equation 4.8 with an infinite sum. Divide the real line into intervals of length $\Delta x$ with boundary points $d_i$, then

$$E[g(x)] = \int_{-\infty}^{\infty} g(x)f(x)\, dx \approx \sum_{i=-\infty}^{\infty} g(d_i)f(d_i)\, \Delta x$$

Notice that a very important assumption is made. All of the sample values of $x$ have the same pdf; that is, they are identically distributed. Since data points may equal boundary values, the intervals are half-open and

$$f(d_i)\, \Delta x \approx P[d_i \leq x < d_i + \Delta x]$$

Using the interpretation of relative frequency, then

$$E[g(x)] \approx \sum_{i=-\infty}^{\infty} g(d_i)P[d_i \leq x < d_i + \Delta x] = \sum_{i=-\infty}^{\infty} g(d_i)\, \frac{N_i}{N}$$

where $N_i$ is the number of measurements in the interval $[d_i \leq x < d_{i+1}]$. Since $g(d_i) \cdot N_i$ approximates the sum of values of $g(x)$ for points within interval $i$, then

$$E[g(x)] \approx \frac{1}{N} \sum_{j=1}^{N} g(x_j) \tag{4.31}$$

where $x_j$ is the $j$th data point. Equation 4.31 is the *estimator* for the average value of $g(x)$. Since no criterion was optimized for the derivation of the estimator, it can be considered an empirical *sample moment*. Notice that this is a commonly used estimator. It is obvious if $g(x) = x$, or

$$E[x] \approx \hat{m} = \frac{1}{N} \sum_{j=1}^{N} x_j \qquad (4.32)$$

This is the estimator for the mean value. The circumflex is used to denote an estimator of the theoretical function. Similarly, an estimator for the variance is

$$\hat{\sigma}^2 = \frac{1}{N} \sum_{j=1}^{N} (x_j - m)^2$$

Typically the mean is not known, and in this case the variance estimator becomes

$$\hat{\sigma}^2 = \frac{1}{N-1} \sum_{j=1}^{N} (x_j - \hat{m})^2 \qquad (4.33)$$

Notice that the coefficient in equation 4.33 is $\frac{1}{(N-1)}$ instead of $\frac{1}{N}$. The reason will be given in Section 5.5. The actual values that are calculated using equations 4.32 and 4.33 are *estimates* of the mean and variance, respectively.

There are other methods for developing estimators. For instance, in Exercise 2.1 the estimator for the sample mean is derived using the least squares criterion and the result is the same as equation 4.32. This is not usually the situation with other estimators; different criteria can produce different estimators for the same parameter.

The estimate of a moment of a random variable is also a random variable. Consider estimating the mean value of the daily river flow plotted in Figure 4.9 from the table in Appendix 4.3. If only the first ten values were available, then $\hat{m} = 2368$. If only the last ten values were available, then $\hat{m} = 1791$. Thus $\hat{m}$ itself is a random variable with a pdf $f(\hat{m})$, and it is necessary to know its relationship to $m$. In general, it is necessary to know the relationship between a sample estimate of a moment or function and its true value. An elegant branch of mathematical statistics called *estimation and sampling theory* has as its definition the development of these relationships (Fisz, 1963). The premises for most of the developments are that the measurements arise from the same underlying distribution and that they are independent. The latter condition can sometimes be relaxed if the number of measurements is large enough to represent the sample space. For many probabilistic parameters, the distribution of values produced by an estimator, the *sampling distribution*, has been derived and relates the estimate to the true value. The usage of sampling distributions is essential in signal analysis; this will be illustrated in detail for the sample mean.

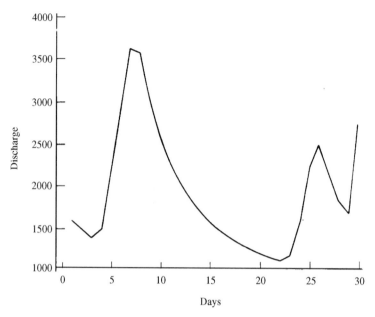

**FIGURE 4.9**   Daily river flow during March, 1981 of the Duck River.

### 4.4.2   Significance of the Estimate

For the sample mean estimator, the *Student's t* distribution relates $m$ and $\hat{m}$ through the $t$ variable if either the random variable $x$ has a normal distribution or $N$ is large. The $t$ variable is a standardized error and is defined as

$$t = \frac{m - \hat{m}}{\hat{\sigma}/\sqrt{N}} \qquad (4.34)$$

where $\hat{\sigma}$ *is the sample standard deviation,* $\sqrt{\hat{\sigma}^2}$. The pdf of the Student's $t$ variable has a complicated form, which is

$$f(t) = \frac{\Gamma\left(\frac{n}{2}\right)}{\sqrt{N-1}\cdot\Gamma\left(\frac{1}{2}\right)\cdot\Gamma\left(\frac{1}{2}(N-1)\right)} \cdot \frac{1}{[1 + t^2/(N-1)]^{N/2}} \qquad (4.35)$$

where $\Gamma(u)$ is the gamma function; $f(t)$ is plotted in Figure 4.10. Notice that $f(t)$ asymptotically approaches zero as the absolute value of $t$ approaches infinity. This means that the error in estimating the mean can be infinitely large; however, judging from the magnitude of $f(t)$ the likelihood is very small.

     This is formalized in the following manner. A bound is set for $t$, $t_c$, such that the probability that the error is larger than this bound is very small; that is,

$$P[|t| \geq t_c] = 0.05 = \alpha \qquad (4.36)$$

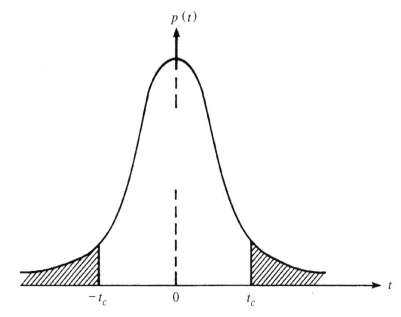

**FIGURE 4.10** Student's $t$ pdf with critical regions shaded.

The range of values of $t$ satisfying equation 4.36 is called the critical or *significance region*. The remaining values of $t$ comprise the *confidence region*. These regions are illustrated also in Figure 4.10. The *probability $\alpha$ is the critical level* and $t_c$ is the critical value. The words *critical* and *significance* are used interchangeably. The utility of this construction is that the critical value can be easily translated into useful bounds for the real value of the mean via equation 4.34.

Rewriting this equation for relating $m$ to its estimate, $\hat{m}$, gives

$$m = \hat{m} - t \frac{\sigma}{\sqrt{N}} \tag{4.37}$$

This is effectively a variable transformation from $t$ space to $m$ space. Inserting the critical values of $t$, $\pm t_c$, into this equation produces the $\alpha$ level *lower, $m_1$, and upper, $m_u$, bounds* for the mean, respectively, as

$$m_1 = \hat{m} - t_c \frac{\sigma}{\sqrt{N}} \tag{4.38}$$

$$m_u = \hat{m} + t_c \frac{\sigma}{\sqrt{N}} \tag{4.39}$$

Thus given the sample mean the confidence interval at level $\alpha$ for the true mean is

$$P[m_1 \leq m \leq m_u] = 1 - \alpha = 0.95 \tag{4.40}$$

The equation for $f(t)$ is complicated to evaluate so the critical values for various significance levels are tabulated in Appendix 4.4. There is one more aspect to the Student's $t$ distribution; notice in equation 4.35 that $f(t)$ depends on the factor $(N - 1)$, called the *degrees of freedom*, $v$. One simply designates a significance level, $\alpha$, and finds the value of $t_c$ for the needed degrees of freedom in the table. The notion of degrees of freedom relates to the number of independent variables that are being summed. The estimate, $\hat{m}$, is a sum of $N$ independent measurements. However, the $t$ variable is not; it is a sum of terms $\{x_i/\hat{\sigma};\ 1 \le i \le N\}$, where $\hat{\sigma}$ is also a function of $x_i$. A sum of any $N - 1$ terms, $x_i/\hat{\sigma}$, are independent. The $N$th term is constrained in value since $\hat{\sigma}$ is calculated. Hence the $t$ variable has only $N - 1$ independent terms; that is, only $N - 1$ degrees of freedom. This concept of the number of independent variables in a summation is used with all estimators.

---

### EXAMPLE 4.9

For the river flow data plotted in Figure 4.9, estimate the mean daily river flow. The data is listed in Appendix 4.3 for the month of March. Use the critical level of 0.05. The degrees of freedom are $v = N - 1 = 29$. The critical value is $t_c = 2.045$. Next the sample mean and variance must be calculated.

$$\hat{m} = \frac{1}{30} \sum_{i=1}^{30} x_i = 1913$$

$$\hat{\sigma} = \sqrt{\frac{1}{29} \sum_{i=1}^{30} (x_i - \hat{m})^2} = 700.8$$

The bounds for the mean are

$$m_1 = \hat{m} - t_c \frac{\hat{\sigma}}{\sqrt{N}} = 1913 - 2.045 \cdot \frac{700.8}{\sqrt{30}} = 1651.3$$

$$m_u = \hat{m} + t_c \frac{\hat{\sigma}}{\sqrt{N}} = 1913 + 2.045 \cdot \frac{700.8}{\sqrt{30}} = 2174.7$$

The confidence interval for the true mean is

$$P[1651.3 \le m \le 2174.7] = 0.95$$

Thus it can be said that based on the 30 observations the mean river flow is within the range stated with a confidence of 0.95.

---

An alternative aspect of the procedures used for estimation is called *hypothesis testing*. In this situation a statement is made about the value of some parameter or moment of a random variable. This is the *null hypothesis*. The null hypothesis is tested for feasibility using the *sampling distribution* of the parameter or moment. The sampling distribution relates the true value of a parameter or moment to a sampled value. The general procedure is similar to the procedure explained in Example 4.9. For instance, a hypothesis is made about the true mean value of the distribution from which a set of data has been measured; that is, the true mean has a value $m_0$. This is symbolized by $H_0$. The alternative to $H_0$ is that given the estimate of the mean, the true mean value is other than $m_0$. This is given the symbol $H_1$. Mathematically, this is concisely stated as

$$H_0 : m = m_0$$
$$H_1 : m \neq m_0$$

(4.41)

In the Example 4.9, the sampling distribution is the Student's $t$ and bounds for a confidence interval are defined in equations 4.38 and 4.39 given $\hat{m}$, $\hat{\sigma}$, $t_c$, and $\alpha$. If $m_0$ lies within the confidence region, then $H_0$ is accepted as true, otherwise it is rejected and $H_1$ is accepted as true. Hypothesis testing is covered in detail in textbooks on introductory statistics.

## 4.5 DENSITY FUNCTION ESTIMATION

### 4.5.1 General Principle

There are situations when knowing the statistical parameters of the data is not sufficient, and it is desired to discover or model its probability distribution. In quality control, for example (as in Example 4.1), does an exponential pdf portray the distribution of failure times accurately, or is another model necessary? This hypothesis can be tested using *Pearson's $\chi^2$ statistic* (Fisz, 1963; Otnes and Enochson, 1972). Let $F(x)$ be the hypothesized cdf for the data. Divide the range of $x$ into $N_b$ disjoint intervals, $S_j$, such that

$$P[S_j] = P[d_{j-1} \leq x < d_j], \qquad 1 \leq j \leq N_b$$

(4.42)

are the theoretical probabilities of occurrence. The test depends on comparing the observed number of ocurrences of values in a set of samples to the number expected from the proposed distribution. Let $o_j$ and $e_j$ represent the number of observed and expected occurrences, respectively, in $S_j$. If $N$ equals the total number of data points, $e_j = NP[S_j]$. The metric for this comparison, the chi-square statistic, is

$$\chi^2 = \sum_{j=1}^{N_b} \frac{(o_j - e_j)^2}{e_j}$$

(4.43)

where

$$\sum_{j=1}^{N_b} o_j = N$$

(4.44)

The $e_j$ are calculated from the proposed model. The chi-square statistic in equation 4.43 has been developed upon the hypothesis that the sample has the proposed cdf and the density function for $\chi^2$ is

$$f(u) = \begin{cases} \dfrac{1}{2^{v/2}\Gamma\left(\dfrac{v}{2}\right)} e^{-u/2} u^{v/2-1}, & u \geq 0 \\ 0, & u < 0 \end{cases} \tag{4.45}$$

where $u = \chi^2$ and $v = $ the degrees of freedom $= N_b - 1$.

This density function is plotted in Figure 4.11. Obviously, $\chi^2$ is positive so that an $\alpha$ level significance region is established only over one tail of the distribution such that

$$P[\chi^2 \geq \chi^2_{v,\alpha}] = \alpha \tag{4.46}$$

The value of $\chi^2_{v,\alpha}$ corresponding to various significance levels is tabulated in Appendix 4.5. If the calculated value of $\chi^2$ lies within the significance region, then the proposed model is rejected as being appropriate and another one must be selected. A simple example will illustrate this procedure.

---

### EXAMPLE 4.10

Cards are drawn at random 20 times from a full deck. The result is 8 clubs, 3 diamonds, 5 hearts, and 4 spades. The hypothesis being tested is the honesty of the deck of cards. Thus the probability for drawing a card of a desired suit is $\frac{1}{4}$ and this is the proposed model. The mapping from the nominal variables to a magnitude or selection level is: clubs $= 1$, diamonds $= 2$, hearts $= 3$, spades $= 4$. The expected numbers for each level is 5; therefore, $e_1 = e_2 = e_3 = e_4 = 5$. The

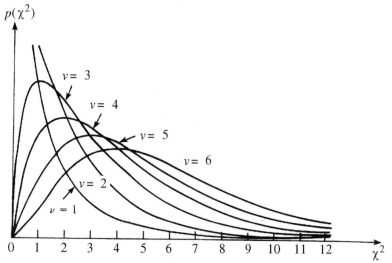

**FIGURE 4.11**   Chi-square density function with various degrees of freedom, $v$.

observed numbers are: $o_1 = 8$, $o_2 = 3$, $o_3 = 5$, $o_4 = 4$. Calculating $\chi^2$ from equation 4.43 produces

$$\chi^2 = \sum_{j=1}^{N_b} \frac{(o_j - e_j)^2}{e_j} = \frac{9}{5} + \frac{4}{5} + \frac{1}{5} + 0 = 2.8$$

The confidence limit is based on a 95% confidence level and $v = 4 - 1 = 3$ degrees of freedom. The entry in Appendix 4.5 shows that $\chi^2_{3,0.05} = 7.81$. Thus the inequality $\chi^2 \geq \chi^2_{v,\alpha}$ is not satisfied and the proposed probability model is accepted as true; that is, the deck of cards is honest.

### 4.5.2   Detailed Procedure

In practice one does not have well-defined categories as in Example 4.10, so one needs to produce a histogram. Consider the collector currents in a sample of 102 manufactured transistors. The current is measured with an accuracy of 0.05 microamperes. The number of transistors with different values is plotted in Figure 4.12a. The production of the histogram requires defining the number and boundaries of these categories.

First define the range of interest for the random variable, $a \leq x \leq b$. Then divide this range into $k$ subintervals or *class intervals*. The width, $W$, of these class intervals is $W = (b - a)/k$. To account for the entire sample space, define two classes for $x < a$ and $x \geq b$. Number the $k + 2$ classes or bins consecutively from 0 to $k + 1$. Let $j$ indicate the bin number and $N_j$ the number in each bin. The histogram is then computed in the following manner.

1. Initialize $N_j$ to 0,   $0 \leq j \leq k + 1$.
2. Sort $x_i$, $1 \leq i \leq N$, into classes, and increment $N_j$ according to the following specifications.
   a. If $x_i < a$, increment $N_0$.
   b. If $x_i \geq b$, increment $N_{k+1}$.
   c. For other $x_i$, calculate $j = \text{INT}\left(\frac{x_i - a}{W}\right) + 1$, and increment $N_j$.

The INT stands for the integerization operation. The sample probability that a measurement lies within a bin is the relative frequency definition, or

$$\hat{P}[d_{j-1} \leq x < d_j] = \frac{N_j}{N} \tag{4.47}$$

To approximate the magnitude of the pdf, remember that the pdf is probability per unit value, and divide equation 4.47 by the bin width, $W$, or

$$\hat{f}(x) = \frac{N_j}{NW}, \qquad d_{j-1} \leq x < d_j \tag{4.48}$$

Figure 4.12b shows estimates of the histogram and the pdf for the collector current data using two different bin widths. As one can surmise from this figure, if $W$ is too large then peak magnitudes of $f(x)$ will be underestimated. It has been shown that there is a bias in estimating a pdf (Shanmugan and Breipohl, 1988). The mean value of the estimator is

$$E[\hat{f}(x_{cj})] = f(x_{cj}) + f''(x_{cj})\frac{W^2}{24}, \qquad d_{j-1} \leq x < d_j \qquad (4.49)$$

where $x_{cj} = (d_{j-1} + d_j)/2$. Thus some knowledge of the general form of the pdf is important for a good histogram representation.

In signal processing since most of the variables are continuous, so are the pdf models that one wishes to test as feasible representations. This is handled in a straightforward manner since, for a continuous variable,

$$e_j = N \cdot P[d_{j-1} \leq x < d_j] \qquad (4.50)$$

Thus once one chooses the boundaries for the histogram, one must integrate the hypothesized pdf over the bin boundaries to calculate the bin probabilities.

---

### EXAMPLE 4.11

In the manufacture of cotton thread, one of the properties being controlled is the tensile strength. Three hundred balls of cotton thread are chosen from a consignment and their breaking tension measured. The range of tensions is found to be $0.5 \text{ kg} \leq x \leq 2.3 \text{ kg}$. It is desired to test whether a normal distribution is a suitable model for this property. A bin width of $0.14 \text{ kg}$ is chosen and results in the creation of 13 bins. A histogram is calculated, and the numbers of occurrences are shown in Table 4.1.

One practical problem that arises is that the mean and the standard deviation of the hypothesized distribution are unknown and must be estimated from the data. These are

$$\hat{m} = 1.41, \qquad s = 0.26$$

The probabilities are found by using the standard normal pdf. For the sixth bin,

$$P[S_6] = P[1.20 \leq x < 1.34] = P\left[-0.81 \leq \frac{x - 1.41}{0.26} < -0.27\right] = 0.1846$$

The probabilities $P[S_j]$ are also listed in the table. The expected number of occurrences are calculated with equation 4.50 and the results are rounded off. An empirical rule used is that the number of expected ocurrences must at least be 2. This is accommodated by combining bins 1, 2, and 3 and bin 13 with bin 12, resulting in 10 bins.

Another practical problem arises: The use of sample moments affects the degrees of freedom. For each estimated parameter in the hypothesized

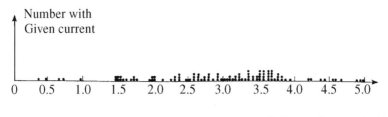

Collecter Current ($\mu$A)

(a)

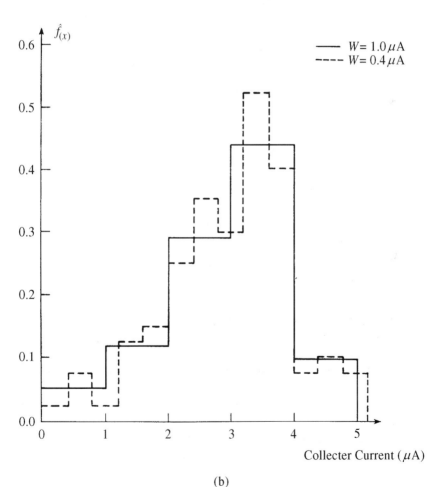

Collecter Current ($\mu$A)

(b)

**FIGURE 4.12** Statistical data on collector currents of 102 transistors. (a) Histogram with $W = 0.05$ microamperes. (b) Histogram with $W = 1$ and estimates of probability density functions with $W = 1.0$ and 0.4 microamperes. [Adapted from Jenkins and Watts, figs. 3.3 and 3.6, with permission.]

**TABLE 4.1**  Histogram of Breaking Tensions (kilograms)

| $j$ | $d_{j-1}-d_j$ | $N_j$ | $P[S_j]$ |
|-----|---------------|-------|----------|
| 1   | 0.5 –0.64     | 1     | 0.0015   |
| 2   | 0.64–0.78     | 2     | 0.0065   |
| 3   | 0.78–0.92     | 9     | 0.0223   |
| 4   | 0.92–1.06     | 25    | 0.0584   |
| 5   | 1.06–1.20     | 37    | 0.1205   |
| 6   | 1.20–1.34     | 53    | 0.1846   |
| 7   | 1.34–1.48     | 56    | 0.2128   |
| 8   | 1.48–1.62     | 53    | 0.1846   |
| 9   | 1.62–1.76     | 25    | 0.1205   |
| 10  | 1.76–1.90     | 19    | 0.0584   |
| 11  | 1.90–2.04     | 16    | 0.0223   |
| 12  | 2.04–2.18     | 3     | 0.0065   |
| 13  | 2.18–2.32     | 1     | 0.0015   |

distribution, the degrees of freedom must be reduced by one. Thus $v = 10 - 1 - 2 = 7$. For a 5% significance level,

$$\chi^2_{7,0.05} = 14.07$$

For the breaking strength of the threads, the chi-square is

$$\chi^2 = \sum_{j=2}^{11} \frac{(o_j - e_j)^2}{e_j} = 300 \sum_{j=2}^{11} \frac{(\hat{P}[S_j] - P[S_j])^2}{P[S_j]} = 22.07$$

Thus the hypothesis is rejected and the thread strengths are not represented by a normal distribution.

---

Many interesting applications arise when a model of the pdf of time intervals between events is needed. One such application arises when studying the kinetics of eye movement. As one tracks a moving object visually, the position of the eye changes slowly and then it moves quickly back to its initial position. These quick movements are called nystagmus, and their times of occurrence are called events. The histogram of time intervals between events (IEI) for one measurement session is shown in Figure 4.13. Two pdf models are proposed—log-normal and gamma—and both are fitted by equating the sample moments of the IEI and the models' moments. The resulting models are also plotted in Figure 4.13. Both demonstrate good "fits" by generating $\chi^2$ values with significance regions less than the 0.05 level (Anderson and Correia, 1977).

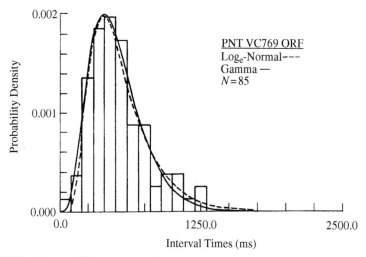

**FIGURE 4.13** An IEI histogram with log-normal and gamma pdf models superimposed. [Adapted from Anderson and Correia, fig. 7, with permission.]

## 4.6 ESTIMATE OF CORRELATION

It is common to need to assess the dependence between two variables or the strength of some cause-and-effect relationship using the correlation coefficient. The correlation measure is used to compare the rainfall patterns in cities, the simularity between electrocardiographic waveforms, the incidences of diseases with pollution, etc. The estimator for the correlation coefficient, $\rho$, is a direct translation of the theoretical definition given in Section 4.3.2. The sample covariance is

$$\hat{\sigma}_{xy} = \frac{1}{N-1} \sum_{i=1}^{N} (x_i - \hat{m}_x)(y_i - \hat{m}_y) \tag{4.51}$$

Using the estimators for the sample variance, the sample correlation coefficient is defined as

$$\hat{\rho} = \frac{\hat{\sigma}_{xy}}{\hat{\sigma}_x \hat{\sigma}_y} \tag{4.52}$$

If both variables, $x$ and $y$, have normal distributions, this estimator is also a maximum likelihood estimator. The sample correlation is calculated to find the strength of a relationship, therefore it is necessary to test it with some hypotheses. The sampling distribution of $\hat{\rho}$ is quite asymmetric and a transformation that creates an approximately normal random variable $z$ is implemented. The transformation is

$$z = \frac{1}{2} \ln\left(\frac{1 + \hat{\rho}}{1 - \hat{\rho}}\right) \tag{4.53}$$

Its mean and variance are, respectively (Fisz, 1963)

$$m_z = \frac{1}{2} \ln\left(\frac{1 + \rho}{1 - \rho}\right) + \frac{\rho}{2(N - 1)}; \qquad \sigma_z^2 = \frac{1}{N - 3} \tag{4.54}$$

When $N$ is not small, the second term for $m_z$ can be ignored.

### EXAMPLE 4.12

For hospital patients suffering circulatory shock it is desired to know (a) if there is a correlation between the blood pH in the venous system ($x$) and the arterial system ($y$); and (b) if so, what is the confidence interval? The plot of the pH values is in Figure 4.14. The sample moments and correlation coefficient calculated from measurements on 108 patients are

$$\hat{m}_x = 7.373, \qquad \hat{m}_y = 7.413, \qquad \hat{\sigma}_x^2 = 0.1253, \qquad \hat{\sigma}_y^2 = 0.1184,$$

$$\hat{\sigma}_{xy} = 0.1101, \qquad \hat{\rho} = 0.9039$$

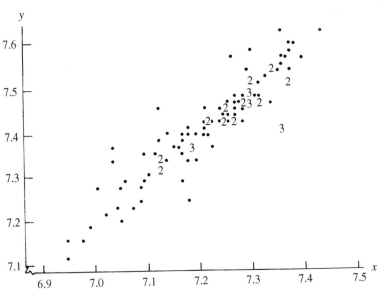

**FIGURE 4.14**   The arterial blood pH ($y$) and the venous blood pH ($x$) for critically ill patients. The numbers indicate multiple points at that coordinate. [Adapted from Afifi and Azen, fig. 3.1.2, with permission.]

The first hypothesis to be tested is whether the variables are correlated. The null hypothesis is $\rho = 0$. Using the $z$ transformation, $m_z = 0.0$ and $\sigma_z^2 = \dfrac{1}{N-3} = 0.00952$. The 95% confidence interval is $\pm 1.96\sigma_z$. That is, if

$$-0.1913 \le z \le 0.1913$$

then $\rho = 0$. Using equation 4.53

$$z_\rho = \frac{1}{2}\ln\left(\frac{1+\hat{\rho}}{1-\hat{\rho}}\right) = \frac{1}{2}\ln\left(\frac{1.9039}{0.0961}\right) = 1.493$$

and the variables are correlated.

Now the confidence interval for $\rho$ must be established. The bounds for the confidence interval of $z$ are $z_\rho \pm 1.96\sigma_z$, or

$$1.3017 \le z \le 1.6843$$

Using equation 4.53 the inverse transformation from $z$ to $\rho$ is

$$\rho = \frac{e^{2z}-1}{e^{2z}+1}$$

and substituting the upper and lower bounds for $z$ yields

$$0.8622 \le \rho \le 0.9334$$

Thus the two pH values are highly correlated.

---

An important relationship between correlation and linear regression is apparent when the conditional pdf, $f(y|x)$, of the normal distribution is considered. This is directly derived from Bayes' rule, equation 4.24, and is reserved as an exercise. The conditional mean, $m_{y|x}$, and standard deviation, $\sigma_{y|x}$, are

$$m_{y|x} = m_y + \rho\frac{\sigma_y}{\sigma_x}(x-m_x) = m_y - \frac{\sigma_{xy}}{\sigma_x^2}m_x + \frac{\sigma_{xy}}{\sigma_x^2}x \qquad (4.55)$$

$$\sigma_{y|x} = \sigma_y\sqrt{1-\rho^2} \qquad (4.56)$$

Now return to the solution of the linear model relating two sets of measurements $\{y_i: 1 \le i \le N\}$ and $\{x_i: 1 \le i \le N\}$. The model is

$$\hat{y}_i = a + bx_i \qquad (2.2)$$

Solving equations 2.6 and 2.7 explicitly for the coefficients yields

$$a = \hat{m}_y - \frac{\hat{\sigma}_{xy}}{s_x^2}\hat{m}_x, \qquad b = \frac{\hat{\sigma}_{xy}}{s_x^2} \qquad (4.57)$$

The solution for these coefficients contains the estimates of the moments used in equation 4.55 in the same algebraic form. Hence this shows that the correlation coefficient is a measure of *linear dependence* between two samples or time series.

## 4.7 GENERAL PROPERTIES OF ESTIMATORS

### 4.7.1 Convergence

As has been studied, an estimator produces estimates that are random variables with a sampling distribution. This distribution cannot always be derived but we need to know if the estimate is close to the true value in some sense. Fortunately, *convergence* can be determined using two very important general properties of estimators: the bias and consistency.

The *bias* property describes the relationship between the function being estimated and the mean value of the estimator. For instance, when estimating the mean value of a random variable using equation 4.32, $E[\hat{m}] = m$ and the estimator is said to be unbiased. If another estimator for mean value, $\hat{m}_2$, were used, perhaps the result would be $E[\hat{m}_2] \neq m$. This estimator is said to be biased. This is generalized for any estimator, $\hat{g}_N(x)$, being a function of the set of $N$ measurements, $\{x_i\}$. If $E[\hat{g}_N(x)] = g(x)$, then the estimator is unbiased; otherwise the estimator is biased. Equation 4.33 computes the sample's variance and is an unbiased estimator because $E[\hat{\sigma}^2] = \sigma^2$. Another estimator, which is more intuitive because the coefficient of the summation is $1/N$, is defined as

$$\hat{\sigma}_2^2 = \frac{1}{N} \sum_{i=1}^{N} (x_i - \hat{m})^2 \tag{4.58}$$

However, this estimator is biased because

$$E[\hat{\sigma}_2^2] = \frac{N-1}{N} \sigma^2 \neq \sigma^2 \tag{4.59}$$

These relationships are derived in many introductory books on probability and statistics and are left for the reader to review. Some references are Hoel (1960) and Larsen and Marx (1986).

The *mean square error* is the mean square difference between the estimator and the true value. It is also the variance, Var[·], of the estimator if the estimator is unbiased. Let $G_N(x)$ equal the sample mean of $g(x)$ using $N$ data points, or

$$G_N(x) = \frac{1}{N} \sum_{i=1}^{N} g(x_i) \tag{4.60}$$

If

$$\lim_{N \to \infty} \text{Var}[G_N(x)] = \lim_{N \to \infty} E[(G_N(x) - E[g(x)])^2] = 0 \tag{4.61}$$

then the estimator is *consistent*. For the sample mean when the random variable $x$ has a variance $\sigma^2$,

$$\text{Var}[\hat{m}] = E[(\hat{m}_N - m)^2] = \frac{\sigma^2}{N} \tag{4.62}$$

and $\hat{m}$ is a consistent estimator. Essentially, *consistency* means that the variance of the estimate approaches zero if an infinite amount of data is available. In other words, the estimate will converge to some value if enough data is used. Proving convergence can be an arduous task and its usage and implementation is reserved for advanced courses in mathematical statistics and random processes. Some treatments of this topic can be found in Papoulis (1984), Ochi (1990), and Fisz (1963).

### 4.7.2 Recursion

The general formula for calculating an estimate of the mean of a function of a random variable was given in equation 4.31. If this is implemented directly, it becomes a batch algorithm. To reduce truncation errors in machine computation, recursive algorithms can easily be developed for the sample mean. Begin again by rewriting the batch algorithm. Now separate the sum into an $(N - 1)$-point summation and the final value of $g(x_i)$. Equation 4.60 becomes

$$G_N(x) = \frac{1}{N} \sum_{i=1}^{N-1} g(x_i) + \frac{1}{N} g(x_N) \qquad (4.63)$$

Change the summation to become an $(N - 1)$-point sample mean and

$$G_N(x) = \frac{N-1}{N} \cdot \frac{1}{N-1} \sum_{i=1}^{N-1} g(x_i) + \frac{1}{N} g(x_N)$$

$$= \frac{N-1}{N} G_{N-1}(x) + \frac{1}{N} g(x_N) \qquad (4.64)$$

Thus with an estimate of the mean, one can update the estimate simply in one step with the addition of another sample point by using the weighted sum of equation 4.64. This is the recursive algorithm for calculating a sample mean of $g(x)$.

## 4.8 RANDOM NUMBERS AND SIGNAL CHARACTERISTICS

Signals with known characteristics are often needed to test signal processing procedures or to simulate signals with specific properties. This section will introduce methods to generate random signals with different first-order probability characteristics and to further exemplify the probability concepts for describing signals.

### 4.8.1 Random Number Generation

Random number generators can be used to generate random number sequences that can be used to simulate signals by simply assuming that successive numbers are also successive time samples. Almost all higher-level languages and environments have random number function calls in the system library. One of the most used algorithms is the *linear congruential generator*, and it will be described briefly here. This algorithm has the recurrence relationship

$$I(n + 1) = aI(n) + c \quad [\text{modulo } m] \quad (4.65)$$

where $I(n)$, $a$, and $c$ are integers; $b$ is the computer's word length; and $m = 2^b$. Therefore the integers range in value from 0 to $m - 1$ inclusive. The sequence is initiated with a seed number, $I(0)$, and the recursion relationship is used to generate subsequent numbers. All system functions return a floating point number, $y(n) = I(n)/m$; that is, $y(n)$ has a magnitude range $0 \le y(n) < 1$ and is uniformly distributed. Thus for these types of algorithms, $m_y = 0.5$ and $\sigma_y^2 = 0.0833$.

The goal is to produce a series of random numbers that are independent of one another. Proper choices of the seed number, multiplier, and increment in equation 4.65 ensure this. For a good sequence, choose $I(0)$ to be an odd integer, $c = 0$, and $a = 8 \cdot \text{INT} \pm 3$, where INT is some integer. The generator will produce $2^{b-2}$ numbers before repetition begins. This type of number generator is called *pseudo-random* because of this repetition and because the same sequence of numbers will be produced every time the same seed number is used. The time series produced are usually very good if one is only concerned with several sample functions. (It is usually advisable to check the uncorrelatedness of any generated sequence when using a generator for the first time).

If it is necessary to generate a large ensemble of functions, an additional shuffling manipulation must be employed to ensure complete representation of the sample space. A subroutine for producing this shuffling can be found in Press et al. (1992). Consult Press et al. (1992) and Schwartz and Shaw (1975) for more details on random number generation.

---

### EXAMPLE 4.13

Figure 4.15a shows a 100-point random sequence generated to simulate a random signal that has a uniform distribution with $m = 0.5$ and $\sigma^2 = 1/12$. Let the sampling interval be 2 seconds. The sample moments are reasonable and are $\hat{m} = 0.51$ and $s^2 = 0.086$. A histogram of amplitude values is shown in Figure 4.15b. For this histogram the bin width is 0.1. The expected values in each bin all have the value of 10, and the observed values seem reasonably close. Figure 4.15c shows an estimation of the pdf using equation 4.48 with $W = 1/10 = 0.1$ and

$$\hat{f}(x) = \frac{N_j}{10}, \qquad 1 \le j \le 10, \qquad 0.1(j - 1) \le x < 0.1j$$

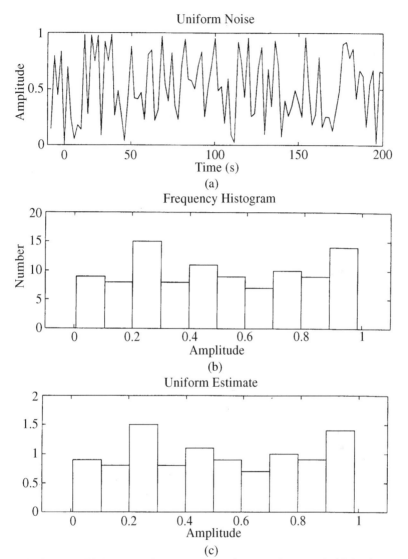

**FIGURE 4.15** (a) A sample function of a uniform random signal. (b) Its frequency histogram. (c) The estimate of its pdf.

### 4.8.2 Change of Mean and Variance

One would like to control the mean and variance of the random number generation to be more useful. For this purpose a linear transformation is very useful. Let $x(n)$ represent the signal produced by the random number generator and $y(n)$ represent the signal resulting from a transformation; that is,

$$y(n) = a + bx(n) \tag{4.66}$$

The mean and variance of $y(n)$ are easily derived as

$$E[y(n)] = E[a + bx(n)] = a + bE[x(n)] = a + bm_x \qquad (4.67)$$

$$
\begin{aligned}
\sigma_y^2 &= E[y^2(n)] - E^2[y(n)] \\
&= E[(a + bx(n))^2] - (a + bm_x)^2 \\
&= E[a^2 + 2abx(n) + b^2x^2(n)] - a^2 - 2abm_x - b^2m_x^2 = b^2\sigma_x^2 \qquad (4.68)
\end{aligned}
$$

Notice that this transformation is independent of the density functions of the random signals $x(n)$ and $y(n)$.

### 4.8.3   Density Shaping

A random number generator produces a number sequence with a uniform pdf. However, the amplitude distribution of a signal can be any pdf. Several techniques can be used to transform a uniform process into another process. The most simple approximation is to use the Central Limit Theorem. Simulation tests have shown that summing 12 uniformly distributed numbers will produce a set of numbers with a Gaussian distribution. The mean and variance must be determined to fully characterize the process. For now, neglecting the time indicator and concentrating on the summing process,

$$y = \sum_{i=1}^{N} x_i$$

$$E[y] = m_y = E\left[\sum_{i=1}^{N} x_i\right] = \sum_{i=1}^{N} E[x_i] = Nm_x \qquad (4.69)$$

Notice that the mean of a sum is the sum of the means for any process. The variance of $y$ is

$$
\begin{aligned}
Var[y] &= E[y^2] - E^2[y] \\
&= E\left[\left(\sum_{i=1}^{N} x_i\right)^2\right] - \left(\sum_{i=1}^{N} E[x_i]\right)^2 \\
&= E\left[\sum_{i=1}^{N}\sum_{j=1}^{N} x_i x_j - E[x_i]E[x_j]\right] \\
&= \sum_{i=1}^{N}\sum_{j=1}^{N} Cov[x_i x_j] \qquad (4.70)
\end{aligned}
$$

For a signal or sequence whose successive numbers are uncorrelated, this becomes

$$Var[y] = \sum_{i=1}^{N} \sigma_x^2 = N\sigma_x^2 \qquad (4.71)$$

## EXAMPLE 4.14

A Gaussian random sequence with zero mean is approximated by summing six uniformly distributed points and using a linear transformation. The process in Example 4.13 but with 600 points is used. For $N = 6$, a random signal is produced with

$$m_y = 6 \cdot \frac{1}{2} = 3; \qquad \sigma_y^2 = 6 \cdot \frac{1}{12} = \frac{1}{2}$$

The linear transformation is simply $z = y - 3$. At 100-point sample function of the process $z(n)$ with $T = 1$ is shown in Figure 4.16. Its sample mean and variance are 0.0687 and 0.47083, respectively. These statistics match very well to the desired process.

---

To produce random processes with other pdfs, single-valued probability transformations are utilized. This topic is covered in detail in textbooks on introductory probability theory and statistics, such as Hoel (1972) and Brownlee (1960). It is assumed that the initial process, $f(x)$, is independent and has the $[0, 1]$ uniform pdf. The transformation is

$$\int_0^x f_x(\alpha)\, d\alpha = \int_{-\infty}^y f_y(\beta)\, d\beta, \qquad 0 \le x \le 1 \tag{4.72}$$

where $x$ is the number in the uniform process and $y$ is the number in the new process. The solution is

$$x = F_y(y) \quad \text{or} \quad y = F_y^{-1}(x) \tag{4.73}$$

where $F_y^{-1}(x)$ indicates the inverse solution of the desired probability distribution function.

---

## EXAMPLE 4.15

Find the transformation to create a random signal with a Rayleigh distribution from a uniformly distributed random sequence

$$x = \int_0^y \frac{\beta}{a^2} e^{-\beta^2/2a^2}\, d\beta = 1 - e^{-y^2/2a^2}$$

or

$$y = \left(2a^2 \ln\!\left(\frac{1}{1-x}\right)\right)^{0.5}$$

Figure 4.17 shows the result of such a transformation with $a = 2.0$ on a signal with a uniform pdf similar to that in Example 4.13. The sample mean and variance are 2.605 and 1.673, respectively. These sample moments correspond well to what is expected theoretically; that is, $m_y = 2.5$ and $\sigma_y^2 = 1.72$.

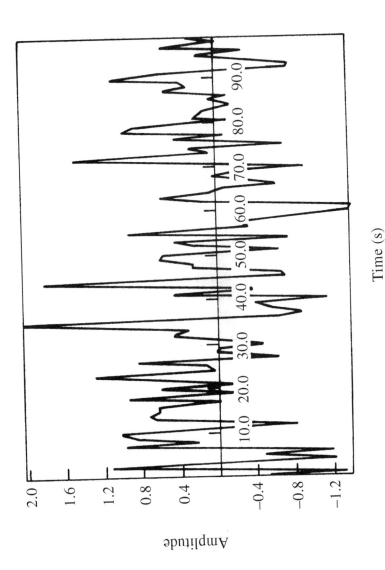

**FIGURE 4.16** A sample function of a Gaussian random signal generated by summing points in the process shown in Figure 4.15.

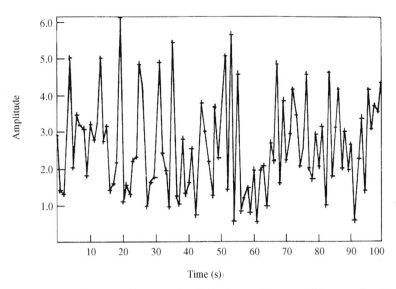

**FIGURE 4.17**   A sample function of a Rayleigh signal generated by transforming a uniform random sequence.

There will be situations in which either the inverse transformation (equation 4.73) cannot be solved or there is no mathematical form for the desired pdf. In these situations a random variable with the desired distribution can be created by using a numerical technique called the rejection method (Press et al., 1992).

## REFERENCES

M. Abramowitz and I. Stegun, *Handbook of Mathematical Functions*, Dover, New York, 1965.

A. Afifi and S. Azen, *Statistical Analysis, A Computer Oriented Approach*, Academic Press, New York, 1979.

D. Anderson and M. Correia, "The Detection and Analysis of Point Processes in Biological Signals," *Proc. IEEE* **65**:773–780 (1977).

K. Brownlee, *Statistical Theory and Methodology in Science and Engineering*, Wiley, New York, 1960.

W. Davenport, *Probability and Random Processes*, McGraw-Hill, New York, 1970.

F. de Coulon, *Signal Theory and Processing*, Artech House, Dedham, MA, 1986.

R. Fante, *Signal Analysis*, Wiley, New York, 1988.

M. Fisz, *Probability Theory and Mathematical Statistics*, Wiley, New York, 1963.

P. Hoel, *Introduction to Probability Theory*, Houghton-Mifflin Co., New York, 1972.

G. Jenkins and D. Watts, *Spectral Analysis and Its Applications*, Holden-Day, New York, 1968.

J. Komo, *Random Signal Analysis in Engineering Systems*, Academic Press, Orlando, FL, 1987.

R. Larsen and M. Marx, *An Introduction to Mathematical Statistics*, Prentice-Hall, Englewood Cliffs, NJ, 1986.

H. Larson and B. Shubert, *Probabilistic Models in Engineering Sciences*, Wiley, New York, 1979.

L. Maisel, *Probability, Statistics, and Random Processes*, Simon and Schuster, New York, 1971.

J. Milton and J. Arnold, *Introduction to Probability and Statistics, Principles and Applications for Engineering and the Computing Sciences*, McGraw-Hill, New York, 1990.

M. Ochi, *Applied Probability and Stochastic Processes*, Wiley, New York, 1990.

M. O'Flynn, *Probabilities, Random Variables, and Random Processes*, Harper & Row, New York, 1982.

R. Otnes and L. Enochson, *Digital Time Series Analysis*, Wiley, New York, 1972.

S. Pandit and S. Wu, *Time Series Analysis and Applications*, Wiley, New York, 1983.

A. Papoulis, *Probability, Random Variables, and Stochastic Processes*, McGraw-Hill, New York, 1984.

P. Peebles, *Probability, Random Variables, and Random Signal Principles*, McGraw-Hill, New York, 1987.

W. Press, B. Flannery, S. Teukolsky, and W. Vetterling, *Numerical Recipes in C— The Art of Scientific Computing*, Cambridge University Press, New York, 1992.

M. Schwartz and L. Shaw, *Signal Processing: Discrete Spectral Analysis, Detection, and Estimation*, McGraw-Hill, New York, 1975.

K. Shanmugan and A. Breipohl, *Random Signals: Detection, Estimation and Data Analysis*, Wiley, New York, 1988.

H. Stark and J. Woods, *Probability, Random Processes, and Estimation Theory for Engineers*, Prentice-Hall, Englewood Cliffs, NJ, 1986.

S. Vardeman, *Statistics for Engineering Problem Solving*, PWS, Boston, 1994.

Water Resources Data–Tennessee–Water Year 1981. U.S. Geological Survey Water—Data Report TN-81-1, p. 277.

## EXERCISES

**4.1** For $f(x) = N(20, 49)$ find the following.
   a. $P[x \leq 30]$
   b. $P[x \geq 30]$
   c. $P[15 \leq x \leq 30]$

**4.2** Prove that the multiplying factor of the exponential density function of equation 4.5 must be $\frac{1}{b}$.

**4.3** Particles from a radioactive device arrive at a detector at the average rate of 3 per second. The time of arrival can be described by an exponential pdf with $b = \frac{1}{3}$.

a. What is the probability that no more than 2 seconds will elapse before a particle is detected?

b. What is the probability that one has to wait between 2 and 5 seconds for a detection?

**4.4** A Rayleigh pdf is defined over the range $x \geq 0$ as

$$f(x) = \frac{x}{\alpha^2} \exp\left(-\frac{x^2}{2\alpha^2}\right)$$

For $\alpha = 2$, what is $P[4 \leq x \leq 10]$?

**4.5** Assume that the radii of the paint droplets in Figure 4.2 can be described by an exponential pdf with $b = 4.1$. What is the probability that the radii are greater than 20 microns?

**4.6** In a certain production line 1000-ohm $(\Omega)$ resistors that have a 10% tolerance are being manufactured. The resistance is described by a normal distribution with $m = 1000 \, \Omega$ and $\sigma = 40 \, \Omega$. What fraction will be rejected?

**4.7** The probability of dying from an amount of radiation is described by a normal random variable with a mean of 500 roentgens and a standard deviation of 150 roentgens.

a. What percentage of people will survive if each receives a dosage of less than 200 roentgens?

b. At what dosage level will only 10% of the exposed individuals survive?

**4.8** Prove equation 4.12, which relates the variance to the mean and mean square.

**4.9** Find the general formulas for the mean and variance of the uniform density function written in equation 4.4

**4.10** Consider the triangular probability function shown in Figure E4.10.

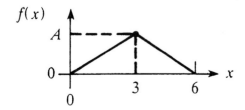

a. For this function to be a pdf, what must $A$ be?

b. Calculate $m$ and $\sigma$.

c. What is $P[x \leq 2]$ and $P[x \leq 5]$?

**4.11** What are the mean and variance of the general exponential density function, equation 4.5?

**4.12** Prove that the variance of the Rayleigh pdf is $\sigma^2 = (2 - \pi/2)\alpha^2$.

**4.13** Derive equation 4.28, the relationship between the covariance and mean product.

**4.14** Consider the bivariate density function in Example 4.7.

a. Verify that the volume is equal to 1 (equation 4.22.2).

b. Find the conditional mean of variable $y$.

**4.15** For the bivariate density function in Example 4.7, do the following.

a. Derive the marginal density function $f(y)$.

b. Derive the conditional density function $f(x|y)$.

**4.16** The lifetime of an electronic safety device is represented by a Rayleigh random variable with $\alpha^2 = 200$ and units in months. Find the conditional pdf that represents the lifetime of the device, given that it has functioned for 15 months.

**4.17** For the conditional normal pdf, derive the mean and variance (equations 4.55 and 4.56).

**4.18** For Example 4.9, verify the critical value of $t$, the estimate of the mean river flow, and its confidence bounds.

**4.19** Sulfur dioxide is a byproduct of burning fossil fuel for energy. It is carried long distances in the air and at some point is converted to acid and falls to the earth in the form of acid rain. A forest suspected of being damaged by acid rain was studied. The concentration of sulfur dioxide, micrograms per cubic meter, was measured at various locations and are listed in Table E4.19.

a. Find the average concentration of sulfur dioxide and the 95% confidence limits for the true mean.

b. In an undamaged forest the average sulfur dioxide concentration was found to be 20 micrograms per cubic meter. Does the data indicate that acid rain caused the damage in the damaged forest?

**TABLE E4.19** Sulfur Dioxide Concentrations

| | | | | | |
|------|------|------|------|------|------|
| 52.7 | 43.9 | 41.7 | 71.5 | 47.6 | 55.1 |
| 62.2 | 56.5 | 33.4 | 61.8 | 54.3 | 50.0 |
| 45.3 | 63.4 | 53.9 | 65.5 | 66.6 | 70.0 |
| 52.4 | 38.6 | 46.1 | 44.4 | 60.7 | 56.4 |

**4.20** Refer to the hospital census data tabulated in Appendix 7.6.

    a. For the first 10 days, calculate the sample mean and confidence limits for the 90% confidence intervals.

    b. Repeat part (a) for the entire table.

    c. Assume that the sample mean and variance from part (b) represent the true theoretical values of $m$ and $\sigma^2$ for the hospital census. Does the average census for the first 10 days reflect what is happening over the entire time period?

**4.21** Prove $-1 \leq \rho \leq 1$. [*Hint*: $E[(x - m_x) + (y - m_y))^2] \geq 0$.]

**4.22** Verify the calculations for testing zero correlation and finding the confidence intervals for the correlation coefficient in Example 4.12.

**4.23** A test was performed to determine whether there is a dependence between the amount of chemical ($y$, grams/liter) in solution and its temperature ($x$, degrees centrigrade) for crystallization. The measurements are listed in Table E4.23.

    a. What is the estimate of the covariance, $\hat{\sigma}_{xy}$?

    b. Normalize the estimate from part (a) to estimate the sample correlation coefficient, $\hat{\rho}$

    c. Does $\rho$ test as being zero?

    d. What are the 95% confidence limits for $\rho$?

**TABLE E4.23** Chemical Density ($y$) *and Temperature* ($x$)

| $x$ | $y$ |
| --- | --- |
| 0.3 | 3.2 |
| 0.4 | 2.4 |
| 1.2 | 4.3 |
| 2.3 | 5.4 |
| 3.1 | 6.6 |
| 4.2 | 7.5 |
| 5.3 | 8.7 |

**4.24** Estimate $\rho$ for the water discharge data in Table E2.3. Are the volume rate and gauge height linearly dependent?

**4.25** Verify the estimates of the probability density function for the transistor collector current data in Figure 4.12b.

**4.26** Verify the values of $\chi^2$ and $\chi^2_{v,\alpha}$ for fitting the normal density function to the thread tension data in Example 4.11.

**4.27** A neurophysiological experiment to investigate neuronal activity in the visual cortex of a cat's brain is being conducted. The time interval, in seconds, between firings in one neuron is measured and is organized into the histogram shown in Table E4.27.

    a. What are the sample mean and variance?

    b. Fit an exponential pdf to the observations.

    c. Test if the exponential distribution is a good model. What is the value of the degrees of freedom?

**TABLE E4.27**  Time Interval

| $N_j$ | |
| --- | --- |
| 0–15 | 63 |
| 15–30 | 25 |
| 30–45 | 14 |
| 45–60 | 7 |
| 60–75 | 6 |
| 75–90 | 3 |
| 90–105 | 2 |

**4.28** Table E4.28 lists the breaking strength, in kilograms, of plastic rods.

    a. Construct a histogram of these observations with a bin width of 0.5 kg and a lessor boundary of the first bin of 6.25 kg.

    b. Scale the histogram to estimate values of a pdf.

    c. What types of models might be suitable for these observations?

**TABLE E4.28**  Breaking Strengths (kg)

| | | | | | | | | | |
| --- | --- | --- | --- | --- | --- | --- | --- | --- | --- |
| 6.70 | 7.04 | 7.21 | 7.29 | 7.35 | 7.45 | 7.59 | 7.70 | 7.72 | 7.74 |
| 7.84 | 7.88 | 7.94 | 7.99 | 7.99 | 8.04 | 8.10 | 8.12 | 8.15 | 8.17 |
| 8.20 | 8.21 | 8.24 | 8.25 | 8.26 | 8.28 | 8.31 | 8.37 | 8.38 | 8.42 |
| 8.49 | 8.51 | 8.55 | 8.56 | 8.58 | 8.59 | 8.60 | 8.66 | 8.67 | 8.69 |
| 8.70 | 8.74 | 8.81 | 8.84 | 8.86 | 8.90 | 9.01 | 9.15 | 9.55 | 9.80 |

**4.29** Verify the derivation of the mean and variance of the linear transformation in Section 4.8.2.

**4.30** We want to produce a Gaussian white noise process, $z(n)$, with a mean of 2 and a variance of 0.5 by summing 12 uniformly distributed numbers from a random number generator. This is similar to the purpose of Example 4.14.

    a. What are the mean and variance of the summing process, $y(n)$?

    b. What linear transformation is needed so that $m_z = 2$ and $\sigma_z^2 = 0.5$?

**4.31** Find the transformation necessary to create random variables with
a. an exponential pdf,
b. a Maxwell pdf with $a = 2$,
from a uniformly distributed random variable.

**4.32** For the Rayleigh random signal produced in Example 4.15, test that the sample moments are consistent with the theoretical moments specified by the transformation.

## Computer Exercises

**4.33** a. Generate and plot independent random time series containing 100 points and having the following characteristics.
(i)  uniform pdf, $m = 0.5$, $\sigma = 1$
(ii) Maxwell pdf, $a = 2$

b. Calculate the sample mean and variance. Are they consistent with what was desired?
c. Statistically test these sample moments.

**4.34** An alternative method for generating pseudo-Gaussian random numbers uses the linear multiplicative congruential generator with the Box–Muller algorithm (Komo, 1987). The method is as follows:
a. Generate two uniform random numbers, $u(1)$ and $u(2)$, as conventionally defined.
b. Make the following angular and magnitude transformations

$$\text{ANG} = 2\pi u(1)$$
$$R = \sqrt{-2\ln(u(2))}$$

c. Two independent pseudo-Gaussian numbers are

$$x(1) = R\cos(\text{ANG}) \quad \text{and} \quad x(2) = R\sin(\text{ANG})$$

Generate 500 random numbers with this method. What are the mean and variance? Plot the histogram. Does it appear Gaussian?

**4.35** Repeat Exercise 4.28 using a computer. What are the sample mean, variance, and mean square?

**4.36** This is an exercise to study a property of the variance of the estimate of a mean value. Use the signal points of the rainfall data in Appendix 8.4.
a. Calculate the mean and variance of the entire signal.
b. Divide the signal into 26 segments containing 4 points and calculate the means of each segment. Calculate the variance of these 26 segment means. How does it compare with the signal variance?
c. Divide the signal into fewer segments (for instance, 10) containing more points and calculate the means of each segment. Then calculate the variance of these segmental means.

    d. How do the segmental variance vary with the number of points in each segment? Is this consistent with theory?

**4.37** This is an exercise to study the properties of the estimator of the mean.

    a. Using a Gaussian random number generator, generate 1000 random numbers with a mean of 0 and a variance of 1.

    b. Calculate the sample mean and sample variance of the entire signal.

    c. Divide the signal into 100 segments containing 10 points and calculate the means of each segment. Calculate the variance of these 100 segment means. How does it compare with the signal variance?

    d. Divide the signal into 75, 50, 25, and 10 segments. Then calculate the variance of each of these segmental means.

    e. How do the segmental variances vary with the number of points in each segment? Is this consistent with theory?

    f. Using the results from the division into 50 segments, calculate the upper and lower bounds for the confidence interval of the true mean. Does the true mean lie between these bounds?

**4.38** An electrocardiogram on a normal healthy individual was measured. The signal was sampled with a computer and the time intervals between heartbeats were measured. In Appendix 4.6, 180 intervals are listed. We want to study the statistical properties of these intervals to provide a basis for evaluating the time intervals from an individual with heart disease.

    a. What are $\hat{m}$ and $\hat{\sigma}^2$?

    b. Using a suitable bin width, $0.004 \le W \le 0.008$, sort these intervals into a histogram.

    c. Consider the symmetry, range, etc. of the histogram. What would be a suitable density function to describe it?

    d. Perform a $\chi^2$ test on this proposal. Is it a good model? If not, what might be a better one?

**4.39** This is an exercise to explore the concept of sampling of correlation coefficients.

    a. Generate 1000 uniform random numbers. What are the sample mean and variance? Are they close to the expected values?

    b. Equate the first 500 values to a variable $x$ and the second 500 values to a variable $y$. Estimate $\rho$. Is it close to the expected value?

    c. Divide the variables $x$ and $y$ into 20 sets of 25 paired observations. Estimate $\rho$ for each set. Test each $\hat{\rho}$ to determine whether it reflects a correlation of zero. Use a 95% confidence interval.

    d. Are any of the values of $\hat{\rho}$ close to or outside of the confidence interval? Are these observations consistent with theory?

# APPENDIX 4.1: PLOTS AND FORMULAS FOR FIVE PROBABILITY DENSITY FUNCTIONS

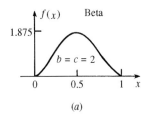

(a)

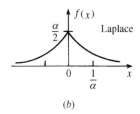

(b)

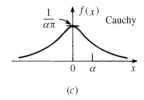

(c)

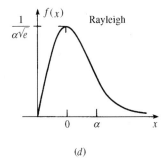

(d)

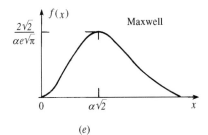

(e)

## APPENDIX 4.2:  VALUES OF THE STANDARDIZED NORMAL CDF $\Phi(z)$

$$\Phi(z) = \begin{cases} 0.5 + \Phi^+(z), & z \geq 0 \\ 1 - \Phi^+(-z), & z \leq 0 \end{cases}$$

$$\Phi^+(z) = \frac{1}{\sqrt{2\pi}} \int_0^z \exp\left(-\frac{t^2}{2}\right)$$

| $z$ | $\Phi^+(z)$ | $z$ | $\Phi^+(z)$ |
|------|------|------|------|
| 0.05 | 0.01994 | 2.05 | 0.47981 |
| 0.10 | 0.03983 | 2.10 | 0.48213 |
| 0.15 | 0.05962 | 2.15 | 0.48421 |
| 0.20 | 0.07926 | 2.20 | 0.48609 |
| 0.25 | 0.09871 | 2.25 | 0.48777 |
| 0.30 | 0.11791 | 2.30 | 0.48927 |
| 0.35 | 0.13683 | 2.35 | 0.49060 |
| 0.40 | 0.15542 | 2.40 | 0.49179 |
| 0.45 | 0.17364 | 2.45 | 0.49285 |
| 0.50 | 0.19146 | 2.50 | 0.49378 |
| 0.55 | 0.20884 | 2.55 | 0.49460 |
| 0.60 | 0.22575 | 2.60 | 0.49533 |
| 0.65 | 0.24215 | 2.65 | 0.49596 |
| 0.70 | 0.25803 | 2.70 | 0.49652 |
| 0.75 | 0.27337 | 2.75 | 0.48701 |
| 0.80 | 0.28814 | 0.85 | 0.49780 |
| 0.85 | 0.30233 | 2.85 | 0.49780 |
| 0.90 | 0.31594 | 2.90 | 0.49812 |
| 0.95 | 0.32894 | 2.95 | 0.49840 |
| 1.00 | 0.34134 | 3.00 | 0.49864 |
| 1.05 | 0.35314 | 3.05 | 0.49884 |
| 1.10 | 0.36433 | 3.10 | 0.49902 |
| 1.15 | 0.37492 | 3.15 | 0.49917 |
| 1.20 | 0.38492 | 3.20 | 0.49930 |
| 1.25 | 0.39434 | 3.25 | 0.49941 |
| 1.30 | 0.40319 | 3.30 | 0.49951 |
| 1.35 | 0.41149 | 3.35 | 0.49958 |
| 1.40 | 0.41924 | 3.40 | 0.49965 |
| 1.45 | 0.42646 | 3.45 | 0.49971 |
| 1.50 | 0.43319 | 3.50 | 0.49976 |
| 1.55 | 0.43942 | 3.55 | 0.49980 |

(*continued*)

| $z$ | $\Phi^+(z)$ | $z$ | $\Phi^+(z)$ |
|------|---------|------|---------|
| 1.60 | 0.44519 | 3.60 | 0.49983 |
| 1.65 | 0.45052 | 3.65 | 0.49986 |
| 1.70 | 0.45543 | 3.70 | 0.49988 |
| 1.75 | 0.45993 | 3.75 | 0.49990 |
| 1.80 | 0.46406 | 3.80 | 0.49992 |
| 1.85 | 0.46783 | 3.85 | 0.49993 |
| 1.90 | 0.47127 | 3.90 | 0.49994 |
| 1.95 | 0.47440 | 3.95 | 0.49995 |
| 2.00 | 0.47724 | 4.00 | 0.49996 |

## APPENDIX 4.3   CHART OF RIVER FLOW

*River Flow in Duck River,*
*March, 1981 (cubic feet*
*per second)*

| Day | Flow |
|------|------|
| 1 | 1590 |
| 2 | 1500 |
| 3 | 1390 |
| 4 | 1490 |
| 5 | 2110 |
| 6 | 2900 |
| 7 | 3610 |
| 8 | 3550 |
| 9 | 2990 |
| 10 | 2550 |
| 11 | 2240 |
| 12 | 2000 |
| 13 | 1810 |
| 14 | 1650 |
| 15 | 1530 |
| 16 | 1470 |
| 17 | 1390 |
| 18 | 1300 |
| 19 | 1230 |
| 20 | 1180 |
| 21 | 1120 |

*(continued)*

*River Flow in Duck River;*
*March, 1981 (cubic feet*
*per second)*

| Day | Flow |
|-----|------|
| 22  | 1100 |
| 23  | 1160 |
| 24  | 1550 |
| 25  | 2190 |
| 26  | 2480 |
| 27  | 2130 |
| 28  | 1810 |
| 29  | 1640 |
| 30  | 2730 |

## APPENDIX 4.4: STUDENTS $t$ DISTRIBUTION

Probabilities, $P$, of Exceeding $t_c$ (table entries) with $v$ degrees of freedom, two-tail test

| d.f. | $P = 0.1$ | 0.05 | 0.02 | 0.01 | 0.002 | 0.001 |
|------|-----------|--------|--------|--------|--------|--------|
| 1  | 6.314 | 12.706 | 31.821 | 63.657 | 318.31 | 636.62 |
| 2  | 2.920 | 4.303  | 6.965  | 9.925  | 22.327 | 31.598 |
| 3  | 2.353 | 3.182  | 4.541  | 5.841  | 10.214 | 12.924 |
| 4  | 2.132 | 2.776  | 3.747  | 4.604  | 7.173  | 8.610  |
| 5  | 2.015 | 2.571  | 3.365  | 4.032  | 5.893  | 6.869  |
| 6  | 1.943 | 2.447  | 3.143  | 3.707  | 5.208  | 5.959  |
| 7  | 1.895 | 2.365  | 2.998  | 3.499  | 4.785  | 5.408  |
| 8  | 1.860 | 2.306  | 2.896  | 3.355  | 4.501  | 5.041  |
| 9  | 1.833 | 2.262  | 2.821  | 3.250  | 4.297  | 4.781  |
| 10 | 1.812 | 2.228  | 2.764  | 3.169  | 4.144  | 4.587  |
| 11 | 1.796 | 2.201  | 2.718  | 3.106  | 4.025  | 4.437  |
| 12 | 1.782 | 2.179  | 2.681  | 3.055  | 3.930  | 4.318  |
| 13 | 1.771 | 2.160  | 2.650  | 3.012  | 3.852  | 4.221  |
| 14 | 1.761 | 2.145  | 2.624  | 2.977  | 3.787  | 4.140  |
| 15 | 1.753 | 2.131  | 2.602  | 2.947  | 3.733  | 4.073  |
| 16 | 1.746 | 2.120  | 2.583  | 2.921  | 3.686  | 4.015  |
| 17 | 1.740 | 2.110  | 2.567  | 2.898  | 3.646  | 3.965  |
| 18 | 1.734 | 2.101  | 2.552  | 2.878  | 3.610  | 3.922  |
| 19 | 1.729 | 2.093  | 2.539  | 2.861  | 3.579  | 3.883  |
| 20 | 1.725 | 2.086  | 2.528  | 2.845  | 3.552  | 3.850  |
| 21 | 1.721 | 2.080  | 2.518  | 2.831  | 3.527  | 3.819  |
| 22 | 1.717 | 2.074  | 2.508  | 2.819  | 3.505  | 3.792  |
| 23 | 1.714 | 2.069  | 2.500  | 2.807  | 3.485  | 3.767  |

*(continued)*

| d.f. | $P = 0.1$ | 0.05 | 0.02 | 0.01 | 0.002 | 0.001 |
|------|-----------|-------|-------|-------|-------|-------|
| 24 | 1.711 | 2.064 | 2.492 | 2.797 | 3.467 | 3.745 |
| 25 | 1.708 | 2.060 | 2.485 | 2.787 | 3.450 | 3.725 |
| 26 | 1.706 | 2.056 | 2.479 | 2.779 | 3.435 | 3.707 |
| 27 | 1.703 | 2.052 | 2.473 | 2.771 | 3.421 | 3.690 |
| 28 | 1.701 | 2.048 | 2.467 | 2.763 | 3.408 | 3.674 |
| 29 | 1.699 | 2.045 | 2.462 | 2.756 | 3.396 | 3.659 |
| 30 | 1.697 | 2.042 | 2.457 | 2.750 | 3.385 | 3.646 |
| 40 | 1.684 | 2.021 | 2.423 | 2.704 | 3.307 | 3.551 |
| 60 | 1.671 | 2.000 | 2.390 | 2.660 | 3.232 | 3.460 |
| 120 | 1.658 | 1.980 | 2.358 | 2.617 | 3.160 | 3.373 |
| $\infty$ | 1.645 | 1.960 | 2.326 | 2.576 | 3.090 | 3.291 |

The last row of the table ($\infty$) gives values of $z$, the standard normal variable.

## APPENDIX 4.5:   CHI-SQUARE DISTRIBUTION

Probabilities, $P$, of Exceeding $\chi^2$ (table entries) with $v$ Degrees of Freedom

| d.f. \ P | 0.995 | 0.975 | 0.050 | 0.025 | 0.010 | 0.005 | 0.001 |
|------|-------|-------|-------|-------|-------|-------|-------|
| 1 | $3.9 \times 10^{-5}$ | $9.8 \times 10^{-4}$ | 3.84 | 5.02 | 6.63 | 7.88 | 10.83 |
| 2 | 0.010 | 0.051 | 5.99 | 7.38 | 9.21 | 10.60 | 13.81 |
| 3 | 0.071 | 0.22 | 7.81 | 9.35 | 11.34 | 12.84 | 16.27 |
| 4 | 0.21 | 0.48 | 9.49 | 11.14 | 13.28 | 14.86 | 18.47 |
| 5 | 0.41 | 0.83 | 11.07 | 12.83 | 15.09 | 16.75 | 20.52 |
| 6 | 0.68 | 1.24 | 12.59 | 14.45 | 16.81 | 18.55 | 22.46 |
| 7 | 0.99 | 1.69 | 14.07 | 16.01 | 18.48 | 20.28 | 24.32 |
| 8 | 1.34 | 2.18 | 15.51 | 17.53 | 20.09 | 21.96 | 26.13 |
| 9 | 1.73 | 2.70 | 16.92 | 19.02 | 21.67 | 23.59 | 27.88 |
| 10 | 2.16 | 3.25 | 18.31 | 20.48 | 23.21 | 25.19 | 29.59 |
| 11 | 2.60 | 3.82 | 19.68 | 21.92 | 24.73 | 26.76 | 31.26 |
| 12 | 3.07 | 4.40 | 21.03 | 23.34 | 26.22 | 28.30 | 32.91 |
| 13 | 3.57 | 5.01 | 22.36 | 24.74 | 27.69 | 29.82 | 34.53 |
| 14 | 4.07 | 5.63 | 23.68 | 26.12 | 29.14 | 31.32 | 36.12 |
| 15 | 4.60 | 6.26 | 25.00 | 27.49 | 30.58 | 32.80 | 37.70 |
| 16 | 5.14 | 6.91 | 26.30 | 28.85 | 32.00 | 34.27 | 39.25 |
| 17 | 5.70 | 7.56 | 27.59 | 30.19 | 33.41 | 35.72 | 40.79 |
| 18 | 6.26 | 8.23 | 28.87 | 31.53 | 34.81 | 37.16 | 42.31 |
| 19 | 6.84 | 8.91 | 30.14 | 32.85 | 36.19 | 38.58 | 43.82 |

(continued)

Probabilities, $P$, of Exceeding $\chi^2$ (table entries) with $v$ Degrees of Freedom

| d.f. \ P | 0.995 | 0.975 | 0.050 | 0.025 | 0.010 | 0.005 | 0.001 |
|---|---|---|---|---|---|---|---|
| 20 | 7.43 | 9.59 | 31.41 | 34.17 | 37.57 | 40.00 | 45.32 |
| 21 | 8.03 | 10.28 | 32.67 | 35.48 | 38.93 | 41.40 | 46.80 |
| 22 | 8.64 | 10.98 | 33.92 | 36.78 | 40.29 | 42.80 | 48.27 |
| 23 | 9.26 | 11.69 | 35.17 | 38.08 | 41.64 | 44.18 | 49.73 |
| 24 | 9.89 | 12.40 | 36.42 | 39.36 | 42.98 | 45.56 | 51.18 |
| 25 | 10.52 | 13.12 | 37.65 | 40.65 | 44.31 | 46.93 | 52.62 |
| 26 | 11.16 | 13.84 | 38.89 | 41.92 | 45.64 | 48.29 | 54.05 |
| 27 | 11.81 | 14.57 | 40.11 | 43.19 | 46.96 | 49.64 | 55.48 |
| 28 | 12.46 | 15.31 | 41.34 | 44.46 | 48.28 | 50.99 | 56.89 |
| 29 | 13.12 | 16.05 | 42.56 | 45.72 | 49.59 | 52.34 | 58.30 |
| 30 | 13.79 | 16.79 | 43.77 | 46.98 | 50.89 | 53.67 | 59.70 |
| 40 | 20.71 | 24.43 | 55.76 | 59.34 | 63.69 | 66.77 | 73.40 |
| 50 | 27.99 | 32.36 | 67.50 | 71.42 | 76.16 | 79.49 | 86.66 |
| 60 | 35.53 | 40.48 | 79.08 | 83.30 | 88.38 | 91.95 | 99.61 |
| 70 | 43.28 | 48.76 | 90.53 | 95.02 | 100.43 | 104.22 | 112.32 |
| 80 | 51.17 | 57.15 | 101.88 | 106.63 | 112.33 | 116.32 | 124.84 |
| 90 | 59.20 | 65.65 | 113.15 | 118.14 | 124.12 | 128.30 | 137.21 |
| 100 | 67.33 | 74.22 | 124.34 | 129.56 | 135.81 | 140.17 | 149.44 |

For degrees of freedom $f > 100$, test $\sqrt{2\chi^2_{(f)}}$ as $N(\sqrt{2f-1}, 1)$.

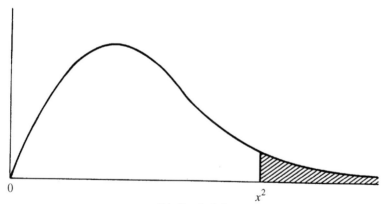

$P$ is the shaded area.

## APPENDIX 4.6: HEARTBEAT MEASUREMENTS

Time Intervals Between Heartbeats, (seconds).

| | | | | | | | | | |
|---|---|---|---|---|---|---|---|---|---|
| 0.796 | 0.803 | 0.803 | 0.810 | 0.810 | 0.813 | 0.813 | 0.816 | 0.816 | 0.816 |
| 0.820 | 0.820 | 0.820 | 0.820 | 0.823 | 0.823 | 0.823 | 0.823 | 0.823 | 0.826 |
| 0.826 | 0.826 | 0.826 | 0.826 | 0.830 | 0.830 | 0.830 | 0.830 | 0.830 | 0.830 |
| 0.830 | 0.833 | 0.833 | 0.833 | 0.833 | 0.833 | 0.833 | 0.836 | 0.836 | 0.836 |
| 0.836 | 0.836 | 0.836 | 0.836 | 0.836 | 0.836 | 0.836 | 0.840 | 0.840 | 0.840 |
| 0.840 | 0.840 | 0.840 | 0.840 | 0.840 | 0.840 | 0.840 | 0.840 | 0.843 | 0.843 |
| 0.843 | 0.843 | 0.843 | 0.843 | 0.843 | 0.843 | 0.843 | 0.843 | 0.843 | 0.843 |
| 0.843 | 0.846 | 0.846 | 0.846 | 0.846 | 0.846 | 0.846 | 0.846 | 0.846 | 0.846 |
| 0.846 | 0.846 | 0.846 | 0.846 | 0.850 | 0.850 | 0.850 | 0.850 | 0.850 | 0.850 |
| 0.850 | 0.850 | 0.850 | 0.850 | 0.850 | 0.850 | 0.850 | 0.850 | 0.853 | 0.853 |
| 0.853 | 0.853 | 0.853 | 0.853 | 0.853 | 0.853 | 0.853 | 0.853 | 0.853 | 0.853 |
| 0.853 | 0.853 | 0.853 | 0.856 | 0.856 | 0.856 | 0.856 | 0.856 | 0.856 | 0.856 |
| 0.856 | 0.856 | 0.856 | 0.856 | 0.856 | 0.856 | 0.856 | 0.856 | 0.856 | 0.860 |
| 0.860 | 0.860 | 0.860 | 0.860 | 0.860 | 0.860 | 0.860 | 0.860 | 0.860 | 0.860 |
| 0.860 | 0.860 | 0.860 | 0.863 | 0.863 | 0.863 | 0.863 | 0.863 | 0.863 | 0.863 |
| 0.863 | 0.863 | 0.863 | 0.863 | 0.863 | 0.866 | 0.866 | 0.866 | 0.866 | 0.866 |
| 0.866 | 0.866 | 0.866 | 0.870 | 0.870 | 0.870 | 0.870 | 0.870 | 0.870 | 0.870 |
| 0.873 | 0.873 | 0.873 | 0.873 | 0.876 | 0.876 | 0.880 | 0.883 | 0.886 | 0.896 |

# V

---

# INTRODUCTION TO RANDOM PROCESSES AND SIGNAL CORRELATION

## 5.1 INTRODUCTION

A random process is a random variable with an additional dimension, time. For each measurement or outcome of an experiment, there exists a time function instead of a single number. This is true for all signal and time series measurements. For example, each time there is a large explosion or sudden movement in the Earth's tectonic plates, seismic waves are produced that travel considerable distances and can be studied. Figure 5.1 shows an example of such a wave. So instead of a single data point for each event, there exists a record of movement over time. This situation is described by assigning two independent arguments, $t$ and $\zeta$, to a random process $x(t, \zeta)$, as depicted in Figure 5.2. The variable $\zeta_n$ indicates the time function, $x(t)$, for the $n$th outcome. Each realization $x(t, \zeta_0), \ldots, x(t, \zeta_n)$ is a *sample function* and the set of sample functions is an *ensemble*. The laws of probability and techniques of statistics are applied by describing the behavior of all the processes at a specific time, $t_0$, as a random variable. This is illustrated in Figure 5.2; $x(t_0, \zeta)$ is a random variable. The behavior of the process at another time, $t_1$, is also a random variable, $x(t_1, \zeta)$. For simplicity, the argument $\zeta$ will be dropped when discussing the ensemble

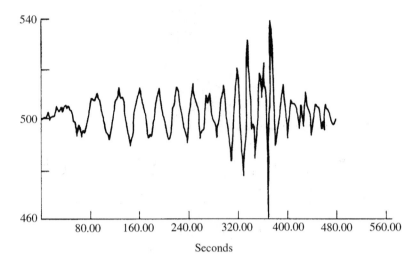

**FIGURE 5.1** Seismic signal, microns/sec, from a nuclear explosion that was recorded at Kongsberg, Norway. [Adapted from Burton, fig. 1, with permission.]

random variable but the time argument will remain explicit; its probability density function is $f_{x(t_1)}(\alpha)$.

The description of the properties of random processes is accomplished by extending the probabilistic description of random variables to include the additional independent variable, time. The presentation of this description will begin with a definition of stationarity, followed by a definition of ensemble moment functions for time domain signals. The estimation of these functions will be considered by using estimators that operate on sample functions. As always, the suitability of the estimators must be evaluated and, most important, the validity of using time domain averages for estimating ensemble averages must be discussed. After this validity is established, several important functions such as the autocorrelation function will be studied. Finally, to develop a better sense of the concept of structure and correlation in a signal, the simulation of signals other than white noise and some methods to create them will be described. The use of simulation methods will be continued in the next chapter.

## 5.2 DEFINITION OF STATIONARITY

The behavior of any random process is not necessarily the same at different times. For example, consider $x(t)$ at times $t_1$ and $t_2$; in general, $f_{x(t_1)}(\alpha) \neq f_{x(t_2)}(\alpha)$. It is highly desirable that this situation be untrue. This leads to the formal definition of *stationarity*. If

$$f_{x(t_1)}(\alpha) = f_{x(t_2)}(\alpha) \qquad (5.1)$$

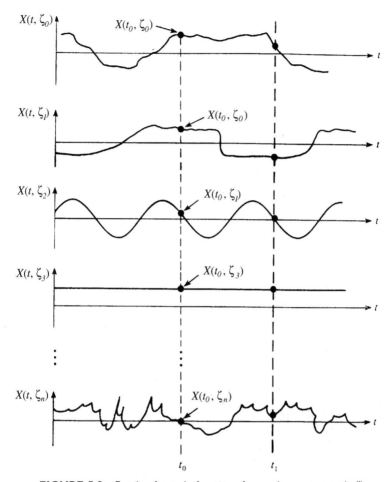

**FIGURE 5.2** Family of sample functions for random process $x(t, \zeta)$.

then the process is *first-order stationary*. This is extended to higher orders of joint probability. In the previous chapter, the joint probability between two random variables was defined. This same concept is used to describe the relationship between values of $x(t)$ at two different times, $t_1$ and $t_2$. The number of variables being related by a pdf is its order. Thus a second-order pdf is $f_{x(t_1)x(t_2)}(\alpha_1, \alpha_2)$. Again, it is desirable that the absolute time not be important. Consider the relationship between $x(t_1)$ and $x(t_2)$ and shift the time by $\tau$ units. If

$$f_{x(t_1+\tau)x(t_2+\tau)}(\alpha_1, \alpha_2) = f_{x(t_1)x(t_2)}(\alpha_1, \alpha_2) \tag{5.2}$$

then the process is *second-order stationary*. Notice that the second-order pdf does not depend on absolute time but only the time difference between $t_1$ and $t_2$. If the pdfs for all orders are equal with a time shift, that is,

$$f_{x(t_1+\tau)\cdots x(t_n+\tau)}(\alpha_1, \ldots, \alpha_n, \ldots) = f_{x(t_1)\cdots x(t_n)}(\alpha_1, \ldots, \alpha_n, \ldots) \tag{5.3}$$

then the process is *strictly stationary*—that is, stationary in all orders. Rarely is it necessary or feasible to demonstrate or have strict stationarity. In fact, it is usually difficult to even ascertain second-order stationarity. It is usually sufficient to have only stationarity of several moments for most applications. A very useful form of stationarity is when the mean and covariance do not change with time. This is called *wide-sense* or *weak stationarity*, which is defined when

$$E[x(t_1)] = E[x(t_2)] \tag{5.4}$$

and

$$\text{Cov}[x(t_1), x(t_2)] = \text{Cov}[x(t_1), x(t_1 + \tau)] \tag{5.5}$$

Equation 5.5 defines the autocovariance function for a stationary process, and $\tau$ is the time difference between the values of the process $x(t)$. Thus, a wide-sense stationary process is stationary in the mean and covariance. Many types of stationarity can be defined, but they are not relevant to the material in this textbook. If one is interested in this topic, consult textbooks such as Papoulis (1984) on stochastic processes.

An omnipresent example of a nonstationary signal is speech. Figure 5.3 shows the sound intensity from the utterance of the phase "should we chase those cowboys?" and the electroglottograph (EGG). The EGG is a measure of the amount of opening in the glottis. Notice how the characteristics of both waveforms, particularly the spread of amplitude values in the speech, change with time. Most signals are stationary over longer periods of time. Figure 5.4 shows an electroencephalogram (EEG) that is stationary over longer time intervals; the stationary periods are separated by the vertical lines.

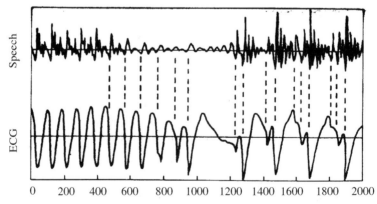

**FIGURE 5.3** The speech and EGG waveforms during the utterance of the phrase "should be chase those cowboys?". A more positive EGG magnitude indicates a greater glottal opening. [Apapted from Childers and Labar, fig. 6, with permission.]

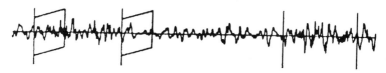

**FIGURE 5.4** EEG signal. Vertical lines indicate period within which the EEG is stationary. [Adapted from Bodenstein and Praetorius, fig. 3, with permission.]

## 5.3 DEFINITION OF MOMENT FUNCTIONS

### 5.3.1 General Definitions

The general definition of an ensemble moment is an extension of the definition of a moment of a random variable with the addition of a time argument. That is

$$E[g(x(t_1))] = \int_{-\infty}^{\infty} g(\alpha) f_{x(t_1)}(\alpha) \, d\alpha \tag{5.6}$$

Several particular ensemble moments are extremely important and must be explicitly defined and studied. The moments now become functions of time. The *ensemble average* is

$$m_x(t_1) = E[x(t_1)] = \int_{-\infty}^{\infty} \alpha f_{x(t_1)}(\alpha) \, d\alpha \tag{5.7}$$

If the process is first-order stationary, then equation 5.4 is true and the time argument can be dropped; that is, $m_x(t_1) = m_x$. As can be easily seen, all first-order moments—such as the variance and mean square—would be independent of time. The converse is not necessarily true. Stationarity of the mean does not imply first-order stationarity. The *ensemble covariance* in equation 5.5 is

$$\text{Cov}[x(t_1), x(t_2)] = \int_{-\infty}^{\infty} \int_{-\infty}^{\infty} (\alpha_1 - m_x(t_1))(\alpha_2 - m_x(t_2)) f_{x(t_1)x(t_2)}(\alpha_1, \alpha_2) \, d\alpha_1 \, d\alpha_2 \tag{5.8}$$

The notation in equation 5.8 is quite cumbersome. The simplified notation is

$$\gamma_x(t_1, t_2) = \text{Cov}[x(t_1), x(t_2)] \tag{5.9}$$

and $\gamma_x(t_1, t_2)$ is more commonly known as the *autocovariance function (ACVF)*. Just as with random variables, with random processes there is a relationship between the covariance and the mean product. It is

$$\gamma_x(t_1, t_2) = E[x(t_1)x(t_2)] - E[x(t_1)]E[x(t_2)] \tag{5.10}$$

and its proof is left as an exercise. The mean product is called the *autocorrelation function (ACF)* and is symbolized by

$$\varphi_x(t_1, t_2) = E[x(t_1)x(t_2)] \tag{5.11}$$

when $t_2 = t_1$, $\gamma_x(t_1, t_1) = \sigma_x^2(t_1)$, the variance as a function of time.

When the process is second-order stationary, all the second-order moments are functions of the time difference $\tau = t_2 - t_1$. This can be proven through the general definition of a moment with the time expressed explicitly. Then

$$E[g(x(t_1), x(t_2))] = \int_{-\infty}^{\infty} \int_{-\infty}^{\infty} g(x(t_1), x(t_2)) f(x(t_1), x(t_2)) \, dx(t_1) \, dx(t_2) \quad (5.12)$$

Let $t_2 = t_1 + \tau$ and reexamine equation 5.12. The integrand is now a function of the time parameters $t_1$ and $\tau$. Because, by definition of stationarity, the moment cannot be a function of $t_1$, it is only a function of the time parameter $\tau$.

### 5.3.2 Moments of Stationary Processes

Several moments similar to the covariance are extremely important in signal processing for stationary situations. A process that is stationary in autocovariance and autocorrelation has

$$\gamma_x(t_1, t_2) = \gamma_x(t_2 - t_1) = \gamma_x(\tau) \quad (5.13)$$
$$\sigma_x^2(t_1) = \sigma_x^2$$

and

$$\varphi_x(t_2, t_1) = \varphi_x(\tau)$$

The autocovariance function indicates any *linear dependence* between the values of the random process, $x(t)$, occurring at different times or, when $x(t)$ is stationary, between values separated by $\tau$ time units. This dependence is common and occurs in the sampled signal plotted in Figure 5.5. Successive points are correlated and

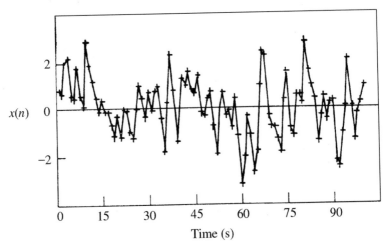

**FIGURE 5.5** A time series containing 100 points. [Adapted from Fuller, fig. 2.1.1, with permission.]

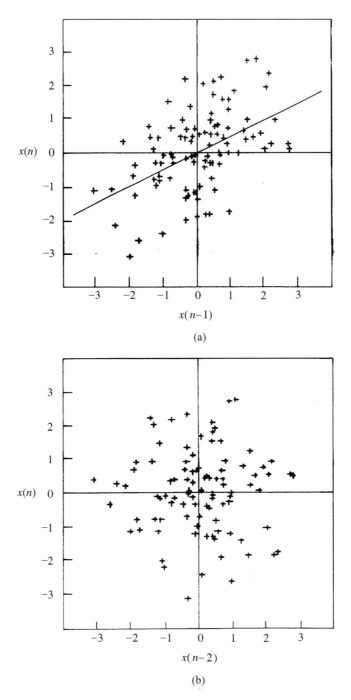

$x(n-1)$

(a)

$x(n-2)$

(b)

**FIGURE 5.6** Scatter plots for the time series, $x(n)$, in Figure 5.5: (a) $x(n)$ vs $x(n-1)$. (b) $x(n)$ vs $x(n-2)$. [Adapted from Fuller, figs. 2.2.2 and 2.1.3, with permission.]

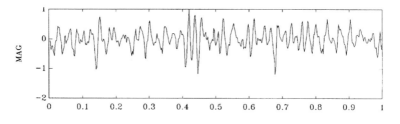

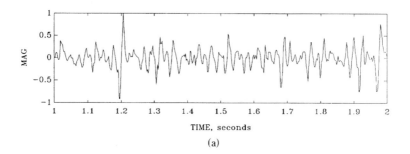

TIME, seconds

(a)

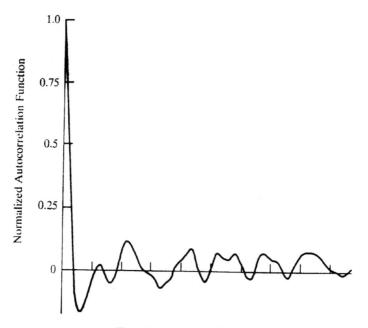

Time Displacement (8 ms per Division)

(b)

**FIGURE 5.7**   (a) A surface EMG during an isometric, constant-force contraction, (b) Its estimated NACF. [Adapted from Kwatney et al., fig. 5, with permission.]

this can be easily ascertained by studying the scatter plot in Figure 5.6. Notice, for points separated by one time unit, that the scatter plot can be described with a regression line indicating a positive correlation. For points separated by two time units, there is no correlation—Figure 5.6b shows a circular distribution of points, which is indicative of zero correlation. The ACVF can show this dependence relationship succinctly. Just as in probability theory, this function needs to be normalized to the range of the correlation coefficient as shown in the previous chapter. The *normalized autocovariance function* (*NACF*) is defined as

$$\xi_x(\tau) = \frac{\varphi_x(\tau) - m_x^2}{\sigma_x^2} = \frac{\gamma_x(\tau)}{\sigma_x^2} \tag{5.14}$$

Figure 5.7 shows an example of a stationary random process and its NACF. The signal is an electromyogram (EMG) measured from the biceps muscle while the elbow is not moving and the muscle's force is constant. Notice how the signal dependence is a function of the time separation or time lag, $\tau$. There seems to be dependence between EMG values that are separated by less than four milliseconds.

The three types of correlation functions are interrelated by equation 5.14 and have some important properties. These will be described for the wide-sense stationary situation since we will be dealing with signals having at least this condition of stationarity. These properties, in terms of the autocovariance, are as follows.

1. $\gamma_x(0) = \sigma_x^2$.
2. $\gamma_x(\tau) = \gamma_x(-\tau)$, even function.
3. If $\gamma_x(\tau) = \gamma_x(\tau + P)$ for all $\tau$, then $x(t)$ is periodic with period $P$.
4. $|\gamma_x(\tau)| \leq \gamma_x(0)$.

A process with no linear dependence among values is a *white-noise* (*WN*) process. It usually has a zero mean and $\gamma_x(\tau) = 0$ for $\tau \neq 0$. More will be stated about correlation functions and their uses in subsequent sections of this chapter.

## 5.4 TIME AVERAGES AND ERGODICITY

In signal analysis, just as in statistics, the goal is to extract information about a random process by estimating the moments and characteristics of its probability description. However, it is much more difficult to measure or often unfeasible to develop a large ensemble of sample functions. It is thus desirable to ascertain properties and parameters of a random process from a single sample function. This means using discrete time averages as estimators for ensembles averages.

For instance, for estimating the mean of a signal, the time average is

$$\hat{m} = \lim_{N \to \infty} \frac{1}{2N + 1} \sum_{n=-N}^{N} x(n) \tag{5.15}$$

where $2N + 1$ is the number of sample points. Two questions immediately arise: (1) Is equation 5.15 convergent to some value? (2) If so, does $\hat{m} = E[x]$? One aspect becomes obvious; The process must be stationary in the mean since only one value of $\hat{m}$ is produced. So these two questions simplify to determining whether $\hat{m} = E[x]$, which is the topic of *ergodicity*. Much of the proof of the ergodic theorems requires knowledge of convergence theorems that are beyond the goals of this textbook. However, some relevant results will be summarized. Good and readable proofs can be found in Papoulis (1984), Davenport (1970), and Gray and Davidson (1986).

A random process is said to satisfy an ergodic theorem if the time average converges to some value. This value is not necessarily the ensemble average. However, for our purposes this discussion will focus on convergence of time averages to ensemble averages in processes that are at least wide-sense stationary. Essentially, we will consider the bias and consistency of time averages. For the time domain, the sample mean is

$$\hat{m}_N = \frac{1}{N} \sum_{n=0}^{N-1} x(n) \tag{5.16}$$

where $x(n)$ is the sampled sequence of the continuous process $x(t)$. The mean value of the sample mean is

$$E[\hat{m}_N] = E\left(\frac{1}{N} \sum_{n=0}^{N-1} x(n)\right) = \frac{1}{N} \sum_{n=0}^{N-1} E[x(n)] = \frac{1}{N} \sum_{n=0}^{N-1} m = m \tag{5.17}$$

and the estimate is unbiased. For convergence it must be shown that equation 5.16 is a consistent estimator. Its variance is

$$\sigma_{\hat{m}}^2 = E[(\hat{m}_N - m)^2]$$

$$= E\left(\left(\left(\frac{1}{N} \sum_{n=0}^{N-1} x(n)\right) - m\right)^2\right)$$

$$= E\left(\left(\frac{1}{N} \sum_{n=0}^{N-1} (x(n) - m)\right)^2\right) \tag{5.18}$$

Using dummy variables for the summing indices that indicate the time sample, equation 5.18 becomes

$$
\begin{aligned}
\sigma_{\hat{m}}^2 &= \frac{1}{N^2} E\left( \sum_{i=0}^{N-1} (x(i) - m) \sum_{j=0}^{N-1} (x(j) - m) \right) \\
&= \frac{1}{N^2} E\left( \sum_{i=0}^{N-1} \sum_{j=0}^{N-1} (x(i) - m)(x(j) - m) \right) \\
&= \frac{1}{N^2} \left( \sum_{i=0}^{N-1} \sum_{j=0}^{N-1} E[(x(i) - m)(x(j) - m)] \right) \\
&= \frac{1}{N^2} \sum_{i=0}^{N-1} \sum_{j=0}^{N-1} \gamma_x(iT, jT)
\end{aligned}
\tag{5.19}
$$

The function $\gamma_x(iT, jT)$ is the discrete time representation of the autocovariance function, equation 5.10. Equation 5.19 can be simplified into a useful form using two properties of the ACVF of a stationary process. First, the time difference is $kT = jT - iT$ and $k$ represents the *lag interval*, the number of sampling intervals within the time difference. Thus, $\gamma_x(iT, jT)$ becomes $\gamma_x(k)$ since the sampling interval is understood. Second, $\gamma_x(k) = \gamma_x(-k)$. The variance of the time series estimator for the mean becomes

$$
\sigma_{\hat{m}}^2 = \frac{1}{N^2} \sum_{i=0}^{N-1} \sum_{j=0}^{N-1} \gamma_x(i - j) = \frac{\sigma_x^2}{N} + \frac{2}{N} \sum_{k=1}^{N-1} \left( 1 - \frac{k}{N} \right) \gamma_x(k)
\tag{5.20}
$$

For the mean value, the convergence in the limit as $N \to \infty$ depends on the ACVF. The first term on the right side of equation 5.20 approaches zero if variance of the process is finite. The second term approaches zero if $\gamma_x(k) = 0$ for $k > M$ and $M$ is finite. In other words, the values of the random process that are far apart in time must be uncorrelated. Both of these conditions are satisfied in almost all engineering applications. Formalizing this situation, it is defined that a weakly stationary process is *ergodic in the mean*, that is,

$$
\lim_{N \to \infty} \hat{m}_N = E[x(t)]
\tag{5.21}
$$

$$
\gamma_x(0) < \infty \quad \text{and} \quad \lim_{k \to \infty} \gamma_x(k) \to 0
\tag{5.22}
$$

Note that because the estimator is unbiased, the variance of the estimate is also the mean squared error. Thus the time average in equation 5.16 is also a consistent estimator and can be used to estimate the mean of a random process under the conditions just stated. All of the concepts discussed in Section 4.4 for establishing confidence intervals and so forth can be applied.

### EXAMPLE 5.1

A simple example of ergodicity can be demonstrated using the ensemble of constant processes with random amplitudes such as sample function $x(t, \zeta_3)$ in Figure 5.2. Assume that the ensemble consists of six sample functions with constant amplitudes. The amplitudes are integers between 1 and 6 inclusive and are chosen by the toss of an honest die. The process is strictly stationary since nothing changes with time and the ensemble moments are

$$m_x(t) = 3.5; \quad \sigma_x^2(t) = 2.92; \quad \gamma_x(\tau) = 2.92$$

However, a time-average estimate of the mean using any sample function will yield integers and does not converge to $m_x(t)$, regardless of the number of samples used. This is because the second condition of equation 5.22 is not satisfied.

---

It is often necessary to estimate other functions or moments of a random process from a sample function besides the mean value. There are also ergodic theorems for them. Again, fortunately for most applications, the conditions naturally exist. In this section only one other function will be mentioned. The probability distribution function of the amplitudes of a random process can be estimated from the histogram of amplitude values of a sample function as in Section 4.5 if and only if the following conditions are met.

    a. The process is stationary,

    b. $F_{x(t)x(t+\tau)}(\alpha_1, \alpha_2) = F_{x(t)}(\alpha_1)F_{x(t+\tau)}(\alpha_2)$ as $\tau \to \infty$.    (5.23)

Condition (b) means that the process values must be independent for large time separations (Papoulis, 1984).

## 5.5   ESTIMATING CORRELATION FUNCTIONS

### 5.5.1   Estimator Definition

The three types of correlation functions are used for a wide range of applications including investigating the structure of a signal, estimating delay times of reflected signals for ranging, and system modeling. Before we discuss the estimation of these functions, consider the use of the normalized autocovariance function for studying the behavior of wind speeds at a site used to generate electric power through the harnessing of the wind's energy. Figures 5.8a and 5.8b show the hourly wind speeds and the estimated NACFs. The wind speeds do vary much over several hours and the NACF shows that the correlation coefficient is at least 0.6 between speeds occurring at time intervals of 10 hours or less. This is a vastly different structure from the average wind speeds considered on a daily basis. The daily average wind speed and its estimated NACF are plotted in

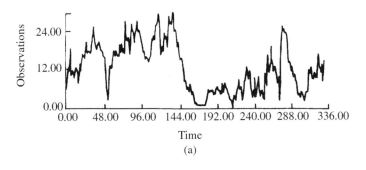

Time

(a)

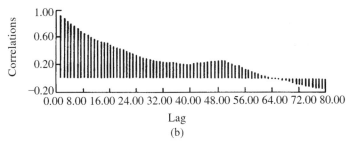

Lag

(b)

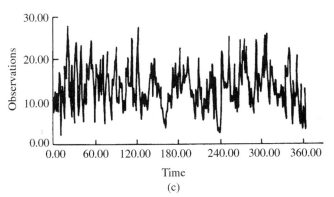

Time

(c)

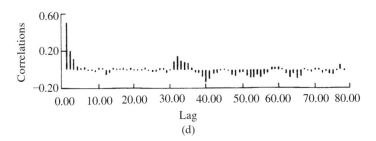

Lag

(d)

**FIGURE 5.8**  Wind speeds (in knots) at Belmullet, Ireland during 1970: (a) Hourly wind speeds during December. (b) The correlation function of (a). (c) Daily average wind speeds during 1970. (d) The autocorrelation function of (c). [Adapted from Raferty et al., figs. 4–7, with permission.]

Figures 5.8c and 5.8d. The signal has many more fluctuations and the NACF shows that only wind speeds of consecutive days have an appreciable correlation.

To obtain accurate estimates of the correlation functions so that good interpretations can be made from them, the statistical properties of their estimators must be understood. Since the properties of the three types of correlation functions are similar, only the properties of the sample autocovariance function will be studied in detail. Usage of these estimators will be emphasized in subsequent sections. Three time domain correlation functions are defined that are analogous to the ensemble correlation functions. These are the time *autocorrelation function*, $R_x(k)$, the time *autocovariance function*, $C_x(k)$, and the time *correlation function*, $\rho_x(k)$. The subscripts are not used when there is no confusion about the random variable being considered.

These functions are defined mathematically as

$$R(k) = \lim_{N \to \infty} \frac{1}{2N+1} \sum_{n=-N}^{N} x(n)x(n+k) \tag{5.24}$$

$$C(k) = R(k) - m^2 \tag{5.25}$$

$$\rho(k) = \frac{C(k)}{C(0)} \tag{5.26}$$

For the present assume that the mean of the signal is zero. Adjustments for a nonzero mean will be mentioned later. Two time-average estimators are commonly used and are finite sum versions of equation 5.24. These are

$$\hat{C}(k) = \frac{1}{N} \sum_{n=0}^{N-k-1} x(n)x(n+k) \tag{5.27}$$

and

$$\hat{C}'(k) = \frac{1}{N-k} \sum_{n=0}^{N-k-1} x(n)x(n+k) \tag{5.28}$$

Notice that the only difference between these two estimators is the divisor of the summation. This difference affects their sampling distribution greatly for a finite $N$. Naturally, it is desired that the estimators be ergodic, and then unbiased and consistent. The bias properties are relatively easy to evaluate. Ergodicity and consistency are more difficult to evaluate but are derived within the same procedure.

## 5.5.2 Estimator Bias

The mean of the estimator in equation 5.27 when $k \geq 0$ is

$$E[\hat{C}(k)] = \frac{1}{N} E\left( \sum_{n=0}^{N-k-1} x(n)x(n+k) \right)$$

$$= \frac{1}{N} \sum_{n=0}^{N-k-1} E(x(n)x(n+k))$$

$$= \frac{1}{N} \sum_{n=0}^{N-k-1} \gamma(k) = \left( 1 - \frac{k}{N} \right)\gamma(k) \qquad (5.29)$$

The estimator is biased but for an infinite sample is unbiased; that is, $\hat{C}(k)$ is an *asymptotically unbiased* estimator of the autocovariance function. For the estimator of equation 5.28, it can be shown that it is an unbiased estimator or that

$$E[\hat{C}'(k)] = \gamma(k) \qquad (5.30)$$

## 5.5.3 Consistency and Ergodicity

The variance of these time-average estimators is examined to determine their ergodicity and convergence for finite time samples. The solution is quite complex. The biased estimator is examined in detail and elaborations of these results are made to study the sampling properties of the unbiased estimator. By definition, the variance of the biased covariance estimator for $k \geq 0$ is

$$\text{Var}[\hat{C}(k)] = E[\hat{C}^2(K)] - E^2[\hat{C}(k)]$$

$$= E\left( \frac{1}{N^2} \sum_{i=0}^{N-k-1} \sum_{j=0}^{N-k-1} x(i)x(i+k)x(j)x(j+k) \right) - \left( 1 - \frac{k}{N} \right)^2 \gamma^2(k)$$

$$= \frac{1}{N^2} \sum_{i=0}^{N-k-1} \sum_{j=0}^{N-k-1} E[x(i)x(i+k)x(j)x(j+k)] - \left( 1 - \frac{k}{N} \right)^2 \gamma^2(k)$$

$$(5.31)$$

The crucial term in the evaluation is the fourth-order moment within the summation. For processes that are fourth-order stationary and Gaussian, this moment can be expressed as a sum of autocovariance functions and is

$$E[x(i)x(i+k)x(j)x(j+k)] = \gamma^2(k) + \gamma^2(i-j) + \gamma(i-j+k)\gamma(i-j-k) \qquad (5.32)$$

For other non-Gaussian processes of practical interest, equation 5.32 is a good approximation (Bendat and Piersol, 1986). The variance expression is simplified by inserting equation 5.32 into equation 5.31 and letting $r = i - j$. The double summation reduces to a single summation and the variance expression becomes

$$\text{Var}\,[\hat{C}(k)] = \frac{1}{N} \sum_{r=-N+k+1}^{N-k-1} \left(1 - \frac{|r|+k}{N}\right)(\gamma^2(r) + \gamma(r+k)\gamma(r-k)) \tag{5.33}$$

This solution is derived in Anderson (1971), Priestley (1981), and in Appendix 5.3. An alternate expression for equatin 5.33 is

$$\text{Var}[\hat{C}(k)] \approx \frac{1}{N} \sum_{r=-\infty}^{\infty} (\gamma^2(r) + \gamma(r+k)\gamma(r+k)) \tag{5.34}$$

for all values of $k$ and large values of $N$. It can easily be seen that

$$\lim_{N\to\infty} \text{Var}[\hat{C}(k)] \to 0$$

iff

$$\lim_{k\to\infty} \gamma(k) \to 0 \quad \text{and} \quad |\gamma(0)| \leq M, \quad \text{where } M \text{ is finite}$$

Thus the random process is ergodic in autocovariance for $\hat{C}(k)$, and $\hat{C}(k)$ is also a consistent estimator. If the unbiased estimator is used, the term $(N - |k|)$ is substituted for $N$ in the variance expression in equation 5.34. Thus the process is ergodic and consistent for this estimator as well, and both are valid estimators for the ensemble autocovariance of a zero mean, fourth-order stationary process.

### 5.5.4 Sampling Properties

We can now proceed to examine the properties of the finite time estimators. For the remainder of this text we will study only estimators of functions that satisfy ergodic properties. Thus the symbols that have been used for time correlation functions will be used since they are commonly used in the engineering literature. The estimator $\hat{C}'(k)$ is unbiased and seems to be the better estimator. The other estimator, $\hat{C}(k)$, has the bias term

$$b(k) = \hat{C}(k) - C(k) = -\frac{|k|}{N}C(k) \tag{5.35}$$

The magnitude of $b(k)$ is affected by the ratio $|k|/N$ and the value of $C(k)$. To keep the bias small, care must be taken to keep the ratio small when values of $C(k)$ are appreciable. Fortunately, for the processes concerned $C(k) \to 0$ as $k \to \infty$ and the bias will be small for large lag times. So the number of samples must be large for small lags.

For a finite $N$, the variance for $\hat{C}(k)$ is equation 5.34, and for $\hat{C}'(k)$ it is

$$\text{Var}[\hat{C}'(k)] \approx \frac{1}{N-|k|} \sum_{r=-\infty}^{\infty} (\gamma^2(r) + \gamma(r+k)(r-k)) \tag{5.36}$$

Note that the variance increases greatly as $|k|$ approaches $N$. In fact, it is greater than the variance of the biased estimator by a factor of $N/(N - |k|)$. This is the reson that so much attention is given to the biased estimator. It is the *estimator of choice* and is implemented in most software algorithms. To compromise with the bias produced for large lag values, the typical *rule of thumb* is to limit the maximum lag such that $|k| \leq 0.1N$. If the mean value must also be estimated so that the estimator is

$$\hat{C}(k) = \frac{1}{N} \sum_{n=0}^{N-k-1} (x(n) - \hat{m}_N)(x(n + k) - \hat{m}_N) \tag{5.37}$$

the bias term is changed approximately by the amount $-\sigma^2(1 - |k|/N)$.

The NACF is estimated by

$$\hat{\rho}(k) = \frac{\hat{C}(k)}{\hat{C}(0)} \tag{5.38}$$

The biased estimator must be used here to ensure that

$$|\hat{\rho}(k)| \leq 1 \quad \text{for all } k \tag{5.39}$$

Equation 5.38 is a biased estimator with the same bias form as that for the autocovariance; that is,

$$E[\hat{\rho}(k)] = \left(1 - \frac{|k|}{N}\right)\rho(k) \tag{5.40}$$

The variance expression is much more complicated (Priestley, 1981). It is

$$\text{Var}[\hat{\rho}(k)] \approx \frac{1}{N} \sum_{r=-\infty}^{\infty} (\rho^2(r) + \rho(r + k)\rho(r - k) + 2\rho^2(k)\rho^2(r) - 4\rho(k)\rho(r)\rho(r - k)) \tag{5.41}$$

### 5.5.5 Asymptotic Distributions

Fortunately, in time series analysis we are almost always concerned with sample functions that have a large number of samples. Therefore, the Central Limit Theorem becomes appropriate because all of the estimators require summing many terms. The sampling distributions for the mean, autocovariance, and normalized autocovariance functions become Gaussian asymptotically with large $N$ and for certain boundary conditions on the random process (Priestley, 1981). For our purposes these conditions are satisfied if the process is weakly stationary and the estimator is ergodic, as discussed in the preceding sections. The major difference between these asymptotic distributions and the exact statistical sampling distributions in Chapter 4 arise from the size of the sample. All of the information is now available to perform hypothesis testing on desired values of the correlation functions. In practice, $\hat{C}(k)$ is used and it is assumed that the sample function is large enough that the extra bias term is negligible.

Another factor arises with correlation functions. These are functions of lag time and it is desired to test all of the values of the estimate simultaneously. This is a complex multivariate problem studied in more advanced courses. The multivariate attribute arises because $\hat{\rho}(k)$ and $\hat{\rho}(k+1)$ are themselves correlated. This attribute disappears for one very useful process, the white-noise process. Thus, in the case of white noise, we can test whether or not a sample function has a structure by knowing only the first-order sampling distribution. From equations 5.40 and 5.41, the mean and variance of the sampling distribution of the correlation function of a white-noise process are

$$E[\hat{\rho}(k)] = 0, \qquad \text{Var}[\hat{\rho}(k)] = \frac{1}{N} \quad \text{for } k \neq 0 \tag{5.42}$$

**EXAMPLE 5.2**

In an industrial process involving distillation, it is desired to know if there is any correlation between batches. Figure 5.9a shows the normalized yields over time. Notice that high values follow low values. This implies that there may be a negative correlation between successive batches. The correlation function is estimated using equation 5.38 and plotted in Figure 5.9b. The magnitude of $\hat{\rho}(1)$ is negative and the magnitudes of $\hat{\rho}(k)$ alternate in sign, which is consistent with our observation of the trends. To test for no structure in the process, the white-noise test is applied. The sampling distribution for $\rho(k)$ of the null hypothesis is Gaussian with the mean and variance given in equation 5.42. Since there are 70 series points in the time series, the variance of $\hat{\rho}(k)$ is 0.0143. The 95% confidence limits for any correlation are

$$|\hat{\rho}(k) - \rho(k)| \leq \frac{1.96}{\sqrt{N}} = 0.234$$

For zero correlation, this becomes

$$-0.234 \leq \hat{\rho}(k) \leq 0.234$$

These limits are indicated in Figure 5.9b by the horizontal dashed lines. The correlations at lags 1 and 2 are nonzero although low in magnitude. This indicates that the initial impressions are correct: Successive yields are weakly negatively correlated and the yields of every second batch are weakly positively correlated. To obtain the confidence limits for the actual value of correlation, the same boundary relationship prevails and

$$\hat{\rho}(k) - 0.234 \leq \rho(k) \leq \hat{\rho}(k) + 0.234$$

For the second lag, $\hat{\rho}(2) = 0.3$, and

$$0.069 \leq \rho(2) \leq 0.534$$

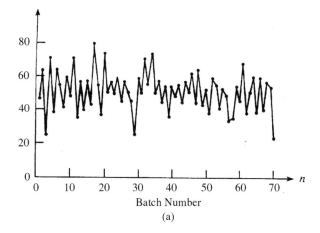

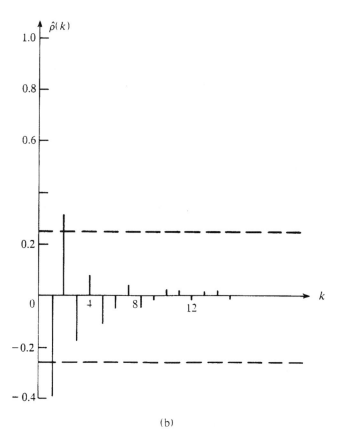

**FIGURE 5.9** (a) The yield of a distillation process over time. (b) Its sample NACF. The dashed lines indicate the zero correlation confidence limits. [Adapted from Jenkins and Watts, figs. 5.2 and 5.6, with permission.]

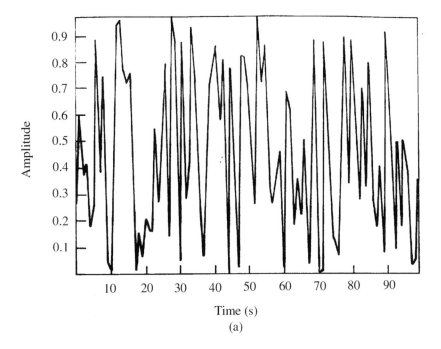

Time (s)

(a)

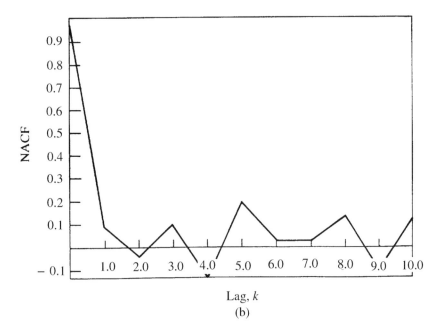

Lag, $k$

(b)

**FIGURE 5.10** (a) A sample function of a uniformly distributed white-noise process. (b) The estimate of its NACF.

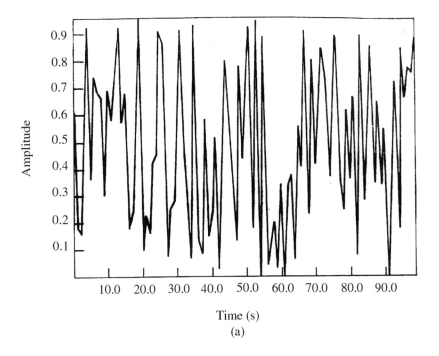

Time (s)

(a)

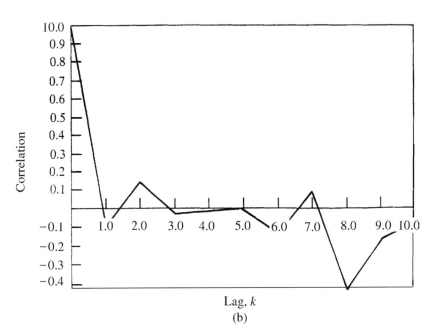

Lag, $k$

(b)

**FIGURE 5.11** (a) A sample function of a uniformly distributed white-noise process. (b) The estimate of its NACF.

## EXAMPLE 5.3

In the previous chapter we discussed using random number generators to produce a sequence of numbers that can be used to simulate a signal. It was too soon then to mention the correlation property between successive numbers. These generators effectively produce a sequence of numbers that are uncorrelated; that is, the signal simulated is white noise. Figure 5.10a shows a 100-point sequence generated using a uniform random number generator. The sample moments, $\hat{m} = 0.453$ and $s^2 = 0.0928$, are reasonable. The estimate of the NACF using equation 5.38 is plotted in Figure 5.10b and is indicative of white noise since $|\hat{\rho}(k)| \leq 0.2$.

## EXAMPLE 5.4

A second uniform white-noise signal is simulated with the same generator used in the previous example. The signal and estimate of the NACF are shown in Figure 5.11. The sample moments are $\hat{m} = 0.5197$ and $s^2 = 0.08127$. Compare these results with those of the previous example. Notice that they differ in specific numbers as expected but that the statistical properties are the same. Something else occurs that at first may seem contradictory. Perform the white-noise test on the sample correlation function. The magnitude of $\hat{\rho}(8)$ exceeds the threshold magnitude of 0.2. Should it then be concluded that this second sequence is indeed correlated? No, it should not be. Remember that a significance level of 5% means that the testing statistic will not adhere to the test inequality for 5% of the tests when the hypothesis is true. Thus, for a white-noise process, it should be expected that one out of twenty values of $\hat{\rho}(k)$ will test as nonzero. The usual indicator of this is when $\hat{\rho}(1)$ is close to zero and another value of $\hat{\rho}(k)$, $k \geq 1$, is not. As has been demonstrated in the previous examples, when an actual signal is not white noise, the correlation functions for low lags tend to have the greater values.

## 5.6   CORRELATION AND SIGNAL STRUCTURE

### 5.6.1   General Moving Average

Many signals encountered are processes that are uncorrelated and have characteristics similar to the white-noise examples. However, most signals measured have a correlation between successive points. Refer again to Figures 5.7 through 5.9. For purposes of simulation, it is necessary to take a white-noise process with zero mean, $x(n)$, and induce a correlation within it. This can be done using algorithms involving moving averages and autoregression. More will be stated about these in subsequent chapters. For the present, the basic concepts will be used to create a structure in a random process.

A *moving-average*, or MA, process, $y(n)$, requires averaging successive points of $x(n)$ using a sliding window. This is different from the procedure for creating a Gaussian white-noise process is in Example 5.5. In the latter procedure successive points of $y(n)$ were not a function of common points of $x(n)$. Remember that for a white-noise process, $R_x(k) = \sigma_x^2\, \delta(k)$. A three-point process is created by the relationship.

$$y(n) = x(n) + x(n-1) + x(n-2) \tag{5.43}$$

Notice that this is simply a sliding or moving sum. For time instant 3,

$$y(3) = x(3) + x(2) + x(1) \tag{5.44}$$

The most general form for a $(q + 1)$-point moving average also has weighting coefficients and is

$$y(n) = b(0)x(n) + b(1)x(n-1) + \cdots + b(q)x(n-q) \tag{5.45}$$

The parameter $q$ is the *order* of the moving average. Since this is a type of system, $x(n)$ and $y(n)$ are usually called the input and output processes, respectively. The mean and variance of this structure for weighting coefficients equaling 1, have already been derived in equations 4.69 and 4.71. For generality, the weighting coefficients need to be included. The mean and variance are

$$m_y = m_x \cdot \sum_{i=0}^{q} b(i) \tag{5.46}$$

$$\sigma_y^2 = \sigma_x^2 \cdot \sum_{i=0}^{q} b(i)^2 \tag{5.47}$$

The derivation is left as an exercise. The output process has an autocovariance function that depends on the order of the MA process and the values of the coefficients.

### 5.6.2 First-Order MA

The autocorrelation function for a two-point MA process will be derived in detail to illustrate the concepts necessary for understanding the procedure. Remember that $m_y = m_x = 0$.

$$R_y(0) = \sigma_y^2 = \sigma_x^2 \cdot \sum_{i=0}^{1} b(i)^2 \tag{5.48}$$

$$
\begin{aligned}
R_y(1) &= E[y(n)y(n+1)] \\
&= E[(b(0)x(n) + b(1)x(n-1))(b(0)x(n+1) + b(1)x(n))] \\
&= E[b(0)x(n)b(0)x(n+1) + b(1)x(n-1)b(0)x(n+1) \\
&\quad + b(0)x(n)b(1)x(n) + b(1)x(n-1)b(1)x(n)]
\end{aligned}
$$

Since the mean of a sum is a sum of the means, the above expectation is a sum of autocorrelations, or

$$R_y(1) = b(0)^2 R_x(1) + b(1)b(0)R_x(2) + b(0)b(1)R_x(0) + b(1)^2 R_x(1) \tag{5.49}$$

Because $x(n)$ is a white-noise process, $R_x(k) = 0$ for $k \neq 0$ and equation 5.49 simplifies to

$$R_y(1) = b(0)b(1)R_x(0) = b(0)b(1)\sigma_x^2 \tag{5.50}$$

Thus the MA procedure has induced a correlation between successive points. For the second lag,

$$
\begin{aligned}
R_y(2) &= E[y(n)y(n+2)] \\
&= E[(b(0)x(n) + b(1)x(n-1))(b(0)x(n+2) + b(1)x(n+1))] \\
&= E[b(0)x(n)b(0)x(n+2) + b(1)x(n-1)b(0)x(n+2) \\
&\quad + b(0)x(n)b(1)x(n+1) + b(1)x(n-1)b(1)x(n+1)]
\end{aligned}
$$

and

$$R_y(2) = b(0)^2 R_x(2) + b(1)b(0)R_x(3) + b(0)b(1)R_x(1) + b(1)^2 R_x(2) \tag{5.51}$$

Again, since $x(n)$ is white noise, $R_y(2) = 0$. All of the other autocorrelation function values have the same form as equation 5.51. Thus,

$$R_y(k) = 0 \quad \text{for } |k| \geq 2 \tag{5.52}$$

Naturally, $R_y(-1) = R_y(1)$. The NACF is $\rho_y(k) = R_y(k)/\sigma_y^2$ and for the two-point MA process in general is

$$
\begin{aligned}
\rho_y(0) &= 1 \\
\rho_y(\pm 1) &= \frac{b(0)b(1)}{b(0)^2 + b(1)^2} \\
\rho_y(k) &= 0 \quad \text{for } |k| \geq 2
\end{aligned}
\tag{5.53}
$$

---

**EXAMPLE 5.5**

Form a two-point moving-average process that is the mean of a Gaussian white-noise process having $m_x = 0$ and $\sigma_x^2 = 0.5$; that is, $b(0) = b(1) = 0.5$. Thus,

$$y(n) = 0.5x(n) + 0.5x(n-1)$$

and $\sigma_y^2 = 0.5\sigma_x^2 = 0.25$, $\rho_y(1) = 0.5$. White noise was used as the input, and the output signal $y(n)$ is shown in Figure 5.12a. Its sample variance is 0.263 and is close to what is expected. The estimate of the NACF is plotted in Figure 5.12b. Here, $\hat{\rho}_y(1) = 0.55$ and tests to be nonzero and is well within the 95% confidence level for the desired correlation.

---

**EXAMPLE 5.6**

The effect of a finite sample on the estimation of the correlation function can be demonstrated through simulating a signal and calculating the parameters of the sampling distribution. The correlated random process generated in the previous example is used. Naturally from the equations, $\hat{\rho}(0) = 1$ always. From equations

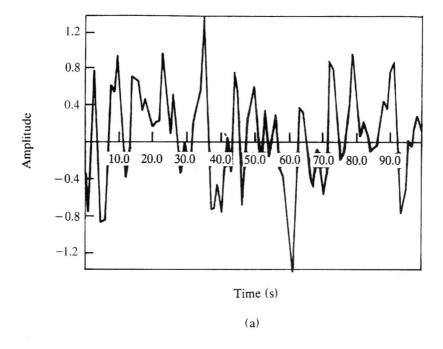

Time (s)

(a)

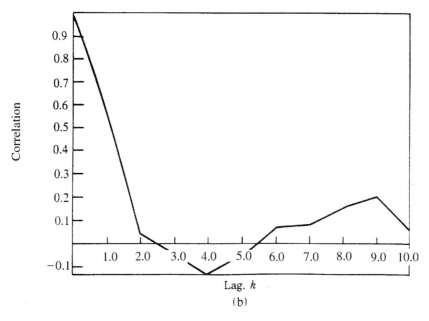

Lag, $k$

(b)

**FIGURE 5.12**  (a) A sample function of a correlated Gaussian process. (b) The estimate of its NACF.

5.40 and 5.41, the sampling mean and variance can be approximated. The sampling mean for lag $= \pm 1$ is

$$E[\hat{\rho}(\pm 1)] = \left(1 - \frac{1}{N}\right)\rho(\pm 1) = 0.5\left(1 - \frac{1}{100}\right) = 0.495$$

For practical purposes this is almost unbiased. The variance expression for $k = 1$ is

$$\text{Var}[\hat{\rho}(1)] \approx \frac{1}{N}\sum_{r=-\infty}^{\infty} (\rho^2(r) + \rho(r+1)\rho(r-1) + 2\rho^2(1)\rho^2(r)$$
$$- 4\rho(1)\rho(r)\rho(r-1))$$

and easily reduces to

$$\text{Var}[\hat{\rho}(1)] \approx \frac{1}{N}\sum_{r=-1}^{1} (\rho^2(r) + \rho(r+1)\rho(r-1) + 2\rho^2(1)\rho^2(r)$$
$$- 4\rho(1)\rho(r)\rho(r-1))$$

On a term-by-term basis, this summation becomes

Term 1: $0.25 + 1 + 0.25 = 1.5$
Term 2: $0 + 0.25 + 0 = 0.25$
Term 3: $2 \cdot 0.25 \cdot (\text{term 1}) = 0.75$
Term 4: $-4 \cdot 0.5 \cdot (0 + 0.5 + 0.5) = -2.0$

and $\text{Var}[\hat{\rho}(1)] \approx \dfrac{1}{100} \cdot 0.5 = 0.005$. The sampling mean and variance for $\hat{\rho}(-1)$ is the same. For $|k| \geq 2$;

$$E[\hat{\rho}(k)] = \left(1 - \frac{|k|}{N}\right)\rho(k) = 0$$

and the estimates are unbiased. The variance for all the lags greater than 1 are the same because only term 1 in the summation is nonzero and is

$$\text{Var}[\hat{\rho}(k)] \approx \frac{1}{N}\sum_{r=-\infty}^{\infty} \rho^2(r) = \frac{1}{100} \cdot 1.5 = 0.015, \quad |k| \geq 2$$

Compared to white noise, the variance of the estimates of the correlation function of the first-order MA process has the following two differences:

1. At lags with no correlation, the variances are larger.
2. At lags of $\pm 1$, the variance is smaller.

### 5.6.3   Second-Order MA

The general formula for the autocorrelation function of a $q$th-order MA process can be derived in a direct manner. Study again the equations in which the ACF is derived for the first-order process. Notice that the only nonzero terms arise from those terms containing $R_x(0)$. Consider a second-order process

$$y(n) = b(0)x(n) + b(1)x(n-1) + b(2)x(n-2) \qquad (5.54)$$

Its value of the ACF at lag 1 is derived from

$$
\begin{aligned}
R_y(1) &= E[y(n)y(n+1)] \\
&= E[(b(0)x(n) + b(1)x(n-1) + b(2)x(n-2))(b(0)x(n+1) \\
&\quad + b(1)x(n) + b(2)x(n-1))]
\end{aligned}
$$

As can be seen from the previous equation, the terms producing nonzero expectations are

$$b(0)x(n) \cdot b(1)x(n) \quad \text{and} \quad b(1)x(n-1) \cdot b(2)x(n-1)$$

Therefore,

$$R_y(1) = (b(0)b(1) + b(1)b(2))\sigma_x^2 \qquad (5.55)$$

Notice that there are two terms and that the indices in the subscripts of the products of the coefficients differ by a value of 1. For lag 2, the autocorrelation value is

$$R_y(2) = b(0)b(2)\sigma_x^2 \qquad (5.56)$$

Notice that there is one term and the indices of the subscript in the coefficient differ by a value of 2.

### 5.6.4  Overview

The ACF for the general $q$th-order MA process is

$$
R_y(k) = 
\begin{cases}
\sigma_x^2 \left( \displaystyle\sum_{i=0}^{q-k} b(i)b(i+k) \right), & |k| \le q \\
0, & |k| > q
\end{cases}
\qquad (5.57)
$$

Its proof is left as an exercise. Does equation 5.57 produce equations 5.55 and 5.56?

There are many types of structure in a time series. The material just presented is actually the simplest manner in which to create a correlation in a time series. More complex correlation functions can be created by using higher-order moving averages or another process called *autoregression*. In this process an output value depends on its previous values. One form is

$$y(n) = -a(1)y(n-1) + b(0)x(n) \qquad (5.58)$$

These processes will be studied in detail in Chapters 6 and 8. Naturally occurring processes usually have a more complex structure than the ones represented by first- and second-order MA processes. Review again several examples. The time series of hourly wind speeds in Figure 5.8 has a correlation function with nonzero

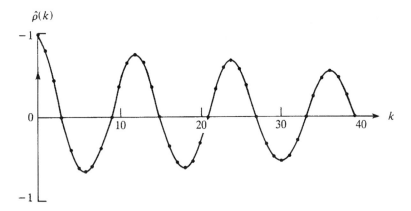

**FIGURE 5.13**   The NACF of monthly measurements of air temperature in a city. The measurements are shown in Figure 1.1. [Adapted from Chatfield, fig. 2.4a, with permission.]

values at many more lags than in the last example. A different structure is the distillation time series in Figure 5.9, which alternates in magnitude; it has an alternating correlation function to reflect this behavior. $\rho(1)$ is negative, which indicates that successive points will have opposite sign. A signal with an oscillation or a periodicity will have a periodic correlation function. An example is the air temperature plotted in Figure 1.1; its correlation function is shown in Figure 5.13.

## 5.7  ASSESSING STATIONARITY OF SIGNALS

Since the techniques for investigating the properties of signals require that the signals be at least weakly stationary, there must be methods to assess signal stationarity. The simple way to assess the stationarity of a sampled signal is to consider the physics and environment that produced it. Again, the speech waveform is an obvious situation. Simple observation of the waveform in Figure 5.3 shows that different words and utterances are composed of different signals called *formants*. If one needs to investigate the properties of a formant, the usage of short time series is necessary. In distinction, many situations are similar to the one reflected in the EEG record in Figure 5.4. Here the signal is stationary for periods greater than one second. What would cause the signals's characteristics to change? Since the EEG is a measurement of brain activity, anything that could affect the activity could cause the EEG to change. It is well known that the state of alertness or the type of mental task causes changes. Atmospheric noise that affects electromagnetic communication is nonstationary and depends on the time of day and cloud conditions. Another situation is the intensity of vibration of an airplane's wing during changes in airspeed.

What if the state of stationarity cannot be assessed by considering physical and environmental conditions? Then one can resort to statistical tests. Fortunately, in many real-world measurements, if the more complex properties such as the ACF or pdf of a signal change, so do some simple properties such as the mean or variance. Thus, testing whether there is a change in these moments over time is adequate for assessing wide-sense stationarity. If the signal also has a Gaussian pdf, testing these moments is a test of complete stationarity. The most direct procedure is to divide a signal's record into multiple segments and to compare the means and variances of each segment with each other. Standard statistical tests are available for testing the equality of sample means and variances. These two tests are summarized in Appendix 5.4. Care must be taken to ensure that the segments are long enough to reflect the signal properties. In particular, if a signal has oscillating components, several cycles of the oscillation must be included in each segment.

Another class of tests called nonparametric tests are also utilized. These are particularly useful when one does not want to make any assumptions about the underlying pdf in the signal being evaluated or when there are a lot of segments. An important and highly utilized nonparametric test is the *runs test*. This test is also summarized in Appendix 5.4. Additional discussions of stationarity testing using these tests and nonparametric tests can be found in Bendat and Piersol (1986), Otnes and Enochson (1972), and Shanmugan and Breipohl (1988).

Again consider the speech signal in Figure 5.3. Divide it into three time segments, $S_i$, of 600 ms such that

$$
\begin{aligned}
S_1 &\Rightarrow 0 \le t \le 600 \text{ ms} \\
S_2 &\Rightarrow 600 \le t \le 1200 \text{ ms} \\
S_3 &\Rightarrow 1200 \le t \le 1800 \ ms
\end{aligned}
\tag{5.59}
$$

Qualitatively compare the means and variances among the three segments. The means of the segments are slightly positive and probably are equal. However, the variances are different. One can assess this by considering the range of amplitudes. The variance of $S_2$ is much less than those of $S_1$ and $S_3$, and most likely the variance of $S_1$ is less than that of $S_3$. Thus the signal is nonstationary.

---

**EXAMPLE 5.7**

The stationarity of the distillation signal used in Example 5.2 and listed in Appendix 5.1 is being assessed using the Student's $t$ and $F$ tests to compare the means and variances, respectively. Refer to Appendix 5.4. The signal is divided into two segments such that

$$
\begin{aligned}
S_1 &\Rightarrow 1 \le n \le 35 \\
S_2 &\Rightarrow 36 \le n \le 70
\end{aligned}
$$

where $N_1 = N_2 = 35$. The sample means and variances for each segment are $\hat{m}_1 = 53.17$, $\hat{\sigma}_1^2 = 176.78$, $m_2 = 49.11$, and $\hat{\sigma}_2^2 = 96.17$.

The $T$ statistic for testing the equality of means has a $t$ distribution with the degrees of freedom being $v = N_1 + N_2 - 2$. It is

$$T = \frac{\hat{m}_1 - \hat{m}_2}{\left(\left(\dfrac{1}{N_1} + \dfrac{1}{N_2}\right)\left(\dfrac{N_1\hat{\sigma}_1^2 + N_2\hat{\sigma}_2^2}{N_1 + N_2 - 2}\right)\right)^{1/2}} = \frac{53.17 - 49.11}{\left(\dfrac{2}{35} \cdot \dfrac{35(176.78 + 96.17)}{68}\right)^{1/2}} = 1.43$$

The significance region is $|T| \geq t_{68;0.025} = 2$. The signal tests as stationary in the mean.

The $F$ statistic for testing the equality of variances uses the unbiased estimates of variances. The number of degrees of freedom are $v_1 = N_1 - 1$ and $v_2 = N_2 - 1 = 34$. The $F$ statistic is

$$F = \frac{\hat{\sigma}_1^2}{\hat{\sigma}_2^2} = 1.84$$

The 95% confidence for a two-sided test are

$$F_{34,34;0.975} \leq F \leq F_{34,34;0.025} \quad \text{or} \quad 0.508 \leq F \leq 1.97$$

Since $F$ falls within this interval, the variances are equal at the 95% confidence level and the signal is stationary in variance.

The tests show that the distillation signal is stationary in the mean and variance at a confidence level of 95%. One of the working assumptions is that any nonstationarity in the covariance function would be reflected in the nonstationarity of the variance. Thus it is concluded that the signal is wide-sense stationary.

---

**EXAMPLE 5.8**

The data for the brightness of a variable star from Appendix 3.2 is plotted in Figure 5.14. Observation of the brightness makes one suspect that there is a fluctuation in amplitude and energy over time. The amplitudes over the period from 80 to 110 days and three other time periods seem much lower than the higher amplitudes in the period from 0 to 80 days and four other time periods. However, the mean values seem equal. Thus, testing for a trend in the mean square value using the runs test would be appropriate. Because the dominant frequency component has a period of approximately 30 days, dividing the signal into 60-day segments is valid. The sequence of estimated mean square values is

[460.0 294.1 464.5 307.3 421.0 343.0 364.1 376.9 336.6 383.0]

and the median value is 370.5. The sequence of observations above and below the median is $[+ - + - + - - + - +]$. The number of runs is 9. For $N = 10$, $N_1 = 5$ and the 95% confidence interval is $[2 < r \leq 9]$. The number of runs

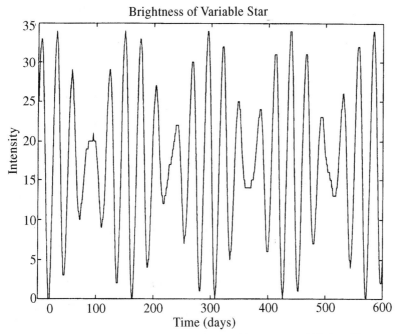

**FIGURE 5.14** Intensity of brightness of a variable star over 600 days. Data are from Appendix 3.2.

observed is exactly on the upper boundary. Thus we would conclude that the high number of runs indicates a fluctuation in signal energy and that the intensity of star brightness is nonstationary over 600 days.

## REFERENCES

T. Anderson, *The Statistical Analysis of Time Series*, Wiley, New York, 1971.

J. Bendat and A. Piersol, *Random Data, Analysis and Measurement Procedures*, Wiley, New York, 1986.

G. Bodenstein and H. Praetorius, "Features Extraction from the Electroencephalogram by Adaptive Segmentation," *Proc. IEEE* **65**:642–652 (1997).

K. Brownlee, *Statistical Theory and Methodology in Science and Engineering*, Wiley, New York, 1960.

P. Burton, "Analysis of a Dispersed and Transient Signal," in *Applications of Time Series Analysis*, Southampton University, September, 1977.

C. Chatfield, *The Analysis of Time Series: Theory and Practice*, Wiley, New York, 1975.

D. Childers and J. Labar, "Electroglottography for Laryngeal Function Assessment and Speech Analysis," *IEEE Trans. Biomed. Eng.* **31**:807–817 (1984).

W. Davenport, *Probability and Random Processes*, McGraw-Hill, New York, 1970.

W. Fuller, *Introduction to Statistical Time Series*, Wiley, New York, 1976.

R. Gray and L. Davisson, *Random Processes, A Mathematical Approach for Engineers*, Prentice-Hall, Englewood Cliffs, NJ, 1986.

P. Hoel, *Introduction to Mathematical Statistics*, Wiley, New York, 1971.

G. Jenkins and D. Watts, *Spectral Analysis and Its Applications*, Holden-Day, San Francisco, 1968.

J. Komo, *Random Signal Analysis in Engineering Systems*, Academic Press, San Diego, CA, 1987.

E. Kwatney, D. Thomas, and H. Kwatney, "An Application of Signal Processing Techniques to the Study of Myoelectric Signals," *IEEE Trans. Biomed. Eng.* **17**:303–312 (1970).

R. Otnes and L. Enochson, *Digital Time Series Analysis*, Wiley, New York, 1972.

A. Papoulis, *Probability, Random Variables and Stochastic Processes*, McGraw-Hill, New York, 1984.

W. Press, B. Flannery, S. Teukolsky, and W. Vetterling, *Numerical Recipes—The Art of Scientific Computing*, Cambridge University Press, Cambridge, 1986.

M. Priestley, *Spectral Analysis and Time Series: Volume 1—Univariate Series*, Academic Press, New York, 1981.

A. Raferty, J. Haslett, and E. McColl, "Wind Power: A Space-Time Approach," in *Time Series Analysis: Theory and Practice 2*, O. Anderson, Ed., North Holland Publishing, Amsterdam, 1982.

M. Schwartz and L. Shaw, *Signal Processing: Discrete Spectral Analysis, Detection, and Estimation*, McGraw-Hill, New York, 1975.

K. Shanmugan and A. Breipohl, *Random Signals: Detection, Estimation and Data Analysis*, Wiley, New York, 1988.

## EXERCISES

**5.1**  Prove $\gamma(t_1, t_2) = E[x(t_1)x(t_2)] - E[x(t_1)]E[x(t_2)]$, starting with equation 5.8.

**5.2**  One of the equations producing the ergodic conditions for the time-average sample mean is equation 5.20.

  a. Derive the equation for $N = 3$.

  b. Derive the equation for any $N$.

**5.3**  Let some signal, $x(n)$, be ergodic in the mean. Suppose it could have a Gaussian pdf. Use the pdf of a bivariate Gaussian random variable from Section 4.3 and let $x \Rightarrow x(n)$ and $y \Rightarrow x(n + k)$. Show that condition 5.23(b) is true and thus the estimate of the pdf from the histogram would be valid.

**5.4**   In Example 5.1, let $x(t, \zeta)$ be the random process described by the probabilities

$$P[x(t, \zeta_i)] = \tfrac{1}{6}, \quad 1 \le i \le 6$$

$$P[x(t, \zeta_i), x(t + \tau, \zeta_j)] = \begin{cases} \tfrac{1}{6}, & i = j \\ 0, & i \ne j \end{cases}$$

Show that $m(t) = 3.5$, $\sigma^2(t) = 2.92$, and $\gamma(\tau) = 2.92$.

**5.5**   Prove that for positive and negative values of lag that the mean of the estimator $\hat{R}(k)$ is $\left(1 - \dfrac{|k|}{N}\right)\varphi(k)$; that is,

$$E[\hat{R}(k)] = \left(1 - \frac{|k|}{N}\right)\varphi(k)$$

Refer to Section 5.5.2.

**5.6**   Prove that the estimator, $\hat{R}'(k)$, of the autocorrelation function is unbiased, where

$$\hat{R}'(k) = \frac{1}{N - k} \sum_{n=0}^{N-k-1} x(n)x(n + k)$$

**5.7**   Study the derivation in Appendix 5.3 of the single summation expression for $\text{Var}[\hat{C}(k)]$, equation 5.33. Combine all of the equations as stated and show that equation 5.33 is true.

**5.8**   Prove that in general for a stationary signal, the variance of a sum is

$$\text{Var}[y] = N\sigma_x^2 + 2N \sum_{k=1}^{N-1} \left(1 - \frac{k}{N}\right)\gamma_x(k)$$

where $y$ is the sum of $N$ values of $x$. Refer to Section 5.4.

**5.9**   Consider the short duration time series in Table E5.9.

**Table E5.9**   Time Series

| $n$ | $x(n)$ |
|---|---|
| 0 | 0.102 |
| 1 | 0.737 |
| 2 | 0.324 |
| 3 | 0.348 |
| 4 | 0.869 |
| 5 | 0.279 |
| 6 | 0.361 |
| 7 | 0.107 |
| 8 | 0.260 |
| 9 | 0.962 |

   a. Plot $x(n)$ vs $x(n+1)$ and guess the approximate value of $\hat{\rho}(1)$.

   b. Calculate the following sample moments: mean, variance, $\hat{R}(1)$, $\hat{\rho}(1)$.

   c. Does $\hat{\rho}(1)$ reflect that $\rho(1) = 0$?

**5.10** Duplicate the results in Example 5.2. The time series is listed in Appendix 5.1.

**5.11** Consider the random process in Example 5.4.

   a. What are the 95% and 99% confidence limits for $\rho(8)$?

   b. Does either level suggest a zero correlation?

   c. What is the maximum significance level for the values of lag 8 to be uncorrelated?

   Use the Gaussian sampling distribution and a two-sided test.

**5.12** For the random process generated in Example 5.5, perform the calculations to show that the process is not white noise.

**5.13** For the three-point moving-average process of equation 5.43, write the algorithms for $y(4)$, $y(5)$, and $y(6)$. How many points of the process $x(n)$ do they have in common with $y(3)$? What is the implication for the correlation among these values of $y(n)$?

**5.14** Derive the general forms for the mean and variance of the moving-average process, equations 5.46 and 5.47.

**5.15** Derive in detail the value of $R_y(2)$, equation 5.56, for the second-order MA process.

**5.16** Find the autocorrelation functions for the following MA processes:

   a. $y(n) = x(n) + x(n-1) + x(n-2)$

   b. $y(n) = x(n) + 0.5x(n-1) - 0.3x(n-2)$

   How do they differ?

**5.17** Prove that the autocorrelation function for the $q$th-order MA process (equation 5.57) is

$$R_y(k) = \begin{cases} \sigma_x^2 \left( \displaystyle\sum_{i=0}^{q-k} b(i)b(i+k) \right), & |k| \le q \\ 0, & |k| > q \end{cases}$$

**5.18** Verify the results of the two tests made on the estimate of $\rho(1)$ in Example 5.5.

**5.19** You are planning to analyze the stationarity of the gas furnace signals listed in Appendix 9.1. Plot both the input gas rate and output carbon dioxide percentage. It is desired to use the runs test. What is the minimum segment duration needed to apply the test appropriately?

## Computer Exercises

**5.20** Use the star brightness data, Appendix 3.2, to exercise some concepts on ensemble statistics and sample function statistics.

  a. Divide the signal into six segments, sample functions, and plot them.

  b. Calculate and plot the ensemble mean and standard deviation for these six sample functions.

  c. Divide the signal into 100 sample functions of 6 points. This is done only to obtain a large number of points for ensemble calculations.

  d. Calculate the ensemble means and standard deviations for the 100 sample functions.

  e. Calculate the mean and standard deviation for the entire signal and compare them to the ensemble values in part (d). How close are they— 5%, 10%?

  f. Using the signals in part (c), calculate the ensemble correlation coefficient between time points 1 and 2. From the original signal, estimate the NACF for the time lag of 1. How close are these ensemble and time correlation values?

**5.21** Generate 100 points of a white-noise process using any random number generator.

  a. Estimate the autocovariance function to a maximum lag of 50 using both biased and unbiased estimators.

  b. Plot the results and qualitatively compare them.

**5.22** The time series of sunspot numbers listed in Appendix 5.2 is going to be tested for stationarity of the mean.

  a. Divide the sample function into seven segments of 25 points. Calculate the sample mean for each segment.

  b. Test to see whether the sample means of each segment are equal to one another. Use the Student's $t$ distribution with a 95% confidence region.

  c. Is the signal stationary in the mean?

**5.23** Perform a correlation analysis of the sunspot numbers in Appendix 5.2.

  a. Estimate the NACF for lags from 0 to 20 and plot it with appropriate labels.

  b. Describe the NACF from part (a) qualitatively and state what it may indicate.

  c. Do any correlation values test as being nonzero? If so, what are their lag times?

## APPENDIX 5.1 YIELDS FROM A DISTILLATION PROCESS

Table A5.1

| 1–18 | 19–36 | 37–54 | 55–70 |
|------|-------|-------|-------|
| 47 | 37 | 45 | 53 |
| 64 | 74 | 54 | 49 |
| 23 | 51 | 36 | 34 |
| 71 | 58 | 54 | 35 |
| 38 | 50 | 48 | 54 |
| 65 | 60 | 55 | 45 |
| 55 | 44 | 45 | 68 |
| 41 | 57 | 57 | 38 |
| 59 | 50 | 50 | 50 |
| 48 | 45 | 62 | 60 |
| 71 | 25 | 44 | 39 |
| 35 | 59 | 64 | 59 |
| 56 | 50 | 43 | 40 |
| 40 | 71 | 52 | 57 |
| 58 | 56 | 38 | 54 |
| 44 | 74 | 60 | 23 |
| 80 | 50 | 55 | |
| 55 | 58 | 41 | |

## APPENDIX 5.2 WOLFER'S SUNSPOT NUMBERS

Table A5.2

| Year | Wolfer's number | Year | Wolfer's number | Year | Wolfer's number | Year | Wolfer's number |
|------|-----------------|------|-----------------|------|-----------------|------|-----------------|
| 1749 | 80.9 | 1793 | 46.9 | 1837 | 138.3 | 1881 | 54.3 |
| 1750 | 83.4 | 1794 | 41.0 | 1838 | 103.2 | 1882 | 59.7 |
| 1751 | 47.7 | 1795 | 21.3 | 1839 | 85.8 | 1883 | 63.7 |
| 1752 | 47.8 | 1796 | 16.0 | 1840 | 63.2 | 1884 | 63.5 |
| 1753 | 30.7 | 1797 | 6.4 | 1841 | 36.8 | 1885 | 52.2 |
| 1754 | 12.2 | 1798 | 4.1 | 1842 | 24.2 | 1886 | 25.4 |
| 1755 | 9.6 | 1799 | 6.8 | 1843 | 10.7 | 1887 | 13.1 |
| 1756 | 10.2 | 1800 | 14.5 | 1844 | 15.0 | 1888 | 6.8 |
| 1757 | 32.4 | 1801 | 34.0 | 1845 | 40.1 | 1889 | 6.3 |
| 1758 | 47.6 | 1802 | 45.0 | 1846 | 61.5 | 1890 | 7.1 |
| 1759 | 54.0 | 1803 | 43.1 | 1847 | 98.5 | 1891 | 35.6 |
| 1760 | 62.9 | 1804 | 47.5 | 1848 | 124.3 | 1892 | 73.0 |
| 1761 | 85.9 | 1805 | 42.2 | 1849 | 95.9 | 1893 | 84.9 |
| 1762 | 61.2 | 1806 | 28.1 | 1850 | 66.5 | 1894 | 78.0 |
| 1763 | 45.1 | 1807 | 10.1 | 1851 | 64.5 | 1895 | 64.0 |

(*continued*)

Table A5.2   (*continued*)

| Year | Wolfer's number | Year | Wolfer's number | Year | Wolfer's number | Year | Wolfer's number |
|------|------|------|------|------|------|------|------|
| 1764 | 36.4 | 1808 | 8.1 | 1852 | 54.2 | 1896 | 41.8 |
| 1765 | 20.9 | 1809 | 2.5 | 1853 | 39.0 | 1897 | 26.2 |
| 1766 | 11.4 | 1810 | 0.0 | 1854 | 20.6 | 1898 | 26.7 |
| 1767 | 37.8 | 1811 | 1.4 | 1855 | 6.7 | 1899 | 12.1 |
| 1768 | 69.8 | 1812 | 5.0 | 1856 | 4.3 | 1900 | 9.5 |
| 1769 | 106.1 | 1813 | 12.2 | 1857 | 22.8 | 1901 | 2.7 |
| 1770 | 100.8 | 1814 | 13.9 | 1858 | 54.8 | 1902 | 5.0 |
| 1771 | 81.6 | 1815 | 35.4 | 1859 | 93.8 | 1903 | 24.4 |
| 1772 | 66.5 | 1816 | 45.8 | 1860 | 95.7 | 1904 | 42.0 |
| 1773 | 34.8 | 1817 | 41.1 | 1861 | 77.2 | 1905 | 63.5 |
| 1774 | 30.6 | 1818 | 30.4 | 1862 | 59.1 | 1906 | 53.8 |
| 1775 | 7.0 | 1819 | 23.9 | 1863 | 44.0 | 1907 | 62.0 |
| 1776 | 19.8 | 1820 | 15.7 | 1864 | 47.0 | 1908 | 48.5 |
| 1777 | 92.5 | 1821 | 6.6 | 1865 | 30.5 | 1909 | 43.9 |
| 1778 | 154.4 | 1822 | 4.0 | 1866 | 16.3 | 1910 | 18.6 |
| 1779 | 125.9 | 1823 | 1.8 | 1867 | 7.3 | 1911 | 5.7 |
| 1780 | 84.8 | 1824 | 8.5 | 1868 | 37.3 | 1912 | 3.6 |
| 1781 | 68.1 | 1825 | 16.6 | 1869 | 73.9 | 1913 | 1.4 |
| 1782 | 38.5 | 1826 | 36.3 | 1870 | 139.1 | 1914 | 9.6 |
| 1783 | 22.8 | 1827 | 49.7 | 1871 | 111.2 | 1915 | 47.4 |
| 1784 | 10.2 | 1828 | 62.5 | 1872 | 101.7 | 1916 | 57.1 |
| 1785 | 24.1 | 1829 | 67.0 | 1873 | 66.3 | 1917 | 103.9 |
| 1786 | 82.9 | 1830 | 71.0 | 1874 | 44.7 | 1918 | 80.6 |
| 1787 | 132.0 | 1831 | 47.8 | 1985 | 17.1 | 1919 | 63.6 |
| 1788 | 130.9 | 1832 | 27.5 | 1876 | 11.3 | 1920 | 37.6 |
| 1789 | 118.1 | 1833 | 8.5 | 1877 | 12.3 | 1921 | 26.1 |
| 1790 | 89.9 | 1834 | 13.2 | 1878 | 3.4 | 1922 | 14.2 |
| 1791 | 66.6 | 1835 | 56.9 | 1879 | 6.0 | 1923 | 5.8 |
| 1792 | 60.0 | 1836 | 121.5 | 1880 | 32.3 | 1924 | 16.7 |

## APPENDIX 5.3   VARIANCE OF AUTOCOVARIANCE ESTIMATE

By definition, the variance of the biased covariance estimator for $k \geq 0$ is

$$\mathrm{Var}[\hat{C}(k)] = E[\hat{C}^2(k)] - E_2[\hat{C}(k)]$$

$$= E\left( \frac{1}{N^2} \sum_{i=0}^{N-k-1} \sum_{j=0}^{N-k-1} x(i)x(i+k)x(j)x(j+k) \right) - \left(1 - \frac{k}{N}\right)^2 \gamma^2(k)$$

$$= \frac{1}{N^2} \sum_{i=0}^{N-k-1} \sum_{j=0}^{N-k-1} E[x(i)x(i+k)x(j)x(j+k)] - \left(1 - \frac{k}{N}\right)^2 \gamma^2(k)$$

$$(A5.1)$$

Using the expression in equation 5.32 for the fourth-order moments, equation A5.1 becomes

$$\text{Var}[\hat{C}(k)] = \frac{1}{N^2} \sum_{i=0}^{N-k-1} \sum_{j=0}^{N-k-1} (\gamma^2(k) + \gamma^2(i-j)$$

$$+ \gamma(i-j+k)\gamma(i-j-k)) - \left(1 - \frac{k}{N}\right)^2 \gamma^2(k) \qquad (A5.2)$$

Equation A5.2 will be resolved into a single summation form by simplifying each term separately.

Term 1: Since $\gamma^2(k)$ is not a function of the summing indices,

$$\sum_{i=0}^{N-k-1} \sum_{j=0}^{N-k-1} \gamma^2(k) = (N-k)^2 \gamma^2(k) \qquad (A5.3)$$

Term 2: The second term

$$\sum_{i=0}^{N-k-1} \sum_{j=0}^{N-k-1} \gamma^2(i-j) \qquad (A5.4)$$

has $(i-j)$ in all of the arguments. All of the values of $\gamma^2(i-j)$ that are summed can be represented in a matrix as

$$\begin{bmatrix} \gamma^2(0) & \gamma^2(1) & \ddots & \gamma^2(p-1) \\ \gamma^2(-1) & \gamma^2(0) & \ddots & \gamma^2(p-2) \\ \vdots & \vdots & \ddots & \vdots \\ \gamma^2(-p+1) & \gamma^2(-p+2) & \ddots & \gamma^2(0) \end{bmatrix} \qquad (A5.5)$$

with $p = N - k$. All of the elements along a diagonal are equal and the equivalent summation can be made by summing over the diagonals. The value of the lag is represented by the substitution $r = i - j$. Now $r$ ranges from $-p + 1$ to $p - 1$ and the number of elements along each diagonal is $N - |r| - k$. Thus

$$\sum_{i=0}^{N-k-1} \sum_{j=0}^{N-k-1} \gamma^2(i-j) = \sum_{r=-N+k+1}^{N-k-1} \gamma^2(r)(N - |r| - k) \qquad (A5.6)$$

Term 3. The resolution for the third term

$$\sum_{i=0}^{N-k-1} \sum_{j=0}^{N-k-1} \gamma(i-j+k)\gamma(i-j-k) \qquad (A5.7)$$

follows the same procedure as that for term 2 with the matrix element now being $\gamma(r + k)\gamma(r - k)$. Its summation is

$$\sum_{r=-N+k+1}^{N-k-1} \gamma(r + k)\gamma(r - k)(N - |r| - k) \tag{A5.8}$$

By combining equations A5.8, A5.6, and A5.3 with equation A5.2, the result is

$$\text{Var}[\hat{C}(k)] = \frac{1}{N} \sum_{r=-N+k+1}^{N-k-1} \left(1 - \frac{|r| + k}{N}\right)(\gamma^2(r) + \gamma(r + k)\gamma(r - k)) \tag{A5.9}$$

## APPENDIX 5.4   STATIONARITY TESTS

### A5.4.1   Equality of Two Means

The hypothesis being tested is whether or not the means in two independent sets of measurements are equal. Since this test is used in conjunction with the test for equality of variances, the variances of the two sets of measurements will be assumed to be equal. The test statistic is a $t$ statistic, which uses a normalized value of the difference of the sample means of the two populations. Let the sample means and variances for populations 1 and 2 be represented by the symbols $\hat{m}_1$ and $\sigma_1^2$, $\hat{m}_2$ and $\sigma_2^2$. The test statistic is

$$T = \frac{\hat{m}_1 - \hat{m}_2}{\left(\left(\frac{1}{N_1} + \frac{1}{N_2}\right)\left(\frac{N_1\hat{\sigma}_1^2 + N_2\hat{\sigma}_2^2}{N_1 + N_2 - 2}\right)\right)^{1/2}} \tag{A5.10}$$

$T$ has a Student's $t$ distribution with $v = N_1 + N_2 - 2$ degrees of freedom (Anderson, 1971). The $(1 - \alpha)$ confidence interval is

$$t_{v,1-\alpha/2} \leq T \leq t_{v,\alpha/2} \tag{A5.11}$$

The table for the $t$ test is in Appendix 4.4.

### A5.4.2   Equality of Variances

The second parametric test for assessing the stationarity of signals requires testing the equality of variances of two sets of measurements. It is assumed that the true mean values, $m_1$ and $m_2$, are unknown. The $F$ distribution is appropriate for this task (Anderson, 1971). The test statistic, $F$, is the ratio of the two unbiased estimates of the variance and is

$$F = \frac{\hat{\sigma}_1^2}{\hat{\sigma}_2^2} \tag{A5.12}$$

## APPENDIX 5.5 F DISTRIBUTION: SIGNIFICANCE LIMIT FOR THE 97.5TH PERCENTILE

97.5th Percentile

| $v_1/v_2$ | 1 | 2 | 3 | 4 | 5 | 6 | 7 | 8 | 9 | 10 | 12 | 15 | 20 | 24 | 30 | 40 | 60 | 120 | $\infty$ |
|---|---|---|---|---|---|---|---|---|---|---|---|---|---|---|---|---|---|---|---|
| 1 | 647.8 | 799.5 | 864.2 | 899.6 | 921.8 | 937.1 | 948.2 | 956.7 | 963.3 | 968.6 | 976.7 | 984.9 | 993.1 | 997.2 | 1001 | 1006 | 1010 | 1014 | 1018 |
| 2 | 38.51 | 39.00 | 39.17 | 39.25 | 39.30 | 39.33 | 39.36 | 39.37 | 39.39 | 39.40 | 39.41 | 39.43 | 39.45 | 39.46 | 39.46 | 39.47 | 39.48 | 39.49 | 39.50 |
| 3 | 17.44 | 16.04 | 15.44 | 15.10 | 14.88 | 14.73 | 14.62 | 14.54 | 14.47 | 14.42 | 14.34 | 14.25 | 14.17 | 14.12 | 14.08 | 14.04 | 13.99 | 13.95 | 13.90 |
| 4 | 12.22 | 10.65 | 9.98 | 9.60 | 9.36 | 9.20 | 9.07 | 8.98 | 8.90 | 8.84 | 8.75 | 8.66 | 8.56 | 8.51 | 8.46 | 8.41 | 8.36 | 8.31 | 8.26 |
| 5 | 10.01 | 8.43 | 7.76 | 7.39 | 7.15 | 6.98 | 6.85 | 6.76 | 6.68 | 6.62 | 6.52 | 6.43 | 6.33 | 6.28 | 6.23 | 6.18 | 6.12 | 6.07 | 6.02 |
| 6 | 8.81 | 7.26 | 6.60 | 6.23 | 5.99 | 5.82 | 5.70 | 5.60 | 5.52 | 5.46 | 5.37 | 5.27 | 5.17 | 5.12 | 5.07 | 5.01 | 4.96 | 4.90 | 4.85 |
| 7 | 8.07 | 6.54 | 5.89 | 5.52 | 5.29 | 5.12 | 4.99 | 4.90 | 4.82 | 4.76 | 4.67 | 4.57 | 4.47 | 4.42 | 4.36 | 4.31 | 4.25 | 4.20 | 4.14 |
| 8 | 7.57 | 6.06 | 5.42 | 5.05 | 4.82 | 4.65 | 4.53 | 4.43 | 4.36 | 4.30 | 4.20 | 4.10 | 4.00 | 3.95 | 3.89 | 3.84 | 3.78 | 3.73 | 3.67 |
| 9 | 7.21 | 5.71 | 5.08 | 4.72 | 4.48 | 4.32 | 4.20 | 4.10 | 4.03 | 3.96 | 3.87 | 3.77 | 3.67 | 3.61 | 3.56 | 3.51 | 3.45 | 3.39 | 3.33 |
| 10 | 6.94 | 5.46 | 4.83 | 4.47 | 4.24 | 4.07 | 3.95 | 3.85 | 3.78 | 3.72 | 3.62 | 3.52 | 3.42 | 3.37 | 3.31 | 3.26 | 3.20 | 3.14 | 3.08 |
| 11 | 6.72 | 5.26 | 4.63 | 4.28 | 4.04 | 3.88 | 3.76 | 3.66 | 3.59 | 3.53 | 3.43 | 3.33 | 3.23 | 3.17 | 3.12 | 3.06 | 3.00 | 2.94 | 2.88 |
| 12 | 6.55 | 5.10 | 4.47 | 4.12 | 3.89 | 3.73 | 3.61 | 3.51 | 3.44 | 3.37 | 3.28 | 3.18 | 3.07 | 3.02 | 2.96 | 2.91 | 2.85 | 2.79 | 2.72 |
| 13 | 6.41 | 4.97 | 4.35 | 4.00 | 3.77 | 3.60 | 3.48 | 3.39 | 3.31 | 3.25 | 3.15 | 3.05 | 2.95 | 2.89 | 2.84 | 2.78 | 2.72 | 2.66 | 2.60 |
| 14 | 6.30 | 4.86 | 4.24 | 3.89 | 3.66 | 3.50 | 3.38 | 3.29 | 3.21 | 3.15 | 3.05 | 2.95 | 2.84 | 2.79 | 2.73 | 2.67 | 2.61 | 2.55 | 2.49 |

| | | | | | | | | | | | | | | | | | | | |
|---|---|---|---|---|---|---|---|---|---|---|---|---|---|---|---|---|---|---|---|
| 15 | 6.20 | 4.77 | 4.15 | 3.80 | 3.58 | 3.41 | 3.29 | 3.20 | 3.12 | 3.06 | 2.96 | 2.86 | 2.76 | 2.70 | 2.64 | 2.59 | 2.52 | 2.46 | 2.40 |
| 16 | 6.12 | 4.69 | 4.08 | 3.73 | 3.50 | 3.34 | 3.22 | 3.12 | 3.05 | 2.99 | 2.89 | 2.79 | 2.68 | 2.63 | 2.57 | 2.51 | 2.45 | 2.38 | 2.32 |
| 17 | 6.04 | 4.62 | 4.01 | 3.66 | 3.44 | 3.28 | 3.16 | 3.06 | 2.98 | 2.92 | 2.82 | 2.72 | 2.62 | 2.56 | 2.50 | 2.44 | 2.38 | 2.32 | 2.25 |
| 18 | 5.98 | 4.56 | 3.95 | 3.61 | 3.38 | 3.22 | 3.10 | 3.01 | 2.93 | 2.87 | 2.77 | 2.67 | 2.56 | 2.50 | 2.44 | 2.38 | 2.2 | 2.26 | 2.19 |
| 19 | 5.92 | 4.51 | 3.90 | 3.56 | 3.33 | 3.17 | 3.05 | 2.96 | 2.88 | 2.82 | 2.72 | 2.62 | 2.51 | 2.45 | 2.39 | 2.33 | 2.27 | 2.20 | 2.13 |
| 20 | 5.87 | 4.46 | 3.86 | 3.51 | 3.29 | 3.13 | 3.01 | 2.91 | 2.84 | 2.77 | 2.68 | 2.57 | 2.46 | 2.41 | 2.35 | 2.29 | 2.22 | 2.16 | 2.09 |
| 21 | 5.83 | 4.42 | 3.82 | 3.48 | 3.25 | 3.09 | 2.97 | 2.87 | 2.80 | 2.73 | 2.64 | 2.53 | 2.42 | 2.37 | 2.31 | 2.25 | 2.18 | 2.11 | 2.04 |
| 22 | 5.79 | 4.38 | 3.78 | 3.44 | 3.22 | 3.05 | 2.93 | 2.84 | 2.76 | 2.70 | 2.60 | 2.50 | 2.39 | 2.33 | 2.27 | 2.21 | 2.14 | 2.08 | 2.00 |
| 23 | 5.75 | 4.35 | 3.75 | 3.41 | 3.18 | 3.02 | 2.90 | 2.81 | 2.73 | 2.67 | 2.57 | 2.47 | 2.36 | 2.30 | 2.24 | 2.18 | 2.11 | 2.04 | 1.97 |
| 24 | 5.72 | 4.32 | 3.72 | 3.38 | 3.15 | 2.99 | 2.87 | 2.78 | 2.70 | 2.64 | 2.54 | 2.44 | 2.33 | 2.27 | 2.21 | 2.15 | 2.08 | 2.01 | 1.94 |
| 25 | 5.69 | 4.29 | 3.69 | 3.35 | 3.13 | 2.97 | 2.85 | 2.75 | 2.68 | 2.61 | 2.51 | 2.41 | 2.30 | 2.24 | 2.18 | 2.12 | 2.05 | 1.98 | 1.91 |
| 26 | 5.66 | 4.27 | 3.67 | 3.33 | 3.10 | 2.94 | 2.82 | 2.73 | 2.65 | 2.59 | 2.49 | 2.39 | 2.28 | 2.22 | 2.16 | 2.09 | 2.03 | 1.95 | 1.88 |
| 27 | 5.63 | 4.24 | 3.65 | 3.31 | 3.08 | 2.92 | 2.80 | 2.71 | 2.63 | 2.57 | 2.47 | 2.36 | 2.25 | 2.19 | 2.13 | 2.07 | 2.00 | 1.93 | 1.85 |
| 28 | 5.61 | 4.22 | 3.63 | 3.29 | 3.06 | 2.90 | 2.78 | 2.69 | 2.61 | 2.55 | 2.45 | 2.34 | 2.23 | 2.17 | 2.11 | 2.05 | 1.98 | 1.91 | 1.83 |
| 29 | 5.59 | 4.20 | 3.61 | 3.27 | 3.04 | 2.88 | 2.76 | 2.67 | 2.59 | 2.53 | 2.43 | 2.32 | 2.21 | 2.15 | 2.00 | 2.03 | 1.96 | 1.89 | 1.81 |
| 30 | 5.57 | 4.18 | 3.59 | 3.25 | 3.03 | 2.87 | 2.75 | 2.65 | 2.57 | 2.51 | 2.41 | 2.31 | 2.20 | 2.14 | 2.07 | 2.01 | 1.94 | 1.87 | 1.79 |
| 40 | 5.42 | 4.05 | 3.46 | 3.13 | 2.90 | 2.74 | 2.62 | 2.53 | 2.45 | 2.39 | 2.29 | 2.18 | 2.07 | 2.01 | 1.94 | 1.88 | 1.80 | 1.72 | 1.64 |
| 60 | 5.29 | 3.93 | 3.34 | 3.01 | 2.79 | 2.63 | 2.51 | 2.41 | 2.33 | 2.27 | 2.17 | 2.06 | 1.94 | 1.88 | 1.82 | 1.74 | 1.67 | 1.58 | 1.48 |
| 120 | 5.15 | 3.80 | 3.23 | 2.89 | 2.67 | 2.52 | 2.39 | 2.30 | 2.22 | 2.16 | 2.05 | 1.94 | 1.82 | 1.76 | 1.69 | 1.61 | 1.53 | 1.43 | 1.31 |
| ∞ | 5.02 | 3.69 | 3.12 | 2.79 | 2.57 | 2.41 | 2.29 | 2.19 | 2.11 | 2.05 | 1.94 | 1.83 | 1.71 | 1.64 | 1.57 | 1.48 | 1.39 | 1.27 | 1.00 |

$F$ has two degrees of freedom, $v_1 = N_1 - 1$ and $v_2 = N_2 - 1$. The $(1 - \alpha)$ confidence interval for a two-sided test is

$$F_{v_1, v_2; 1-\alpha/2} \leq F \leq F_{v_1, v_2; \alpha/2} \qquad (A5.13)$$

The tables for the $F$ distribution are quite voluminous and the values are usually only tabulated for the upper confidence limit. Since $F$ is a ratio of positive numbers, the lower limit can be determined from the tables because

$$F_{v_1, v_2; 1-\alpha/2} = \frac{1}{F_{v_2, v_1; \alpha/2}} \qquad (A5.14)$$

The values of the 97.5% confidence limits of the $F$ distribution are listed in Appendix 5.5. For tables of values of the $F$ distribution, please consult either the references Shanmugan and Breipohl (1988) and Jenkins and Watts (1968) or any book of statistical tables.

### A5.4.3 Runs Test

The situation being considered is whether or not there is some type of trend in the data or signal. The data or signal is divided into $N$ segments. For each segment one estimates a moment, such as the mean or variance, and produces a sequence of these measures. The actual data magnitudes can be used, in which case each segment has one point. One classifies each measure as being above $(+)$ or below $(-)$ the median of this set of measures. The test statistic is the number of runs $(r)$ in the sequence of signs. Below is a sequence of 12 measures, $N = 12$.

$$[5.5 \ 5.7 \ 4.8 \ 5.0 \ 5.4 \ 6.8 \ 4.9 \ 5.9 \ 6.8 \ 5.5 \ 5.1 \ 5.2]$$

The median value is $(5.4 + 5.5)/2 = 5.45$. The sequence of signs is

$$[+ + - - - + - + + + - -].$$

Then one counts the number of runs; a run is a sequence of like signs. The number of runs is 6, so $r = 6$. It is hypothesized that there is no trend in the data and that the sequence of $N$ measures are independent observations from the same random variable. Then assuming that the number of $(+)$ observations equals the number of $(-)$ observations, the number of runs in the sequence will have the sampling distribution as given in Appendix 5.6. The mean and variance of the number of runs are

$$m_r = \frac{N}{2} + 1; \qquad \sigma_r^2 = \frac{N(N - 2)}{4(N - 1)}$$

For this data the 95% confidence interval for the number of runs is $3 < r \leq 10$. Since $r = 6$ and falls within the confidence interval, the hypothesis is accepted. For additional information on the runs test, refer to Bendat and Piersol (1986).

# APPENDIX 5.6 PERCENTAGE POINTS OF RUN DISTRIBUTION

Values of $r_{n;\alpha}$ such that $\text{Prob}[r_n > r_{n;\alpha}] = \alpha$, where $n = N/2$

| | | | | $\alpha$ | | |
|---|---|---|---|---|---|---|
| $n = N/2$ | 0.99 | 0.975 | 0.95 | 0.05 | 0.025 | 0.01 |
| 5 | 2 | 2 | 3 | 8 | 9 | 9 |
| 6 | 2 | 3 | 3 | 10 | 10 | 11 |
| 7 | 3 | 3 | 4 | 11 | 12 | 12 |
| 8 | 4 | 4 | 5 | 12 | 13 | 13 |
| 9 | 4 | 5 | 6 | 13 | 14 | 15 |
| 10 | 5 | 6 | 6 | 15 | 15 | 16 |
| 11 | 6 | 7 | 7 | 16 | 16 | 17 |
| 12 | 7 | 7 | 8 | 17 | 18 | 18 |
| 13 | 7 | 8 | 9 | 18 | 19 | 20 |
| 14 | 8 | 9 | 10 | 19 | 20 | 21 |
| 15 | 9 | 10 | 11 | 20 | 21 | 22 |
| 16 | 10 | 11 | 11 | 22 | 22 | 23 |
| 18 | 11 | 12 | 13 | 24 | 25 | 26 |
| 20 | 13 | 14 | 15 | 26 | 27 | 28 |
| 25 | 17 | 18 | 19 | 32 | 33 | 34 |
| 30 | 21 | 22 | 24 | 37 | 39 | 40 |
| 35 | 25 | 27 | 28 | 43 | 44 | 46 |
| 40 | 30 | 31 | 33 | 48 | 50 | 51 |
| 45 | 34 | 36 | 37 | 54 | 55 | 57 |
| 50 | 38 | 40 | 42 | 59 | 61 | 63 |
| 55 | 43 | 45 | 46 | 65 | 66 | 68 |
| 60 | 47 | 49 | 51 | 70 | 72 | 74 |
| 65 | 52 | 54 | 56 | 75 | 77 | 79 |
| 70 | 56 | 58 | 60 | 81 | 83 | 85 |
| 75 | 61 | 63 | 65 | 86 | 88 | 90 |
| 80 | 65 | 68 | 70 | 91 | 93 | 96 |
| 85 | 70 | 72 | 74 | 97 | 99 | 101 |
| 90 | 74 | 77 | 79 | 102 | 104 | 107 |
| 95 | 79 | 82 | 84 | 107 | 109 | 112 |
| 100 | 84 | 86 | 88 | 113 | 115 | 117 |

*Source:* J. Bendat and A. Piersol.

# VI

---

# RANDOM SIGNALS, LINEAR SYSTEMS, AND POWER SPECTRA

## 6.1 INTRODUCTION

This chapter is devoted to some basic principles and definitions that are essential for understanding additional approaches for analyzing random signals. Direct Fourier transformation is no longer sufficient for performing frequency analysis. Other concepts must be used and are based on the definition of signal power. Systems concepts are presented in the context of random signals. They are important because systems can modify the properties of any random input signal and because discrete time systems can be models for signals. In fact, this latter capability is the basis for the contemporary approaches to signal analysis. Comprehensive treatments of discrete time systems can be found in books such as Jong (1982), Ziemer et al. (1989), and DeFatta et al. (1988). Please consult them for additional information.

## 6.2 POWER SPECTRA

Harmonic analysis for deterministic waveforms and signals requires only windowing and Fourier transformation of the desired time series. However, implementing *spectral analysis*, frequency analysis for random signals, requires understanding the probabilistic properties of the spectral estimates. Equation 6.1 shows the Fourier transform of a discrete time random signal, $x(n)$, having a sampling interval of $T$ units.

$$X(f) = T \sum_{n=-\infty}^{\infty} x(n)e^{-j2\pi fnT} \tag{6.1}$$

What is implied, and not shown directly, is that $X(f)$ is also a random variable. So equation 6.1 alone will not suffice as a mechanism for performing the frequency analysis. The major problem is that the Fourier transform of a wide-sense stationary random signal does not exist.

Let us examine this. For $X(f)$ to exist the energy must be finite or

$$\text{energy} = T \sum_{n=-\infty}^{\infty} x(n)^2 < \infty \tag{6.2}$$

Since $x(n)$ is at least wide-sense stationary, the energy is infinite for every sample function (Priestley, 1981). In fact, the average energy is also infinite; that is,

$$E[\text{energy}] = T \sum_{n=-\infty}^{\infty} E[x(n)^2] = \infty \tag{6.3}$$

However, equation 6.3 suggests that if average power is considered, it would be a finite quantity on which to base a definition of a frequency transformation. The average power is defined as

$$E[\text{power}] = \lim_{N \to \infty} \sum_{n=-N}^{N} \frac{TE[x(n)^2]}{(2N+1)T} = E[x(n)^2] < \infty \tag{6.4}$$

The methodology for incorporating a frequency variable into equation 6.4 requires an additional definition. Define a signal, $x_p(n)$, that equals a finite duration portion of $x(n)$; that is,

$$x_p(n) = \begin{cases} x(n), & |n| \leq N \\ 0, & |n| > N \end{cases} \tag{6.5}$$

such that $E[x(n)^2] < \infty$. Frequency is introduced by using *Parseval's theorem*,

$$T \sum_{n=-\infty}^{\infty} x_p(n)^2 = \int_{-1/2T}^{1/2T} X_p(f)X_p^*(f) \, df \tag{6.6}$$

Since the $x_p(n)$ sequence is finite, the summation limits can be changed to $-N$ and $N$ and equation 6.6 can be inserted directly into equation 6.4. The order of mathematical operations is changed as shown. Now,

$$
\begin{aligned}
E[\text{power}] &= \lim_{N\to\infty} E\left(\sum_{n=-N}^{N} \frac{T x_p(n)^2}{(2N+1)T}\right) \\
&= \lim_{N\to\infty} E\left(\frac{\int_{-1/2T}^{1/2T} X_p(f)X_p^*(f)\, df}{(2N+1)T}\right) \\
&= \int_{-1/2T}^{1/2T} \lim_{N\to\infty} E\left(\frac{X_p(f)X_p^*(f)}{(2N+1)T}\right) df
\end{aligned}
\tag{6.7}
$$

The integrand defines how the power is distributed over frequency and is the *power spectral density* (*PSD*) function, $S(f)$:

$$
S(f) = \lim_{N\to\infty} E\left(\frac{X_p(f)X_p^*(f)}{(2N+1)T}\right)
\tag{6.8}
$$

The PSD is the Fourier transform of the autocorrelation function. The proof of this relationship is called the *Weiner-Khintchin Theorem* for continuous time processes and *Wold's Theorem* for discrete time processes. Their proofs and contributions are summarized in detail by Priestley (1981) and Koopmans (1974). Since only wide-sense stationary signals are being considered, as stated in Section 5.5, $R(k) = \varphi(k)$ and $C(k) = \gamma(k)$. The transformation pair formed using the DTFT is

$$
S(f) = T \sum_{k=-\infty}^{\infty} R(k)e^{-j2\pi f k T}
\tag{6.9}
$$

$$
R(k) = \int_{-1/2T}^{1/2T} S(f)e^{j2\pi f k T}\, df
\tag{6.10}
$$

This relationship is logical since $R(0) = E[x(n)^2]$ and $R(k)$ contains information about frequency content. One of the properties of $R(k)$, stated in Chapter 5, is that if $R(k)$ is periodic, then $x(n)$ is a periodic random process. Since the ACF is a moment function, hence deterministic, and asymptotically approaches zero for large lag values, it is Fourier-transformable.

An alternative name for $S(f)$ is the *variance spectral density* function. This definition is best understood if one considers the inverse transform in equation 6.10 with $k = 0$ and the mean value of the process being zero. Under those conditions,

$$
R(0) = \int_{-1/2T}^{1/2T} S(f)\, df = \sigma^2
\tag{6.11}
$$

and the area under the PSD function is equal to the variance of the signal. The PSD function has the following three important properties:

   a. It is real function; $S(f) = S^*(f)$.
   b. It is an even function; $S(f) = S(-f)$.
   c. It is nonnegative; $S(f) \geq 0$.

The proofs of properties (a) and (b) are left as exercises for the reader (see Exercise 6.1). Notice that because of property (a), all phase information is lost.

## 6.3   SYSTEM DEFINITION REVIEW

### 6.3.1   Basic Definitions

Linear time-invariant discrete time systems are studied in courses that treat filters, systems, and deterministic signals. These systems have an important role in the framework of random signals as well since random signals are subject to filtering and are inputs to measurement systems, etc. The general concept is the same as with deterministic signals. Figure 6.1 shows the conventional block diagram with $x(n)$ as the *input signal*, $y(n)$ as the *output signal*, and $h(n)$ as the *unit impulse response*. The fundamental relationship between all of these is given by the *convolution sum* and is

$$y(n) = \sum_{i=-\infty}^{\infty} h(n-i)x(i) = \sum_{i=-\infty}^{\infty} x(n-i)h(i) \qquad (6.12)$$

For the system to be *stable* it is necessary that

$$\sum_{n=-\infty}^{\infty} |h(n)| < \infty \qquad (6.13)$$

A system is *causal* if

$$h(n) = 0, \quad n < 0 \qquad (6.14)$$

The complementary relationship in the frequency domain among these signals and the impulse response is also well known from the *convolution theorem* and is

$$Y(f) = H(f)X(f), \quad -\frac{1}{2T} \leq f \leq \frac{1}{2T} \qquad (6.15)$$

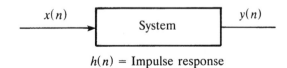

$$h(n) = \text{Impulse response}$$

**FIGURE 6.1**   System block diagram $x(n)$ is the input signal, $y(n)$ is the output signal, and $h(n)$ is the impulse response.

where $Y(f)$, $H(f)$, and $X(f)$ are the DTFT of $y(n)$, $h(n)$, and $x(n)$, respectively. The function $H(f)$ describes the frequency domain properties of the system and is called the *transfer function* or *frequency response*. As with most Fourier transforms, it is a complex function and can be written in polar form,

$$H(f) = |H(f)|e^{j\phi(f)} \tag{6.16}$$

where $|H(f)|$ and $\phi(f)$ are the *magnitude* and *phase responses*, respectively.

The impulse response has two general forms. One is when the convolution sum has finite limits; that is,

$$y(n) = \sum_{i=-s}^{q} x(n-i)h(i) \tag{6.17}$$

where $s$ and $q$ are finite. This structure represents a *nonrecursive* system and has the same form as the moving-average signal model. If $s \leq 0$, the system is causal. When a nonrecursive system is noncausal, usually the impulse response is an even function; that is, $s = q$ and $h(i) = h(-i)$. If either $s$ or $q$ are infinite, the system can take a more parsimonious representation. Most often in this latter situation the systems are causal and therefore $s = 0$ and $q = \infty$. The structure becomes

$$y(n) = b(0)x(n) - \sum_{i=1}^{p} a(i)y(n-i) \tag{6.18}$$

This is a *recursive* structure and is identical mathematically to the autoregressive signal model. A system can have recursive and nonrecursive components simultaneously and takes the structure

$$y(n) = \sum_{l=0}^{q} b(l)x(n-l) - \sum_{i=1}^{p} a(i)y(n-i) \tag{6.19}$$

The relationship among the system coefficients, $a(i)$ and $b(l)$, and the components of the impulse response are complicated except for the nonrecursive system.

The general frequency response is obtained from the DTFT of equation 6.19. After rewriting the equation as

$$\sum_{i=0}^{p} a(i)y(n-i) = \sum_{l=0}^{q} b(l)x(n-l)$$

with $a(0) = 1$ and taking the DTFT, it becomes

$$\sum_{i=0}^{p} a(i)Y(f)e^{-j2\pi fiT} = \sum_{l=0}^{q} b(l)X(f)e^{-j2\pi flT}$$

and finally

$$H(f) = \frac{Y(f)}{X(f)} = \frac{\sum_{l=0}^{q} b(l)e^{-j2\pi f l T}}{\sum_{i=0}^{p} a(i)e^{-j2\pi f i T}} \tag{6.20}$$

The general signal model, represented by equation 6.19 and whose transfer function is represented by equation 6.20, is the *autoregressive moving-average* (*ARMA*) model of order $(p, q)$. Sometimes this is written as ARMA($p, q$). Two very useful models are simplifications of the ARMA model. When $a(0) = 1$ and $a(i) = 0$ for $i \geq 1$, the moving-average model of order $q$ is produced. This was used in Chapter 5. Since the denominator portion of its transfer function is 1, the MA model, MA($q$), is represented by an *all-zero* system. When $b(0) = 1$ and $b(1) = 0$ for $1 \geq 1$, the model is the *autoregressive* (AR) model of order $p$, also written as AR($p$). Since the numerator portion of the transfer function is 1, the AR model is represented by an *all-pole* system.

Because the input and output signals are random, these deterministic relationships *per se* de not describe all the relationships between $x(n)$ and $y(n)$. They must be used as a framework to find other relationships among the probabilistic moments and nonrandom functions that describe the signals. The purpose of this chapter is to define some of the more important relationships and to define some functional forms for the system. For the presentations in this textbook, *it shall be assumed that the signals are at least wide-sense stationary and that all systems generating signals are causal, stable, and minimum phase* (Oppenheim and Willsky, 1983).

### 6.3.2   Relationships between Input and Output

The mean value of the output is found directly from equation 6.12 and for causal systems is

$$E[y(n)] = E\left[ \sum_{i=0}^{\infty} x(n-i)h(i) \right] = \sum_{i=0}^{\infty} E[x(n-i)]h(i)$$

Since $x(n)$ is stationary, let $m_x = E[x(n-i)]$ and

$$E[y(n)] = m_y = m_x \sum_{i=0}^{\infty} h(i) \tag{6.21}$$

Thus the mean values have a simple relationship and if $m_x = 0$ and $\sum_{i=0}^{\infty} |h(i)| < \infty$, then $m_y = 0$. If the summation is considered from the DTFT perspective, then

$$\sum_{i=0}^{\infty} h(i) = \sum_{i=0}^{\infty} h(i)e^{-j2\pi f i T} \bigg|_{f=0} = \frac{H(0)}{T} \tag{6.22}$$

and

$$m_y = m_x \frac{H(0)}{T}$$

The simplest relationship between correlation functions is derived also from the convolution relationship. Now multiply both sides of equation 6.12 by $x(n - k)$ and take the expectation or

$$R_{xy}(k) = E[x(n - k)y(n)] \qquad R_{yx}(k) = E[x(n) \cdot y(n-k)]$$

$$= E\left[ \sum_{i=-\infty}^{\infty} h(i)x(n - k)x(n - i) \right] = R_{xy}(-k)$$

$$= \sum_{i=-\infty}^{\infty} h(i)E[x(n - k)x(n - i)] \qquad (6.23)$$

The term $R_{xy}(k)$ in equation 6.23 is the mean of a cross product and defines the *cross correlation function* (*CCF*) between signals $y(n)$ and $x(n)$ for a time difference $(kT)$. The variable $k$ defines the number of time units that the signal $y(n)$ is delayed or lagged with respect to $x(n)$. Hence $kT$ is also called the *lag time*. The more conventional definition is

$$R_{xy}(k) = E[x(n)y(n + k)] \qquad (6.24)$$

The ordering of the terms within the expectation brackets is very important. The expectation within the summation of equation 6.23 is an autocorrelation function. Since only time differences are important, $(n - i) - (n - k) = (k - i)$ and

$$R_x(k - i) = E[x(n - k)x(n - i)] \qquad (6.25)$$

Equation 6.23 is rewritten as

$$R_{xy}(k) = \sum_{i=-\infty}^{\infty} R_x(k - i)h(i) = R_x(k)*h(k) \qquad (6.26)$$

That is, the CCF between the output and input signals is equal to the convolution of the ACF of the input signal with the system's impulse response. The Fourier transform of equation 6.26 is simple and produces several new functions. Now

$$DTFT[R_{xy}(k)] = DTFT[R_x(k)] \cdot H(f) \qquad (6.27)$$

and

$$DTFT[R_x(k)] = S_x(f) \qquad (6.28)$$

is the PSD of the signal $x(n)$, and

$$DTFT[R_{xy}(k)] = S_{xy}(f) \qquad (6.29)$$

is the *cross power spectral density function* (*CPSD*) between signals $y(n)$ and $x(n)$. A detailed explanation of the meaning of these functions is contained in subsequent sections in this chapter.

A more complicated but also much more useful relationship is that between the PSD of input and output signals. The basic definition is

$$R_y(k) = E[y(n)y(n+k)] = E\left[\left(\sum_{i=-\infty}^{\infty} x(n-i)h(i)\right) \cdot \left(\sum_{l=-\infty}^{\infty} x(n+k-l)h(l)\right)\right]$$

$$= \sum_{i=-\infty}^{\infty} \sum_{l=-\infty}^{\infty} h(i)h(l)E[x(n-i)x(n+k-l)]$$

$$= \sum_{i=-\infty}^{\infty} \sum_{l=-\infty}^{\infty} h(i)h(l)R_x(k+i-l) \qquad (6.30)$$

Reordering the terms in equation 6.30 and using equation 6.26 yields

$$R_y(k) = \sum_{i=-\infty}^{\infty} h(i) \sum_{l=-\infty}^{\infty} h(l)R_x(\{k+i\}-l) \qquad (6.31)$$

$$R_y(k) = \sum_{i=-\infty}^{\infty} h(i)R_{xy}(k+i) \qquad (6.32)$$

Substituting $u = -i$ in equation 6.32, it becomes

$$R_y(k) = \sum_{u=-\infty}^{\infty} h(-u)R_{xy}(k-u) \qquad (6.33)$$

The relationships between PSDs are found by taking the DTFT of equation 6.33

$$S_y(f) = H^*(f)S_{xy}(f) \qquad (6.34)$$

and substituting from equation 6.29 and the two preceding equations

$$S_y(f) = H^*(f)H(f)S_x(f) = |H(f)|^2 S_x(f) \qquad (6.35)$$

The term $|H(f)|^2$ is the *power transfer function*.

## 6.4 SYSTEMS AND SIGNAL STRUCTURE

In the previous chapter it was seen that a moving average of a sequence of uncorrelated numbers can produce a correlation in the resulting sequence. The concept of using a discrete time system to induce structure in a sequence will be formalized in this section. Essentially, this is accomplished by using white noise as the input to the system and the structured sequence as the output of the system.

### 6.4.1 Moving-Average Process

The general form for a moving-average process of order $q$ is

$$y(n) = \sum_{l=0}^{q} b(l)x(n-l) \qquad (6.36)$$

The mean, variance, and correlation structure of the output process, $y(n)$, can be derived in general from this system and the properties of the input process. The mean value of the output is

$$E[y(n)] = m_y$$

$$= E\left( \sum_{l=0}^{q} b(l)x(n - l) \right)$$

$$= \sum_{l=0}^{q} b(l)E[x(n - l)]$$

$$= E[x(n)] \sum_{l=0}^{q} b(l) = m_x \sum_{l=0}^{q} b(l) \qquad (6.37)$$

If $m_x = 0$, then $m_y = 0$. The derivation of the ACF and variance of $y(n)$ is more complex. Fortunately, the general form has already been developed in Section 6.3.2. Compare the form of the MA process, equation 6.36, to the general definition of a system's output, equation 6.12. It can be seen that $b(l) = h(l)$. Thus, from equation 6.30

$$R_y(k) = E[y(n)y(n + k)] = \sum_{i=0}^{q} \sum_{l=0}^{q} b(i)b(l)R_x(k + i - l) \qquad (6.38)$$

The variance becomes

$$\sigma_y^2 = R_y(0) - m_y^2 = \sum_{i=0}^{q} \sum_{l=0}^{q} b(i)b(l)R_x(i - l) - \left( m_x \sum_{l=0}^{q} b(l) \right)^2 \qquad (6.39)$$

For signal modeling, the input process, $x(n)$, is often a zero mean white-noise process, $R_x(k) = \sigma_x^2 \delta(k)$. For this situation, let $u = k + i$ in equation 6.38 and the ACF and variance simplify to

$$R_y(k) = \sigma_x^2 \sum_{l=k}^{q} b(l)b(l - k) = \sigma_x^2 \sum_{u=0}^{q-k} b(u)b(u + k) \qquad (6.40)$$

$$\sigma_y^2 = \sigma_x^2 \sum_{l=0}^{q} b(l)^2 \qquad (6.41)$$

One can ascertain that a $q$th-order MA process has an autocorrelation function with magnitudes of zero for lags greater than $q$ and less than $-q$. This is left as an exercise. Examples and exercises are provided in Chapter 5.

## 6.4.2 Structure with Autoregressive Systems

The autoregressive model with stationary inputs can also be used to develop structure in an uncorrelated sequence and produce a sequence with stationary signal-like properties. The derivation of the correlational structure of the output process is more complex than those produced through MA processes. Thus we will study the first-order process in great detail. From the essential concepts covered, extensions to higher-order AR processes is straightforward, but the manipulations are more complicated as the order is increased.

For a first-order model

$$y(n) = -a(1)y(n-1) + b(0)x(n) \tag{6.42}$$

The mean of the output signal is

$$
\begin{aligned}
E[y(n)] &= m_y \\
&= E[-a(1)y(n-1) + b(0)x(n)] \\
&= -a(1)E[y(n-1)] + b(0)E[x(n)] \\
&= -a(1)m_y + b(0)m_x
\end{aligned}
\tag{6.43}
$$

Solving for $m_y$ yields

$$m_y = \frac{b(0)}{1 + a(1)} m_x \tag{6.44}$$

This is a more complex relationship between the means of the input and output than equation 6.37. Again $m_x$, and thus $m_y$, is usually zero. The variance is

$$
\begin{aligned}
\sigma_y^2 &= E[y(n)^2] \\
&= E[(-a(1)y(n-1) + b(0)x(n))^2] \\
&= E[a(1)^2 y(n-1)^2 - 2a(1)b(0)y(n-1)x(n) + b(0)^2 x(n)^2] \\
&= a(1)^2 E[y(n-1)^2] - 2a(1)b(0)E[y(n-1)x(n)] + b(0)^2 E[x(n)^2]
\end{aligned}
\tag{6.45}
$$

The middle term of equation 6.45, $E[y(n-1)x(n)]$, represents the cross correlation between the present input and previous output values, $R_{yx}(1)$. Since this is a causal system, there can be no correlation and $R_{yx}(1) = m_y m_x$. For $m_x = 0$, equation 6.45 becomes

$$\sigma_y^2 = \frac{b(0)^2}{1 - a(1)^2} \sigma_x^2 \tag{6.46}$$

The autocorrelation function is by definition

$$
\begin{aligned}
R_y(k) &= E[y(n)y(n+k)] \\
&= E[(-a(1)y(n-1) + b(0)x(n))(-a(1)y(n-1+k) + b(0)x(n+k))]
\end{aligned}
\tag{6.47}
$$

Several lag values will be considered. For $k = 1$,

$$
\begin{aligned}
R_y(1) = E[a(1)^2 y(n-1)y(n) &- a(1)b(0)y(n-1)x(n+1) - a(1)b(0)x(n)y(n) \\
&+ b(0)^2 x(n)x(n+1)]
\end{aligned}
\tag{6.48}
$$

Notice that when the expectation is distributed the following will occur.

a. The first and fourth terms will result in autocorrelation function values of signals $y(n)$ and $x(n)$, respectively.
b. The second and third terms will result in cross correlation function values between $x(n)$ and $y(n)$.

Writing this explicitly yields

$$R_y(1) = a(1)^2 R_y(1) - a(1)b(0)R_{yx}(2) - a(1)b(0)R_{xy}(0) + b(0)^2 R_x(1) \quad (6.49)$$

As we seek to simplify this expression, it is apparent that $R_x(1) = 0$. The values of the cross correlations depend on the lag. Consider the third term.

$$R_{xy}(0) = E[x(n)y(n)] = -a(1)E[x(n)y(n-1)] + b(0)E[x(n)^2] \quad (6.50)$$

The first term on the right side of equation 6.50 contains $R_{xy}(-1)$. For this model it is known from our previous discussion that $R_{xy}(-1) = R_{yx}(1) = m_y m_x = 0$. Strictly, one could carry this one more step and obtain

$$
\begin{aligned}
E[x(n)y(n-1)] &= -a(1)E[x(n)y(n-2)] + b(0)E[x(n)x(n-1)] \\
&= -a(1)E[x(n)y(n-2)] \quad (6.51)
\end{aligned}
$$

Continuing this procedure, it terminates with an expression

$$E[x(n)y(0)] = y(0)m_x = 0 \quad (6.52)$$

Thus $R_{xy}(0) = b(0)\sigma_x^2$. The same reasoning shows the value of the second term of equation 6.49 is zero. Demonstrating this is left as an exercise. Therefore,

$$R_y(1) = a(1)^2 R_y(1) - a(1)b(0)^2 \sigma_x^2$$

and

$$R_y(1) = -\frac{a(1)b(0)^2 \sigma_x^2}{1 - a(1)^2} \quad (6.53)$$

Referring to the expression for the variance of $y(n)$, then

$$R_y(1) = -a(1)\sigma_y^2$$

and

$$\rho_y(1) = -a(1) \quad (6.54)$$

This procedure can be continued to provide a closed-form solution for the ACF:

$$R_y(k) = \sigma_y^2 (-a(1))^{|k|} \quad (6.55)$$

Plots of several NACFs for different values of $a(1)$ are shown in Figure 6.2. Notice that in contrast to the correlation functions for an MA process that these NACFs have nonzero values at all lags and approach zero only asymptotically. The ACF in equation 6.55 can also be derived through a frequency domain approach using equation 6.35 and the IDFT of $S_y(f)$.

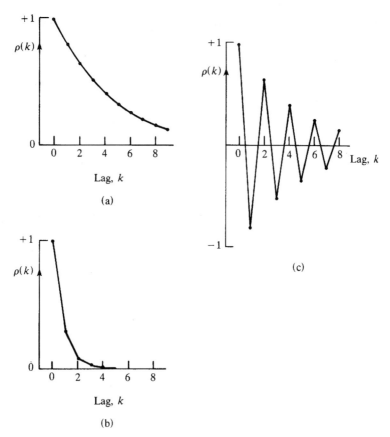

**FIGURE 6.2**   Three examples of NACFs for a first-order AR process: (a) $a(1) = -0.8$. (b) $a(1) = -0.3$. (c) $a(1) = 0.8$. [Adapted from Chatfield, fig. 3.1, with permission.]

### EXAMPLE 6.1

Because the ACF reflects structure in a signal, interpreting these should indicate some qualitative properties of the signal. Compare Figure 6.3a and 6.3b. One ACF has a monotonically decreasing trend in magnitude while the other has values that oscillate positively and negatively. The first ACF indicates that signal points that are one, two, etc. time units apart are positively correlated to one another. Thus it should be expected that the signal should have some short-term trends over time. A signal containing 100 points was simulated and is plotted in Figure 6.3c. Examining it reveals that groupings of four to five consecutive points are either positive or negative. The other ACF indicates that signal points that are odd time units apart are negatively correlated. Thus any positive signal values should be followed by negative ones and vice versa. Figure 6.3d shows a simulated time series from such a system. Again, its qualitative properties correspond to what is expected from the ACF.

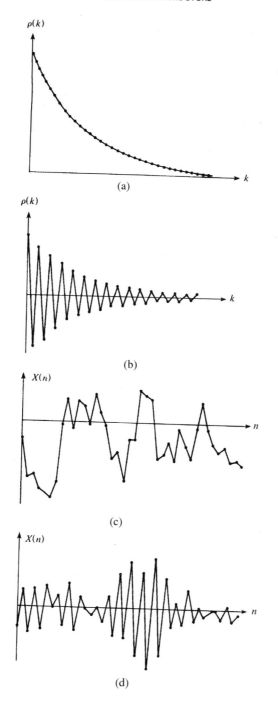

**FIGURE 6.3**   The ACFs from a first-order AR process with (a) $a(1) = -0.9$, (b) $a(1) = 0.9$, and (c), (d) realizations of these processes. [Adapted from Jenkins and Watts, figs. 6.4 and 6.5, with permission.]

### 6.4.3   Higher-Order AR Systems

The description of the ACF for higher-order AR models is much more complex, and the derivation of closed-form solutions is the subject of the study of difference equations (Pandit and Wu, 1983). However, a recursive solution for $R_y(k)$ can be derived in a simple manner. Begin with the equation for a $p$th-order AR model

$$y(n) + a(1)y(n-1) + a(2)y(n-2) + \cdots + a(p)y(n-p) = x(n) \qquad (6.56)$$

The usual convention of letting $a(0) = b(0) = 1$ has been adopted. The ACF of $y(n)$ is obtained by simply premultiplying equation 6.56 by $y(n-k)$ and taking the expectations; that is,

$$
\begin{aligned}
R_y(k) &= E[y(n-k)y(n)] \\
&= E[-a(1)y(n-k)y(n-1) - \cdots - a(p)y(n-k)y(n-p) + y(n-k)x(n)] \\
&= -a(1)R_y(k-1) - a(2)R_y(k-2) - \cdots - a(p)R_y(k-p) \\
&\quad + E[y(n-k)x(n)]
\end{aligned}
\qquad (6.57)
$$

It is known from the previous paragraph that for $m_x = 0$ and $k > 0$, then $R_{yx}(k) = 0$ and equation 6.57 becomes

$$R_y(k) + a(1)R_y(k-1) + a(2)R_y(k-2) + \cdots + a(p)R_y(k-p) = 0 \qquad (6.58)$$

This is the recursive expression for the ACF of process $y(n)$. For the situation when $k = 0$, equation 6.57 becomes

$$R_y(0) + a(1)R_y(1) + a(2)R_y(2) + \cdots + a(p)R_y(p) = \sigma_x^2 \qquad (6.59)$$

The proof of the latter expression is left as an exercise. Equations 6.59 and 6.58 will also be very important also in the study of signal modeling.

---

### EXAMPLE 6.2

There is a second-order AR process

$$y(n) - 1.0y(n-1) + 0.5y(n-2) = x(n)$$

By inspection, the recursive equations for the ACF and NACF are

$$R_y(k) - 1.0R_y(k-1) + 0.5R_y(k-2) = 0$$

and

$$\rho_y(k) - 1.0\rho_y(k-1) + 0.5\rho_y(k-2) = 0$$

By simply using the latter equation, one can generate the NACF. However, one must have the initial conditions since to begin

$$\rho_y(2) = 1.0\rho_y(1) - 0.5\rho_y(0)$$

The subscript of $\rho(k)$ is dropped for now. Since $\rho(0) = 1$, $\rho(1)$ must be ascertained. Being clever, let $k = 1$ and

$$\rho(1) - 1.0\rho(0) + 0.5\rho(-1) = 0$$

Solving for $\rho(1)$ yields $\rho(1) = 0.667$ and the NACF can be found. A sample function and the NACF for this signal are plotted in Figure 6.4.

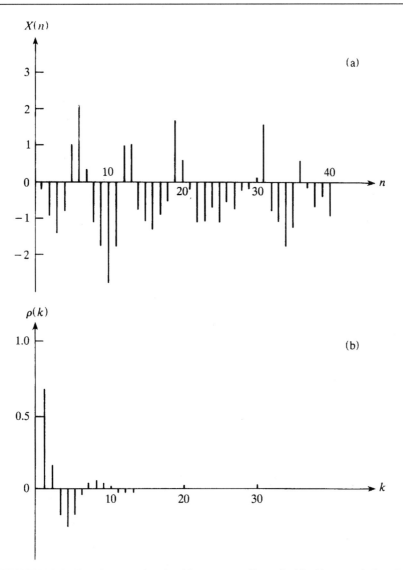

**FIGURE 6.4** For the second-order AR process in Example 6.2, (a) a sample function, $N = 40$, and (b) the ACF are ploted. [Adapted from Jenkins and Watts, fig. 5.9, with permission.]

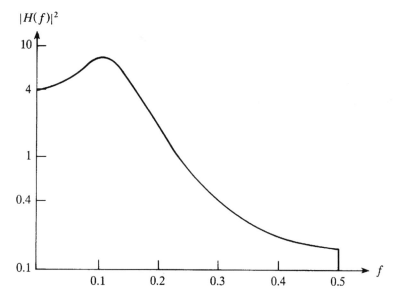

**FIGURE 6.5**   Power transfer function for a second-order AR system with $a(1) = -1.0$, $a(2) = 0.5$, and $T = 1$.

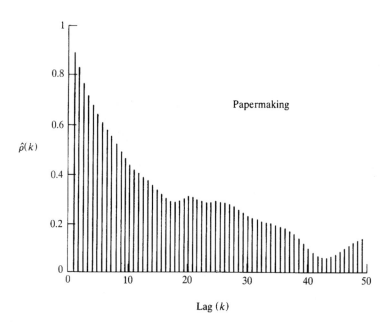

**FIGURE 6.6**   The NACF for an industrial papermaking process represented by an ARMA(2, 1) process. $a(1) = -1.76$, $a(2) = 0.76$, and $b(1) = -0.94$. [Adapted from Pandit and Wu, fig. 3.15, with permission.]

EXAMPLE 6.3

The power transfer function for the AR process in Example 6.2 is to be derived. The transfer function is

$$H(f) = \frac{1}{1 - e^{-j2\pi fT} + 0.5e^{-j4\pi fT}}$$

Therefore

$$|H(f)|^2 = \frac{1}{1 - e^{-j2\pi fT} + 0.5e^{-j4\pi fT}} \frac{1}{1 - e^{j2\pi fT} e^{j4\pi fT}}$$

$$= \frac{1}{1 - e^{-j2\pi fT} + 0.5e^{-j4\pi fT} - e^{j2\pi fT} + 1 - 0.5e^{-j2\pi fT}}$$
$$+ 0.5e^{j4\pi fT} - 0.5e^{j2\pi fT} + 0.25$$

$$= \frac{1}{2.25 - 3\cos(2\pi fT) + \cos(4\pi fT)}$$

This is plotted in Figure 6.5 for $T = 1$.

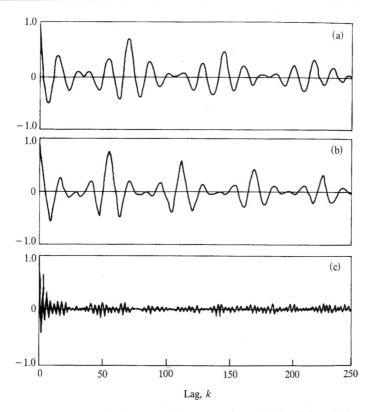

Lag, $k$

**FIGURE 6.7** The NACFs for three different speech signals, $T = 0.1$ ms. [Adapted from Rabiner and Schafer, fig. 4.24, with permission.]

As it might be expected, AR models with different parameters generate signals with different characteristics. An industrial papermaking process can be represented by an ARMA(2, 1) model,

$$y(n) - 1.76y(n-1) + 0.76y(n-2) = x(n) = 0.94x(n-1) \qquad (6.60)$$

Its NACF, plotted in Figure 6.6, indicates some long-duration positive correlations in the output time series. Other processes can be very complex. Models for speech generation can have orders of fourteen or higher. The NACFs of several speech sounds are plotted in Figure 6.7. They are definitely different from each other and have different model parameters. Their interpretation is not easy as with the first- and second-order models. However, it can be appreciated that the autocorrelation function represents some of the time domain characteristics of a signal in a compact manner.

## 6.5   TIME SERIES MODELS FOR SPECTRAL DENSITY

Although one of the main goals of signal analysis is to learn techniques to estimate the PSD of a measured time series, it is instructive to derive PSDs of several models of theoretical signals and to study them. In this manner one can develop a practical understanding of the concepts involved. Also, the evaluation of many estimation techniques are based on the estimation of PSD from sample functions generated from model processes. At this point we consider the white-noise process and several MA processes.

### EXAMPLE 6.4

A zero mean white-noise process is defined by the ACF $R(k) = \sigma^2 \delta(k)$. Its PSD is found using the DTFT defined in Section 6.2 and is

$$S(f) = T \sum_{k=-\infty}^{\infty} R(k)e^{-j2\pi fkT} = T \sum_{k=-\infty}^{\infty} \sigma^2 \delta(k)e^{-j2\pi fkT} = \sigma^2 T$$

This is plotted in Figure 6.8a. The spectral density is constant for all frequencies. This only occurs for white noise. Notice that the area under $S(f)$ equals $\sigma^2$.

### EXAMPLE 6.5

A first-order MA process was studied in Chapter 5, Example 5.5. The process is

$$y(n) = 0.5x(n) + 0.5x(n-1)$$

with $\sigma_x^2 = 0.5$ and autocorrelation function values

$$R_y(0) = 0.25, \qquad R_y(1) = 0.125, \qquad R_y(k) = 0 \quad \text{for } |k| \geq 2$$

The PSD is

$$S_y(f) = T(0.125e^{j2\pi fT} + 0.25 + 0.125e^{-j2\pi fT})$$
$$= T(0.25 + 0.25 \cos(2\pi fT))$$

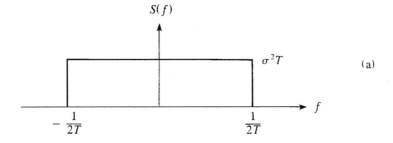

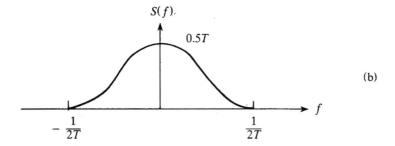

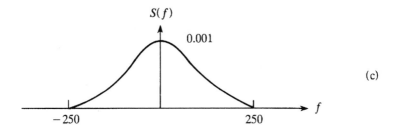

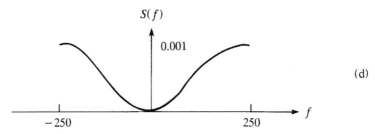

**FIGURE 6.8** The PSDs of several moving-average processes: (a) White noise. (b) First-order, low-frequency in parametric form. (c) First-order, low-frequency with sampling interval = 2 ms. (d) First-order, high-frequency with sampling interval = 2 ms.

and is plotted in Figure 6.8b. Notice that most of the power is concentrated in the lower frequency range. This process is in general a lowpass process.

The spectrum in Example 6.5 was left in parametric form for the illustration of two observations. The first is that in most of the theoretical presentations of PSD, $T = 1$. The second is that the shape of the PSD is independent of the sampling frequency. Figure 6.8c shows the spectrum of Figure 6.8b with $T = 2$ ms. Using equation 6.11, the variance is

$$\sigma_y^2 = \int_{-1/2T}^{1/2T} S(f) \, df = \int_{-1/2T}^{1/2T} T(0.25 + 0.25 \cos(2\pi f T)) \, df$$

$$= 2T \left( 0.25f + 0.25 \frac{\sin(2\pi f T)}{2\pi T} \right)_0^{1/2T} = 0.25 \qquad (6.61)$$

In both cases, the area under the PSD functions is 0.25. The major difference is the range of the frequency axis.

The complement of the low-frequency process is the highpass process where the majority of the variance is concentrated in the higher frequency range. This is easily formed from the first-order MA process of Example 6.5 by changing the sign of the coefficient of the term with lag 1; that is,

$$y(n) = 0.5x(n) - 0.5x(n - 1) \qquad (6.62)$$

Its PSD is shown in Figure 6.8d. The derivation is left as an exercise.

Another type of process that is frequently encountered is the *bandpass* process, which contains power in the middle range of frequencies. A hypothetical PSD for such a process is shown in Figure 6.9. The *lower and upper frequency bounds* are designated $f_l$ and $f_u$ respectively. Second- and higher-order MA processes can have this general type of PSD. In fact, both MA and AR processes of high order can produce spectra with multiple bands of frequencies. Figure 6.10 shows the PSD of the output of an AR(20) system used to simulate speech in a person with muscle tremors (Gath and Yair, 1987). The relative magnitude of the coefficients of the process determine whether the process is a lowpass, highpass,

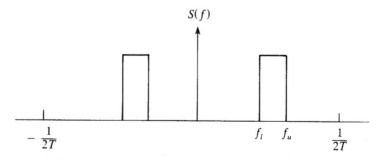

**FIGURE 6.9**   General schematic of the PSD of a bandpass process.

or bandpass process. A direct method to derive autocorrelation functions that will produce a PSD with a certain form is to rewrite equation 6.9 as a cosine series. It is

$$S(f) = T\left(R(0) + 2\sum_{k=1}^{\infty} R(k)\cos(2\pi fkT)\right) \qquad (6.63)$$

Derivation of this equation is straightforward and is left as an exercise. Equation 6.63 is a Fourier cosine series with coefficients

$$[Z(0), \dots, Z(k), \dots] = [TR(0), \dots, 2TR(k), \dots]$$

The integral equations are

$$Z(0) = \frac{1}{2f_N} \int_{-f_N}^{f_N} S(f)\, df \qquad (6.64)$$

$$Z(k) = \frac{1}{f_N} \int_{-f_N}^{f_N} S(f)\cos(2\pi fkT)\, df, \quad k \geq 1 \qquad (6.65)$$

As with any Fourier series, the accuracy of the approximation depends on the number of coefficients $Z(k)$. Let us now derive some autocorrelation functions that are appropriate for a bandpass process.

---

### EXAMPLE 6.6

Develop the autocorrelation function of a second-order process that has the general properties of the bandpass spectrum in Figure 6.9. Using the fact that the PSD is an even function, the first term is

$$Z(0) = TR(0) = \frac{1}{2f_N} \int_{-f_N}^{f_N} S(f)\, df$$

$$= \frac{1}{f_N} \int_{0}^{f_N} S(f)\, df$$

$$= 2T \int_{f_i}^{f_u} \frac{\sigma^2}{2(f_u - f_i)}\, df$$

$$= T\sigma^2 \quad \text{or} \quad R(0) = \sigma^2$$

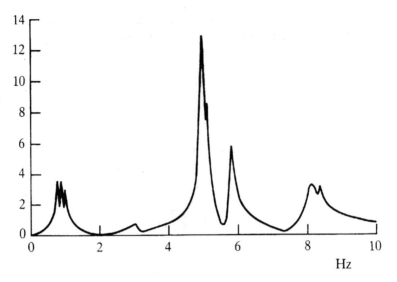

**FIGURE 6.10**   The PSD function of a tremorous speech signal represented by a 20th-order AR process. [Adapted from Gath and Yair, fig. 5, with permission.]

The other coefficients are

$$Z(k) = \frac{1}{f_N} \int_{-f_N}^{f_N} S(f) \cos(2\pi fkT) \, df$$

$$= \frac{2}{f_N} \int_{0}^{f_N} S(f) \cos(2\pi fkT) \, df$$

$$= 4T \int_{f_l}^{f_u} \frac{\sigma^2}{2(f_u - f_l)} \cos(2\pi fkT) \, df$$

$$= \frac{2T\sigma^2}{(f_u - f_l)} \int_{f_l}^{f_u} \cos(2\pi fkT) \, df$$

$$\frac{2T\sigma^2}{(f_u - f_l)} \left( \frac{\sin(2\pi fkT)}{2\pi kT} \right)_{f_l}^{f_u} = \frac{\sigma^2}{\pi k(f_u - f_l)} (\sin(2\pi f_u kT) - \sin(2\pi f_l kT))$$

For the sake of the example let the upper and lower frequency bounds equal 50% and 25% of the folding frequency, respectively, as shown in Figure 6.11a. Then

$$Z(k) = \frac{2\sigma^2}{\pi k f_N} (\sin(0.50\pi k) - \sin(0.25\pi k))$$

and

$$R(k) = \frac{2\sigma^2}{\pi k} (\sin(0.50\pi k) - \sin(0.25\pi k)), \quad k \geq 1$$

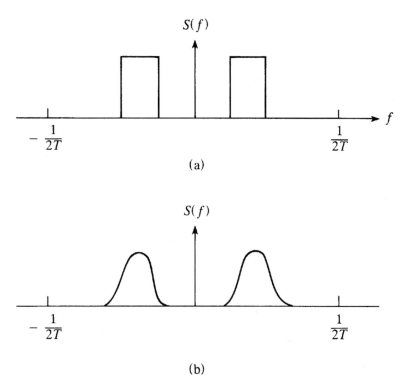

**FIGURE 6.11** Ideal and model PSD: (a) General bandpass. (b) Second-order MA model with bandpass PSD.

or

$$R(k) = [0.19\sigma^2, -0.32\sigma^2, -0.36\sigma^2, 0.0, 0.22\sigma^2, 0.11\sigma^2, \ldots]$$

If we had a second-order process with a variance of 1.0, the autocorrelation function would be

$$R(k) = [1.0, 0.19, -0.32]$$

The PSD is

$$S(f) = T\left(R(0) + 2\sum_{k=1}^{2} R(k)\cos(2\pi fkT)\right)$$
$$= T(1.0 + 2.0 \cdot 0.19\cos(2\pi fT) - 2.0 \cdot 0.32 \ \cos(4\pi fT))$$

as shown schematically in Figure 6.11b. Notice that it only approximates the ideal model but nonetheless is a bandpass process.

## REFERENCES

J. Cadzow, *Foundations of Digital Signal Processing and Data Analysis*, Macmillan, New York, 1987.

C. Chatfield, *The Analysis of Time Series: Theory and Practice*, Halstead Press— Wiley, New York, 1975.

D. Defatta, J. Lucas, and W. Hodgkiss, *Digital Signal Processing: A System Design Approach*, Wiley, New York, 1988.

I. Gath and E. Yair, "Comparative Evaluation of Several Pitch Process Models in the Detection of Vocal Tremor," *IEEE Trans. Biomed. Eng.* **34**:532–538 (1987).

G. Jenkins and D. Watts, *Spectral Analysis and Its Applications*, Holden-Day, San Francisco, 1968.

M. Jong, *Methods of Discrete Signal and Systems Analysis*, McGraw-Hill, New York, 1982.

L. Koopmans, *The Spectral Analysis of Time Series*, Academic Press, New York, 1974.

A. Oppenheim and A. Willsky, *Signals and Systems*, Prentice-Hall, Englewood Cliffs, NJ, 1983.

S. Pandit and S. Wu, *Time Series Analysis with Applications*, Wiley, New York, 1983.

M. Priestley, *Spectral Analysis and Time Series: Volume 1—Univariate Series*, Academic Press; New York, 1981.

L. Rabiner and R. Schafer, *Digital Processing of Speech Signals*, Prentice-Hall, Englewood Cliffs, NJ, 1978.

R. Ziemer, W. Tranter, and D. Fannin, *Signals and Systems—Continuous and Discrete*, Macmillan, New York, 1989.

## EXERCISES

**6.1** Properties of the PSD are listed in Section 6.2. Prove properties (a) and (b), that is, that the PSD is an even and real function.

**6.2** Prove that $R_{yx}(k) = R_{xy}(-k)$.

**6.3** Using

$$R_y(k) = E[y(n)y(n + k)] = \sum_{i=0}^{q} \sum_{l=0}^{q} b(i)b(l)R_x(k + i - l)$$

show that $R_y(k) = 0$ for $|k| > q$.

**6.4** For the following systems, find the transfer functions and express them in polar form.

a. $y(n) = x(n) - 1.4x(n-1) + 0.48x(n-2)$

b. $y(n) + 0.25y(n-1) - 0.5y(n-2) + 0.7y(n-3) = x(n)$

c. $y(n) = 0.89y(n-1) + 0.61y(n-2) = x(n) + 0.54x(n-1)$

**6.5** Using the basic definition of an MA process, equation 6.36, prove that the variance of the output is

$$\sigma_y^2 = \sigma_x^2 \sum_{l=0}^{q} b(l)^2$$

if $x(n)$ is zero mean white noise.

**6.6** Develop the power transfer function relationship, equation 6.35, from the equation

$$S_y(f) = T \sum_{k=-\infty}^{\infty} \sum_{i=-\infty}^{\infty} \sum_{l=-\infty}^{\infty} h(i)h(l)R_x(k+i-l)e^{-j2\pi fkT}$$

[*Hint:* Sum over $k$ first.]

**6.7** Derive the relationship between the autocorrelation function of the system output and the cross correlation between the input and output. [*Hint:* Start with the convolution relationship and multiply by $y(n-k)$.]

**6.8** A boxcar or rectangular moving average is described by the MA system relationship

$$y(n) = \frac{1}{q+1} \sum_{l=0}^{q} x(n+l)$$

For $q = 3$ and $R_x(k) = 4\delta(k)$, show that

$$R_y(k) = \left(1 - \frac{|k|}{4}\right) \quad \text{for } |k| \leq 3$$

**6.9** Start with a general first-order AR system with a white-noise input having a zero mean. Prove that $E[y(3)x(4)] = 0$.

**6.10** Show in detail that for a first-order AR system with a white-noise input that $\rho(2) = a(1)^2$.

**6.11** Show that the second term on the right-hand side of equation 6.49 is zero.

**6.12** For the third-order AR model,

$$y(n) + 0.25y(n-1) - 0.5y(n-2) - 0.7y(n-3) = x(n)$$

show that

$$R_y(0) + 0.25R_y(1) - 0.5R_y(2) - 0.7R_y(3) = \sigma_x^2$$

Remember that $R_y(k)$ is an even function.

**6.13** For a general fourth-order AR system, find the equations that must be solved to use the recursion relationship for finding the values of the correlation function.

**6.14** Consider the second-order AR system in Example 6.2.
  a. Find the variance of the output signal if the variance of the input signal is 22 and its mean is zero.
  b. Plot $R_y(k)$ for $-10 \leq k \leq 10$.

**6.15** Consider the third-order system in Exercise 6.12.
  a. Find the recursion relationship for the autocorrelation function.
  b. Using the equations for $k = 1$ and $k = 2$, solve for the initial conditions for $\rho_y(k)$.
  c. Generate the values of $\rho_y(k)$ for $0 \leq k \leq 10$.

**6.16** Prove that the normalized PSD, $S_N(f) = S(f)/\sigma^2$, and the correlation function, $\rho(k)$, are a Fourier transform pair. Assume $m = 0$.

**6.17** Prove that $S(f) = TR(0) + 2T \sum_{k=1}^{\infty} R(k) \cos(2\pi f kT)$.
  [*Hint:* Remember that $R(k)$ is an even function.]

**6.18** Resketch Figure 6.5 with the sampling interval having the values $T = 5.0$ sec, 0.5 sec, and 0.5 ms.

**6.19** Derive the PSD, shown in Figure 6.8d, of the first-order, high-frequency MA process in equation 6.62; $T = 2$ ms, and $\sigma_x^2 = 0.50$.

**6.20** For the highpass process

$$y(n) = 0.5x(n) - 0.5x(n - 1)$$

with $\sigma_x^2 = 0.50$, calculate and plot the PSD for $T = 1$ sec and $T = 10$ sec.

**6.21** Consider the following first-order MA processes.
  a. $y(n) = 0.7x(n) + 0.3x(n - 1)$; $T = 1$ sec, $\sigma_x^2 = 2$
  b. $y(n) = 0.4x(n) - 0.6x(n - 1)$; $T = 1$ ms, $\sigma_x^2 = 1$

  Calculate and plot the PSD. Are they highpass or lowpass processes? Verify that the area under $S_y(f)$ equals $\sigma_y^2$.

**6.22** Consider the following second-order MA processes.
  a. $y(n) = x(n) + x(n - 1) + x(n - 2)$; $T = 1$ sec, $\sigma_x^2 = 1$
  b. $y(n) = x(n) + 0.5x(n - 1) - 0.3x(n - 2)$; $T = 1$ ms, $\sigma_x^2 = 5$

  Derive and sketch the PSD.

**6.23** Show that that process

$$y(n) = x(n) + 0.8x(n - 1) + 0.5x(n - 2); T = 1 \text{ sec}, \sigma_x^2 = 1$$

has the normalized power density spectrum

$$S*(f) = 1 + 1.27\cos(2\pi f) + 0.53\cos(4\pi f)$$

**6.24** Develop the ACF for a third-order model of a lowpass MA process with $f_u = 0.75 f_N$ and $\sigma^2 = 25$.
   a. What are the ACF and PSD?
   b. Plot them for parametric $T$ and $T = 20$ s.

**6.25** Develop the ACF for a third-order model of a highpass MA process with $f_l = 0.35 f_N$ and $\sigma^2 = 100$.
   a. What are the ACF and PSD?
   b. Plot them for parametric $T$ and $T = 20$ ms.

## Computer Exercises

**6.26** Consider the second-order MA process

$$y(n) = x(n) - 1.4x(n-1) + 0.48x(n-2)$$

   a. Derive $\sigma_y^2$ and $R_y(k)$ when $\sigma_x^2 = 1, 5$.
   b. Generate 50 points of $y(n)$ with $\sigma_x^2 = 5$ and plot them.
   c. Make the scatter plots for $y(n)$ vs $y(n-2)$ and $y(n)$ vs $y(n-4)$. Are these plots consistent with the theory?

**6.27** Model a signal using the second-order AR system

$$y(n) - 0.79y(n-1) + 0.22y(n-2) = x(n)$$

and let $\sigma_x^2 = 10$ with $x(n)$ being a Gaussian white-noise sequence. Generate and plot 100 points of the output signal and label it $y_1(n)$. Use another sequence of values for $x(n)$ and generate another output signal, $y_2(n)$. Are $y_1(n)$ and $y_2(n)$ exactly the same? Are their qualitative characteristics similar?

# VII

---

# SPECTRAL ANALYSIS FOR RANDOM SIGNALS: CLASSICAL ESTIMATION

## 7.1 SPECTRAL ESTIMATION CONCEPTS

The determination of the frequency bands that contain energy or power in a sample function of a stationary random signal is called *spectral analysis*. The mode in which this information is presented is an energy or power spectrum. It is the counterpart of the Fourier spectrum for deterministic signals. The shape of the spectrum and the frequency components contain important information about the phenomena being studied. Two applications will be shown before the methodologies are studied. The first application is in machinery. Vibrations are an intrinsic component of rotating machinery. Excessive levels indicate malfunction of some component. Consider Figure 7.1, which is the power spectrum of the vibration measurements from a $\frac{1}{15}$ hp electric motor (Noori et al., 1988). Vibrations at different frequencies are normal and are created by various moving parts. Some of the frequency components and their sources are indicated in the figure. A malfunction, such as worn-out bearings or loose components, is indicated when its corresponding frequency component has too large a magnitude. Thus the spectrum is a valid tool for machine diagnosis.

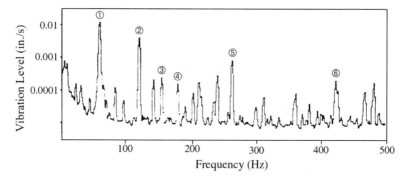

**FIGURE 7.1** Vibration spectrum of a $\frac{1}{15}$ horsepower electric motor. The frequencies of the peaks in the spectrum correspond to various motions in the motor: (1) 60 Hz, rotor unbalance; (2) 120 Hz, electrically induced vibration; (3) 155 Hz, bearing outer race; (4) 180 Hz, bent shaft; (5) 262 Hz, bearing inner race; (6) 420 Hz, bearing rolling elements. [Adapted from Noori et al., fig 18.39, with permission.]

Another application is in the investigation of the behavior of the smooth muscle of the intestines by measuring the electrical activity generated during contraction with an electrogastrogram (EGG). The smooth muscle plays an important role in the regulation of gastrointestinal motility, and the EGG reflects the mode of muscular contraction. Figure 7.2 shows a measurement from the colon of a experimental dog (Reddy et al., 1987). The activity is irregular. However, when a specific stimuli is impressed, such as eating, the activity pattern changes dramatically and becomes regular. Spectra from consecutive two-minute periods are shown in Figure 7.3. Before the dog eats, the peaks of the power spectra occur at different frequencies over time. Notice that after the dog eats, the peaks occur at approximately the same frequency, indicating a change in function.

These are only a sample of the many existing applications. The task now is to learn how to estimate these spectra accurately.

In general, there are two basic methods that are considered classical approaches. (The modern methods, which now have been used extensively for several decades, will be studied in the next chapter.) One classical approach is to develop an estimator from the Fourier transform of the sample function. This is called the direct or *periodogram* method and is based on Parseval's Theorem. The name evolved in the context of the first applications of spectral analysis in astronomy and geophysics. The frequencies were very low—on the order of cycles per week—and it was more convenient to use the period or $\frac{1}{f}$ as the independent variable. For example, Figure 7.4 shows a signal of earthquake occurrences and the estimated spectra. More will be stated about these spectra in subsequent sections. The other classical approach is based on the fact that the spectral density function is the Fourier transform of the autocovariance function,

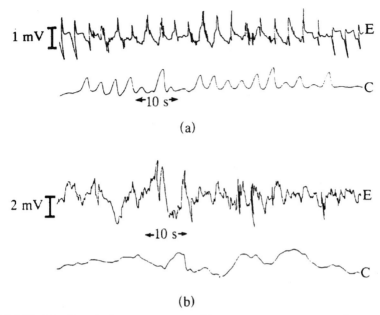

1 mv

C

◄10 s►

(a)

2 mv

E

◄10 s►

C

(b)

**FIGURE 7.2**  Typical electrogastrograms (E) and contractile activity (C) from the duodenum (a) and colon (b) in a dog. [Adapted from Reddy et al., fig. 1, with permission.]

as discussed in Section 6.2. With this method an estimator of the PSD is derived from the Fourier transform of the estimator of the autocorrelation function. This is called the indirect or *Blackman-Tukey* (*BT*) method.

### 7.1.1  Developing Procedures

The classical estimators and their properties are developed directly from the definition of the PSD, and both methods have the same statistical properties. Their mathematical equivalence will be derived from the definition of the Blackman-Tukey estimator. Complete derivations are mathematically intense but are understandable for someone knowledgeable in probability and statistics. Some properties are derived in this chapter, and the expressions for the variance are derived in the appendices. Comprehensive presentations can be found in references such as Geckinli and Yavuz (1983), Jenkins and Watts (1968), and Priestley (1981).

Recall from Chapter 5 that the biased estimator of the autocorrelation function, equation 5.27, is

$$\hat{R}(k) = \begin{cases} \dfrac{1}{N} \displaystyle\sum_{n=0}^{N-|k|-1} x(n)x(n+k), & |k| \le M \\ 0, & |k| > M \end{cases} \qquad (7.1)$$

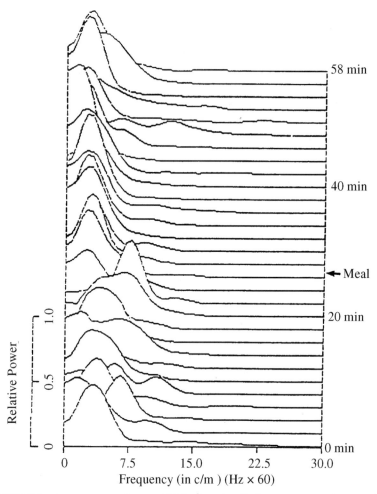

**FIGURE 7.3**   Power spectra from an EGG of a colon. Each spectrum is calculated from successive 2-minute time periods. [Adapted from Reddy et al., fig. 7b, with permission.]

The bounds on the maximum lag, $M$, are typically $0.1N \leq M \leq 0.3N$. The BT estimator is

$$\hat{S}(f) = T \sum_{k=-M}^{M} \hat{R}(k)w(k)e^{-j2\pi fkT} \qquad (7.2)$$

where $w(k)$ is a window function. The other estimator, the periodogram, can be obtained without first estimating $\hat{R}(k)$. This is done by incorporating the

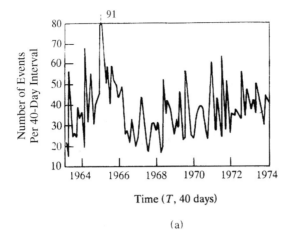

Time (T, 40 days)

(a)

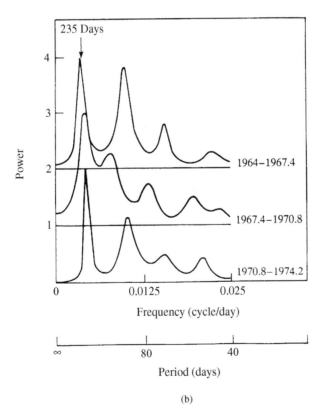

Frequency (cycle/day)

Period (days)

(b)

**FIGURE 7.4**  Earthquake occurrences: (a) Earthquake occurrence rate from 1964 to 1974. (b) Spectra for three nonoverlapping time periods. [Adapted from Landers and Lacoss, figs. 4 and 5, with permission.]

rectangular data window, $d_R(n)$, into equation 7.1. Recall from Section 3.5 that the rectangular data window is defined as

$$d_R(n) = \begin{cases} 1, & 0 \le n \le N - 1 \\ 0, & \text{elsewhere} \end{cases} \tag{7.3}$$

Now equation 7.1 is rewritten as

$$\hat{R}(k) = \frac{1}{N} \sum_{n=-\infty}^{\infty} x(n)d_R(n) \cdot x(n+k)d_R(n+k) \tag{7.4}$$

Its Fourier transform is

$$\hat{S}(f) = T \sum_{k=-\infty}^{\infty} \left( \frac{1}{N} \sum_{n=-\infty}^{\infty} x(n)d_R(n) \cdot x(n+k)d_R(n+k) \right) e^{-j2\pi fkT} \tag{7.5}$$

The equation will be rearranged to sum over the index $k$ first, or

$$\hat{S}(f) = \frac{T}{N} \sum_{n=-\infty}^{\infty} x(n)d_R(n) \cdot \sum_{k=-\infty}^{\infty} (n+k)d_R(n+k)e^{-j2\pi fkT} \tag{7.6}$$

Recognizing that the second summation will be a Fourier transform with the substitution $l = n + k$, equation 7.6 becomes

$$\hat{S}(f) = \frac{T}{N} \sum_{n=-\infty}^{\infty} x(n)d_R(n)e^{j2\pi fnT} \cdot \sum_{l=-\infty}^{\infty} (l)d_R(l)e^{-j2\pi f \, lT} \tag{7.7}$$

The second summation of the above equation is proportional to the Fourier transform of the sample function, $x(l)d_R(l)$, a signal with finite energy since it is a truncated version of $x(n)$. Its DTFT is

$$\hat{X}(f) = T \sum_{l=-\infty}^{\infty} (l)d_R(l)e^{-j2\pi f \, lT} \tag{7.8}$$

The first summation is also proportional to a Fourier transform with $f = -(-f)$, or

$$\hat{S}(f) = \frac{1}{NT}\hat{X}(-f)\hat{X}(f) = \frac{1}{NT}\hat{X}^*(f)\hat{X}(f) \tag{7.9}$$

where $\hat{X}^*(f)$ is the complex conjugate of the estimate of the Fourier transform. Equation 7.9 defines the periodogram estimate. Since the periodogram is calculated with a different procedure than the BT estimator, it is also given a different symbol and is represented as

$$I(f) = \frac{1}{NT}\hat{X}^*(f)\hat{X}(f) \tag{7.10}$$

## 7.1.2 Sampling Moments of Estimators

Knowledge of the sampling properties of the PSD estimators is essential and provides the framework for additional steps that will improve the estimation procedure. The mean of the spectral estimate is defined as

$$E[\hat{S}(f)] = E\left[ T \sum_{k=-\infty}^{\infty} \hat{R}(k)e^{-j2\pi fkT} \right] \tag{7.11}$$

Substituting for $\hat{R}(k)$ from equation 7.4 yields

$$
\begin{aligned}
E[\hat{S}(f)] &= \frac{T}{N} \sum_{k=-\infty}^{\infty} \left( E\left[ \sum_{n=-\infty}^{\infty} x(n)d_R(n) \cdot x(n+k)d_R(n+k) \right] \right) e^{-j2\pi fkT} \\
&= \frac{T}{N} \sum_{k=-\infty}^{\infty} \left( \sum_{n=-\infty}^{\infty} [x(n)x(n+k)]d_R(n)d_R(n+k) \right) e^{-j2\pi fkT} \\
&= \frac{T}{N} \sum_{k=-\infty}^{\infty} R(k)e^{-j2\pi fkT} \sum_{n=-\infty}^{\infty} d_R(n)d_R(n+k) \tag{7.12}
\end{aligned}
$$

The second summation results from the implied data window and is a correlation of the rectangular data window with itself. It is defined separately as

$$w(k) = \frac{1}{N} \sum_{n=-\infty}^{\infty} d_R(n)d_R(n+k)$$

$$w(k) = \begin{cases} 1 - \dfrac{|k|}{N}, & |k| \le N-1 \\ 0, & \text{elsewhere} \end{cases} \tag{7.13}$$

It has a triangular shape in the lag domain and is plotted in Figure 7.5a. Equation 7.11 is rewritten as

$$E[\hat{S}(f)] = T \sum_{k=-\infty}^{\infty} w(k)R(k)e^{-j2\pi fkT} \tag{7.14}$$

Thus the mean value of the PSD differs from the actual PSD because $R(k)$ is multiplied by a window called a *lag window*. Its effect is appreciated directly by invoking the convolution theorem. In the frequency domain, equation 7.14 becomes

$$E[\hat{S}(f)] = \int_{-1/2T}^{1/2T} S(g)W(f-g)\, dg \tag{7.15}$$

The function $W(f)$, the Fourier transform of $w(k)$, is called the *spectral window* and is plotted in Figure 7.5b. This particular spectral window is called *Fejer's kernel* and has the formula

$$W(f) = \frac{T}{N} \left( \frac{\sin(\pi f NT)}{\sin(\pi f T)} \right)^2 \tag{7.16}$$

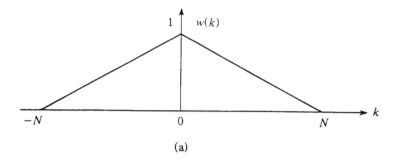

(a)

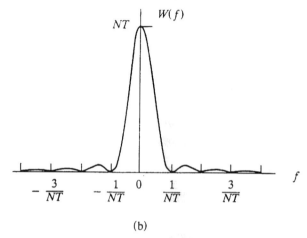

(b)

**FIGURE 7.5**    Implicit window function, Fejer's kernel: (a) Triangular lag window. (b) Corresponding spectral window.

It has a main lobe and side lobes similar to the other windows studied in Chapter 3. The convolution operation in equation 7.15 produces, in general, a biased estimate of $S(f)$. As $N$ approaches infinity, Fejer's kernel approaches a delta function, and the estimator becomes asymptotically unbiased.

## EXAMPLE 7.1

For the purpose of demonstrating the implicit effect of a finite signal sample on the bias, presume that the data window produces the spectral window shown in Figure 7.5 and that the actual PSD has a triangular shape as shown in Figure 7.6. The convolution results in the spectra plotted with the dashed lines in Figure 7.6. Compare these spectra. Notice that some magnitudes are underestimated and others are overestimated.

More quantitative knowledge of the bias is available if advanced theory is considered. It is sufficient to state for now that the bias term is on the order of

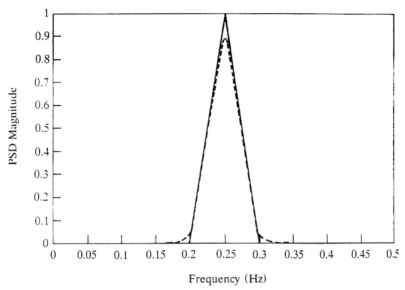

FIGURE 7.6 Bias in Power Spectral Density Estimates

**FIGURE 7.6** Schematic representation of bias caused by finite sample function; actual PSD (–), biased estimate of PSD (- - -).

$\dfrac{(\log N)}{N}$ if the first derivative of $S(f)$ is continuous. Fortunately, this is the situation for most phenomena. Mathematically this is expressed as

$$E[\hat{S}(f)] = S(f) + \mathbb{O}\left(\frac{\log N}{N}\right) \qquad (7.17)$$

It is consistent with intuition that if the sample function has a large number of sample points, then the bias is small or negligible.

The derivation of the variance of the sample PSD is very complicated mathematically and is usually studied at advanced levels. A complete derivation for large $N$ is described by Jenkins and Watts (1968) and is summarized in Appendix 7.1. Several sources derive the variance of a white-noise PSD (e.g., Kay, 1988). For now let it suffice to state that the variance is

$$\text{Var}[\hat{S}(f)] = S^2(f)\left(1 + \left(\frac{\sin(2\pi f\, NT)}{N \sin(2\pi f\, T)}\right)^2\right) \qquad (7.18)$$

when $N$ is large. Study equation 7.18 closely; notice that only the second term on the right side is a function of $N$. The variance is inconsistent and equals the square of the magnitude of the actual PSD for large values of $N$. Essentially this is not a good situation. The estimators developed from basic definitions are biased

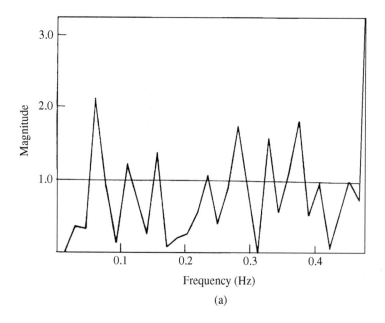

(a)

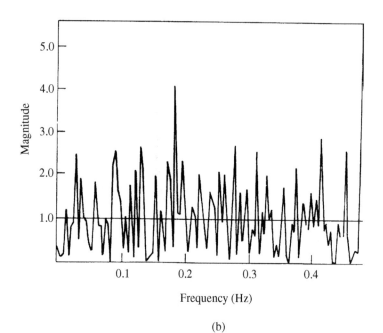

(b)

**FIGURE 7.7** Estimates of the PSD of a uniform white-noise process with $\sigma^2 = 1$ and $T = 1$. (a) $N = 64$. (b) $N = 256$.

and inconsistent—both undesired properties. Techniques that improve the properties of these spectral estimators will be studied after considering their sampling distribution. The sampling distribution will be derived from the periodogram representation. From now on, the calculated version of the spectral estimators will also be used. Thus the periodogram will be calculated at discrete frequencies $f = \dfrac{m}{NT}$. The change in notation is $I(m) = I\left(\dfrac{m}{NT}\right)$ and

$$I(m) = \frac{1}{NT}\hat{X}(m)\hat{X}^*(m) \qquad (7.19)$$

---

**EXAMPLE 7.2**

The sampling theory for power spectra shows that their estimates are inconsistent. This can be easily demonstrated through simulation. A random number generator was used to produce two uniform white-noise processes $x_1(n)$ and $x_2(n)$ with 64 and 256 points, respectively. The signal amplitudes were scaled so that $\sigma^2 = 1$ and the sampling interval, $T$, is 1 second. The periodograms were produced using equation 7.19. The theoretical spectra and the periodograms—their estimates— are plotted in Figure 7.7. Notice that there is great variation in the magnitude of the estimated spectra. They do not resemble the white-noise spectra, and the increase in $N$ does not improve the quality of the estimate. Only the frequency spacing has been reduced.

---

## 7.2 SAMPLING DISTRIBUTION FOR SPECTRAL ESTIMATORS

The sampling distribution of the spectral density function and the relationships among its harmonic components are very important for understanding the properties of PSD estimators. The derivations of some of these properties are presented and the derivations of others are left as exercises.

### 7.2.1 Spectral Estimate for White Noise

Let $x(n)$ represent an $N$-point sample of a Gaussian white-noise process with zero mean and variance $\sigma^2$. Its DFT is defined as usual, and for the purposes of this derivation is represented in complex Cartesian form with real and imaginary components $A(m)$ and $B(m)$, respectively, such that

$$\frac{X(m)}{\sqrt{NT}} = A(m) - jB(m)$$

with

$$A(m) = \sqrt{\frac{T}{N}} \sum_{n=0}^{N-1} x(n)\cos\left(2\pi\frac{mn}{N}\right)$$

and

$$B(m) = \sqrt{\frac{T}{N}} \sum_{n=0}^{N-1} x(n) \sin\left(2\pi \frac{mn}{N}\right) \tag{7.20}$$

The periodogram as defined in equation 7.19 then becomes

$$I(m) = A^2(m) + B^2(m), \quad m = 0, 1, \ldots, \left[\frac{N}{2}\right] \tag{7.21}$$

where $\left[\dfrac{N}{2}\right]$ represents the integer value of $\dfrac{N}{2}$. It can be shown that the estimates of the magnitudes of the PSD have a $\chi^2$ distribution and that the harmonic components are uncorrelated with one another (Jenkins and Watts, 1968). (They are also independent since they also have a Gaussian distribution.) The first- and second-order moments will be discussed first.

### 7.2.1.1   Moments

Since $A(m)$ and $B(m)$ are linear combinations of a Gaussian random variable (equations 7.20), they are also Gaussian random variables. The mean value of the real part of the harmonic components is

$$\begin{aligned}
E[A(m)] &= E\left(\sqrt{\frac{T}{N}} \sum_{n=0}^{N-1} x(n) \cos\left(2\pi \frac{mn}{N}\right)\right) \\
&= \sqrt{\frac{T}{N}} \sum_{n=0}^{N-1} E[x(n)] \cos\left(2\pi \frac{mn}{N}\right) \\
&= 0
\end{aligned} \tag{7.22}$$

The mean value of the imaginary part, $B(m)$, of the harmonic components is also zero. The variance is now written as the mean square and for the real part is

$$\begin{aligned}
\text{Var}[A(m)] &= E[A^2(m)] \\
&= E\left[\left(\sqrt{\frac{T}{N}} \sum_{n=0}^{N-1} x(n) \cos\left(2\pi \frac{mn}{N}\right)\right)^2\right]
\end{aligned} \tag{7.23}$$

This operation requires evaluating the mean values of all the cross products. Because $x(n)$ is white noise, by definition they are zero and

$$\begin{aligned}
E[A^2(m)] &= \frac{T}{N} \sum_{n=0}^{N-1} E[x^2(n)] \left(\cos\left(2\pi \frac{mn}{N}\right)\right)^2 \\
&= \frac{T\sigma^2}{N} \sum_{n=0}^{N-1} \cos^2\left(2\pi \frac{mn}{N}\right)
\end{aligned} \tag{7.24}$$

After some extensive but straightforward algebra with complex variables, it can be shown that

$$E[A^2(m)] = \frac{T\sigma^2}{N}\left(\frac{N}{2} + \cos\left(\frac{2\pi m(N-1)}{N}\right) \cdot \frac{\sin(2\pi m)}{2\sin(2\pi m/N)}\right)$$

$$E[A^2(m)] = \begin{cases} T\sigma^2/2, & m \neq 0 \quad \text{and} \quad \left[\dfrac{N}{2}\right] \\ T\sigma^2, & m = 0 \quad \text{and} \quad \left[\dfrac{N}{2}\right] \end{cases} \tag{7.25}$$

The derivation is left as an exercise. The term $B^2(m)$ is also a random variable with the same properties as $A^2(m)$. The next task is to define the correlational properties. The covariance of the real and imaginary parts of the DFT can be derived using the trigonometric or complex exponential forms. The complex exponentials provide a shorter derivation. Start with the expression for the covariance of different harmonics of the DFT,

$$E[X(m)X^*(p)] = E\left(T\sum_{n=0}^{N-1} x(n)e^{-j2\pi mn/N} \cdot T\sum_{l=0}^{N-1}(l)e^{j2\pi pl/N}\right)$$

$$= T^2 \sum_{n=0}^{N-1}\sum_{l=0}^{N-1} E[x(n)x(l)]e^{-j2\pi mn/N} \cdot e^{j2\pi pl/N} \tag{7.26}$$

$$= T^2 \sum_{n=0}^{N-1} \sigma^2 e^{-j2\pi(m-p)n/N}$$

Using the geometric sum formula, equation 7.26 becomes

$$E[X(m)X^*(p)] = T^2\sigma^2 \frac{1 - e^{-j2\pi(m-p)}}{1 - e^{-j2\pi(m-p)/N}}$$

$$E[X(m)X^*(p)] = \begin{cases} NT^2\sigma^2, & m = p \\ 0, & \text{otherwise} \end{cases} \tag{7.27}$$

Similarly, it can be shown that

$$E[X(m)X(p)] = \begin{cases} NT^2\sigma^2, & m = p = 0 \quad \text{and} \quad m = p = \left[\dfrac{N}{2}\right] \\ 0, & \text{otherwise} \end{cases} \tag{7.28}$$

Thus the harmonic components of the Fourier transform are uncorrelated with one another. Using the last two relationships, it can be shown that the real and imaginary components at different frequencies are also uncorrelated. First express them in real and imaginary parts:

$$E[X(m)X(l)] = E[(A(m) - jB(m))(A(l) - jB(l))] \tag{7.29}$$

and

$$E[X(m)X^*(l)] = E[(A(m) - jB(m))(A(l) + jB(l))] \tag{7.30}$$

Solving these two equations simultaneously yields

$$\text{Cov}[A(m), A(l)] = \text{Cov}[B(m), B(l)] = 0, \quad m \neq l$$
$$\text{Cov}[A(m), B(l)] = \text{Cov}[B(m), A(l)] = 0, \quad m \neq l \tag{7.31}$$

### 7.2.1.2 Sample Distribution

The density function for the periodogram needs to be known so that confidence intervals can be developed for the estimators. The pdf can be defined because it is known that the sum of squared independent Gaussian variables with a unit variance form a chi-square, $\chi_\nu^2$, random variable. The number of degrees of freedom, $\nu$, is equal to the number of independent terms summed. Through standardization, a function of $I(m)$ can be made to have unit variance. Thus standardizing and squaring $A(m)$ and $B(m)$ yields

$$\frac{A^2(m)}{T\sigma^2/2} + \frac{B^2(m)}{T\sigma^2/2} = \frac{I(m)}{T\sigma^2/2} = \chi_2^2, \quad m \neq 0 \quad \text{and} \quad \left[\frac{N}{2}\right] \tag{7.32}$$

Since the zero and folding frequency terms have only real parts

$$\frac{I(m)}{T\sigma^2} = \frac{A^2(m)}{T\sigma^2} = \chi_1^2, \quad m = 0 \quad \text{and} \quad \left[\frac{N}{2}\right] \tag{7.33}$$

The mean and variance of the periodogram can now be derived from the chi-square random variable. It is known that

$$E[\chi_\nu^2] = \nu \quad \text{and} \quad \text{Var}[\chi_\nu^2] = 2\nu \tag{7.34}$$

Thus for all frequency components except for $m = 0$ and $\left[\dfrac{N}{2}\right]$

$$E\left(\frac{I(m)}{T\sigma^2/2}\right) = 2 \quad \text{or} \quad E[I(m)] = T\sigma^2 \tag{7.35}$$

and

$$\text{Var}\left(\frac{I(m)}{T\sigma^2/2}\right) = 4 \quad \text{or} \quad \text{Var}[I(m)] = T^2\sigma^4 \tag{7.36}$$

It can be similarly shown that for the frequencies $m = 0$ and $\left[\dfrac{N}{2}\right]$,

$$E[I(m)] = T\sigma^2 \quad \text{and} \quad \text{Var}[I(m)] = 2T^2\sigma^4 \tag{7.37}$$

This verifies the general results observed in the previous section. The estimation procedure is inconsistent and the variance of the estimate is equal to the square of the actual PSD magnitude. Let us now study several examples of results from estimating the PSD of a known independent process.

## 7.2.2 Sampling Properties for General Random Processes

The theoretical derivations of the sampling properties of a white-noise process can be extended to the general random process with the use of signal models. Any random process can be expressed as a weighted sum of white-noise values. If the sum has a finite number of terms, then the process is a moving-average one; if the sum is infinite, then the process is an AR or ARMA one. The general method is

$$y(n) = \sum_{l=-\infty}^{\infty} h(l)x(n-l) \tag{7.38}$$

where $x(n)$ is a white-noise process. Referring back to Chapters 3 and 6, this also is the form of a convolution between a sequence of coefficients, $h(n)$, and $x(n)$. The Fourier transform of equation 7.38 is

$$Y(m) = H(m)X(m) \tag{7.39}$$

with

$$H(m) = T \sum_{n=-\infty}^{\infty} h(n)e^{-j2\pi mn/N} \tag{7.40}$$

For a sample function of finite duration for the process $y(n)$, the periodogram is

$$I_y(m) = \frac{Y(m)Y^*(m)}{NT} = \frac{H(m)X(m)H^*(m)X^*(m)}{NT} = |H(m)|^2 I_x(m) \tag{7.41}$$

All of the statistical variations are contained in the $x(n)$ process. We want to have an expression that relates the periodogram of $y(n)$ to its actual PSD, $S_y(m)$. This can be accomplished by again using a model for the random process with $S_x(m) = T\sigma^2$. For a system model, we know from Chapter 6 that

$$S_y(m) = |H(m)|^2 S_x(m) \tag{7.42}$$

Substituting equations 7.42 and 7.41 into equation 7.32 yields

$$\frac{2I_y(m)}{S_y(m)} = \chi_2^2, \quad m \neq 0 \quad \text{and} \quad \left[\frac{N}{2}\right]$$

and

$$\frac{I_y(m)}{S_y(m)} = \chi_1^2, \quad m = 0 \quad \text{and} \quad \left[\frac{N}{2}\right] \tag{7.43}$$

Equations 7.43 define the relationship between the periodogram estimate of a PSD and the actual spectrum. Using the moments of the $\chi_2^2$ random variable, it can be seen that the estimate is unbiased

$$E\left[\frac{2I_y(m)}{S_y(m)}\right] = v = 2 \quad \text{or} \quad E[I_y(m)] = S_y(m) \tag{7.44}$$

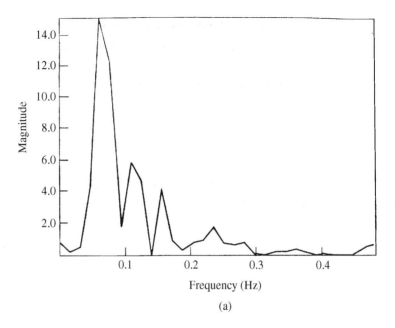

(a)

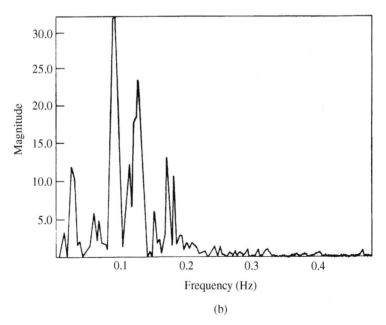

(b)

**FIGURE 7.8** Estimates of the PSD of a Gaussian second-order AR process with $\sigma_x^2 = 1$ and $T = 1$. (a) $N = 64$. (b) $N = 256$.

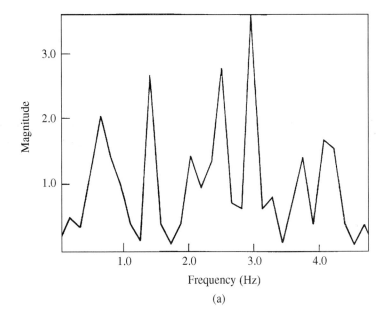

Frequency (Hz)

(a)

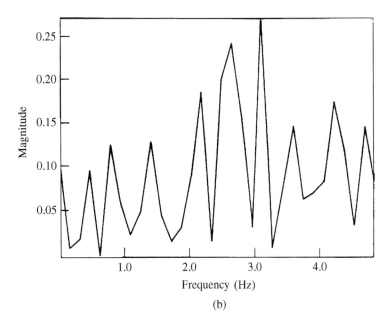

Frequency (Hz)

(b)

**FIGURE 7.9**   Estimates of the PSD of a uniform white-noise process with 64 points: (a) $T = 0.1$, $\sigma^2 = 10$. (b) $T = 0.1$, $\sigma^2 = 1$.

and inconsistent

$$\text{Var}\left[\frac{2I_y(m)}{S_y(m)}\right] = 2\nu = 4 \quad \text{or} \quad \text{Var}[I_y(m)] = S_y^2(m) \tag{7.45}$$

The same results apply for $m = 0$ and $\left[\dfrac{N}{2}\right]$. At first these results seem to agree with the bias and variance results stated in Section 7.1.2 for large $N$. This is true because in the convolution operation of equation 7.38, it is presumed that enough points of $x(n)$ are available to accurately represent $y(n)$. For signals with small $N$, additional processing must be done to make $I_y(m)$ unbiased so that equation 7.44 is appropriate. This situation will be addressed in the next section.

Equation 7.45 shows that the periodogram is not only inconsistent but also has a very large variance, similar to the situation with the periodogram of the white-noise process. Several examples will serve to illustrate this point, and then methods to reduce the variance will be discussed.

---

### EXAMPLE 7.3
Let us now consider the random process with the AR model

$$y(n) = y(n - 1) - 0.5y(n - 2) + x(n)$$

with $T = 1$ and $x(n)$ being zero mean white-noise with $\sigma^2 = 1$. The power transfer function was derived in Example 6.3 and the PSD for $y(n)$ is

$$S_y(m) = \frac{1}{2.25 - 3\cos(2\pi m/N) + \cos(4\pi m/N)} \, S_x(m)$$

For this process the spectrum, $S_y(m)$, equals the power transfer function that is plotted in Figure 6.5. There is appreciable power in the range of 0.05 to 0.15 Hz. The periodogram for this process is estimated from sample functions with $N = 64$ and $N = 256$. These are plotted in Figure 7.8. Notice that these estimates are very erratic in magnitude and that increasing $N$ does not improve the accuracy of the estimation. Again, this is showing that the variance of the estimate is large and inconsistent.

---

### EXAMPLE 7.4
Very seldom in real applications does $T = 1$. If, for a white-noise process, $T = 0.1\,\text{s}$ and $T\sigma^2 = 1$, the estimated PSD would only change in the frequency scale, as plotted in Figure 7.9a. Now obviously $\sigma^2 = 10$. If the variance remained 1, then the PSD would appear as in Figure 7.9b. Notice the additional change in scale to keep the area equal to the variance.

---

## 7.3  CONSISTENT ESTIMATORS: DIRECT METHODS

During the 1950s and 1960s, much research was devoted to improving the properties of spectral estimators. The results were successful and produced useful techniques that are still used. As with most estimators, there is a compromise. Any technique that reduces the variance of an estimate also increases its bias and vice versa. This trade-off will be discussed simultaneously with the explanation of the techniques.

## 7.3.1 Spectral Averaging

The simplest technique for developing a consistent estimator is called *periodogram averaging* or the *Bartlett approach*. It is based on the principle that the variance of a $K$-point sample average of independent random variables with variance $\sigma^2$ is $\dfrac{\sigma^2}{K}$. Thus an ensemble average of $K$ periodogram estimates produces an estimate whose variance is reduced by the factor $K$. In this situation an $N$-point sample function is divided into $K$ segments, each containing $M$ points. This is illustrated in Figure 3.23a; the signal is divided into four segments of 60 points. The periodogram for each segment, $I_i(m)$, is estimated; the integer $i$ is the index for the time series. The averaged periodogram is formed by averaging over all the periodograms at each frequency or

$$I_K(m) = \frac{1}{K} \sum_{i=1}^{K} I_i(m) \tag{7.46}$$

Since the periodogram estimates are independent among each other, the summation in equation 7.46 is a summation of $2K$ squared terms if $m \neq 0$ or $\dfrac{M}{2}$; that is,

$$KI_K(m) = \sum_{i=1}^{K} A_i^2(m) + B_i^2(m) \tag{7.47}$$

and $I_K(m)$ has $2K$ degrees of freedom. Its relationship to the true PSD is still found by transforming it to be a $\chi^2$ random variable. Thus,

$$\frac{2KI_K(m)}{S(m)} = \chi_{2K}^2, \quad m \neq 0 \quad \text{and} \quad \left[\frac{M}{2}\right] \tag{7.48}$$

and

$$E\left(\frac{2KI_K(m)}{S(m)}\right) = v = 2K \quad \text{or} \quad E[I_K(m)] = S(m) \tag{7.49}$$

and the estimator is unbiased. Remember that this bias property assumes a large $N$. The expectation of equation 7.46 is

$$E[I_K(m)] = \frac{1}{K} \sum_{i=1}^{K} E[I_i(m)] = E[I(m)], \quad |m| \leq \left[\frac{M}{2}\right] \tag{7.50}$$

The bias, which is essentially the result of leakage error, Section 3.5, is now defined exactly as in equations 7.14 and 7.15 with $M = N$. This bias is minimized in the same manner as the leakage was reduced for deterministic functions; that is, multiply the observed signal by a data window before the periodogram is produced. Now we have

$$y(n) = x(n)d(n), \quad 0 \le n \le M - 1 \tag{7.51}$$

and

$$I_i(m) = \frac{1}{MT} Y_i(m)Y_i^*(m) \tag{7.52}$$

The Hamming or Hanning data windows are usually used for this purpose. The leakage error has been resolved, but another source of constant bias has been introduced but is easily corrected. Remember that the variance is the area under the PSD. However, if equation 7.51 is considered with the fact that the amplitude of a data window is between 0 and 1, the sample variance of $y(n)$ is

$$\hat{\sigma}_y^2 = E\left[\frac{1}{M} \sum_{n=0}^{M-1} x^2(n)d^2(n)\right] = \sigma_x^2 \frac{1}{M} \sum_{n=0}^{M-1} d^2(n) \le \sigma_x^2 \tag{7.53}$$

Thus the variance of signal $y(n)$ is less than that of $x(n)$.

The *variance reduction factor* (*VR*) is the average square value of the data window, as expressed in equation 7.53, and is called *process loss* (*PL*). To correct for PL, simply divide the periodogram estimate by the process loss factor. The PL factors of several windows are listed in Appendix 7.3. The corrected estimator with the windowed signal is approximately unbiased and is

$$I_i(m) = \frac{1}{PL} \frac{1}{MT} Y_i(m)Y_i^*(m) \tag{7.54}$$

Then it can be stated that

$$E[I_K(m)] = S(m), \quad |m| \le \left[\frac{M}{2}\right] \tag{7.55}$$

and

$$\text{Var}\left[\frac{2KI_K(m)}{S(m)}\right] = 2\nu = 2 \cdot 2K \quad \text{or} \quad \text{Var}[I_K(m)] = \frac{S^2(m)}{K} \tag{7.56}$$

The periodogram averaging reduces the variance by a factor of $K$. This is not without a cost. Remember that only $M$ points are used for each $I_i(m)$ and the frequency spacing becomes $\frac{1}{MT}$. Hence the induced leakage before windowing is spread over a broader frequency range as $K$ is increased.

## EXAMPLE 7.5

The estimate of the white-noise spectrum from Example 7.2 will be improved with spectral averaging. The time series with 256 points is subdivided into four segments, and each segment is multiplied by a Hamming window. The spectral resolution becomes $\dfrac{1}{MT} = \dfrac{1}{64}$ Hz. Each individual spectrum has properties similar to those shown in Figure 7.7a. After correcting for process loss, the ensemble is shown in Figure 7.10a. It is much improved and resembles what is expected for a spectrum of white noise. The same procedure is repeated with $K = 8$ and the spectral estimate is shown in Figure 7.10b.

## EXAMPLE 7.6

The spectral estimate for any random process can be improved with averaging. Consider a second-order AR process with $a_1 = a_2 = 0.75$, $T = 0.1$, and $\sigma_x^2 = 20$. A sample function is generated with $N = 256$ and is divided into four segments. Figure 7.11a and b show the entire signal and the first segment after applying a Hamming data window. The four periodogram estimates, $I_1(m)$, $I_2(m)$, $I_3(m)$, and $I_4(m)$ are calculated and plotted in Figures 7.11c through 7.11f. The frequency resolution is $\frac{1}{6.4}$ Hz. Notice that the individual spectra are still very erratic. The ensemble average is plotted in Figure 7.11g. It definitely resembles the theoretical spectrum in Figure 7.11h. As an exercise, verify some of the values of the periodogram from the individual spectra at several of the harmonics.

### 7.3.2 Confidence Limits

The chi-square relationship between the actual PSD and its estimate can be used to establish confidence limits. These limits are boundaries for the actual PSD. Given the estimate, the actual PSD lies within the boundaries with a desired probability. This is a two-tailed test since an upper and a lower bound are necessary. Let the significance level be $\alpha$ and designate one limit as L1 and let it be the right-hand bound on the distribution. Then

$$\text{Prob}[\chi_\nu^2 \geq L1] = \alpha/2 \tag{7.57}$$

As is usual, the other limit is designated L2, is the left-hand bound, and

$$\text{Prob}[\chi_\nu^2 \geq L2] = 1 - \alpha/2 \quad \text{or} \quad \text{Prob}[\chi_\nu^2 \leq L2] = \alpha/2 \tag{7.58}$$

These establish the limits for a confidence level of $1 - \alpha$. These probabilities and boundaries are illustrated in Figure 7.12. For equation 7.57, the lower bound is established with equation 7.48 and

$$\chi_{2K}^2 = \frac{2KI_K(m)}{S(m)} \leq L1 \quad \text{or} \quad S(m) \geq \frac{2KI_K(m)}{L1} \tag{7.59}$$

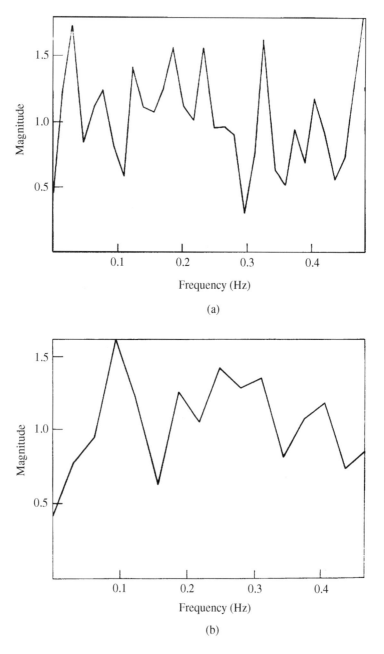

**FIGURE 7.10** Estimate of a spectra of the white-noise process in Figure 7.7 using spectral averaging: (a) $N = 256$, $K = 4$. (b) $N = 256$, $K = 8$.

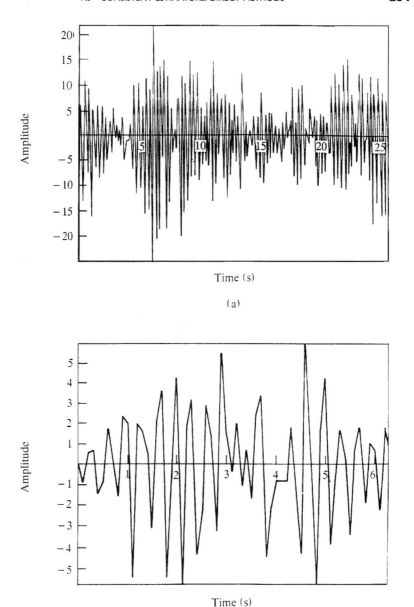

Time (s)

(a)

Time (s)

(b)

**FIGURE 7.11** The spectra of a Gaussian second-order AR process, $a_1 = a_2 = 0.75$, is estimated using segment averaging: (a) Sample function with $N = 256$. (b) Segment 1 multiplied by a Hamming window.

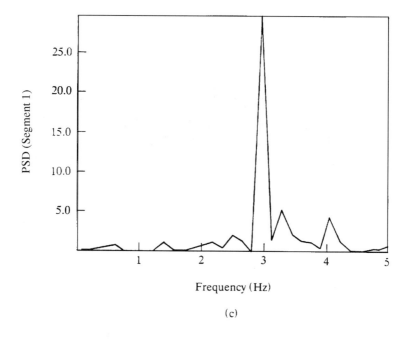

(c)

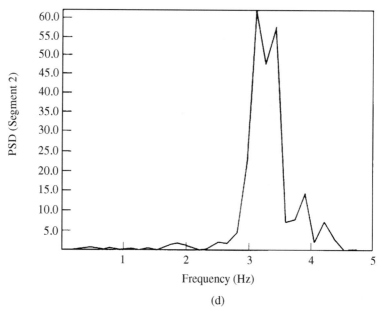

(d)

**FIGURE 7.11** The spectra of a Gaussian second-order AR process, $a_1 = a_2 = 0.75$, is estimated using segment averaging: (c) through (f) Periodogram spectra of segments 1 through 4.

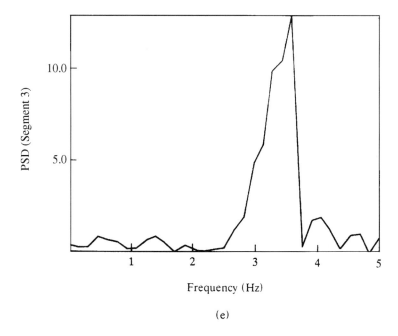

(e)

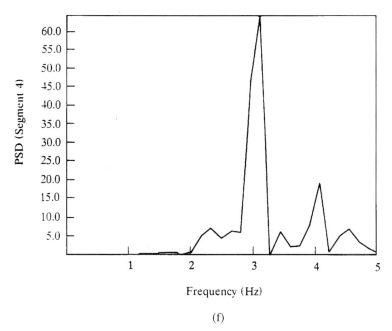

(f)

**FIGURE 7.11**    The spectra of a Gaussian second-order AR process, $a_1 = a_2 = 0.75$, is estimated using segment averaging: (c) through (f) Periodogram spectra of segments 1 through 4.

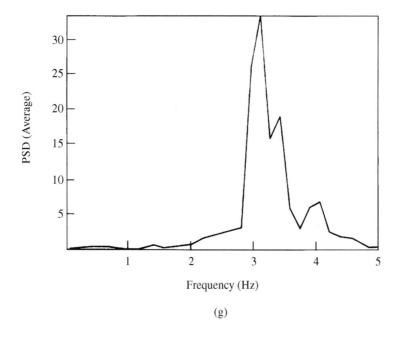

(g)

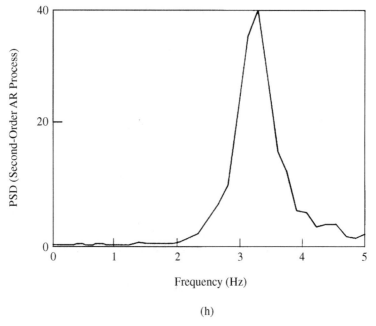

(h)

**FIGURE 7.11** The spectra of a Gaussian second-order AR process, $a_1 = a_2 = 0.75$, is estimated using segment averaging: (g) The average PSD estimate. (h) The ideal spectrum.

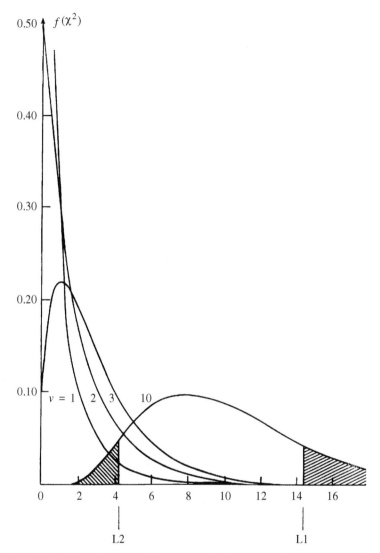

**FIGURE 7.12** Plots of several chi-square probability density functions, $v = 1, 2, 3, 10$, and the confidence limits, L1 and L2, for the pdf with $v = 10$.

Similarly, the upper bound is established for equation 7.58 and is

$$\chi_{2K}^2 = \frac{2KI_K(m)}{S(m)} \geq L2 \quad \text{or} \quad S(m) \leq \frac{2KI_K(m)}{L2} \tag{7.60}$$

The values for L1 and L2 are read directly from the chi-square distribution function table for $v = 2K$ degrees of freedom. Often these limits are written

$$L1 = \chi_{2K,\alpha/2}^2 \quad \text{and} \quad L2 = \chi_{2K,1-\alpha/2}^2 \tag{7.61}$$

### EXAMPLE 7.7

What is the relationship between the averaged periodogram and a signal's PSD for a 95% confidence interval if 6 segments, $K = 6$, are used? From equation 7.61 with $\alpha = 0.05$,

$$L1 = \chi^2_{12,0.025} = 23.34 \quad \text{and} \quad L2 = \chi^2_{12,0.975} = 4.40$$

From equations 7.59 and 7.60, the bounds on the spectrum are

$$\frac{2KI_K(m)}{L1} \le S(m) \le \frac{2KI_K(m)}{L2}$$

or

$$\frac{12I_K(m)}{23.34} = 0.51I_6(m) \le S(m) \le \frac{12I_K(m)}{4.40} = 2.73I_6(m)$$

Thus, with a probability of 0.95 the actual spectrum lies somewhere between a factor of 0.51 and 2.73 times the estimated spectrum.

### EXAMPLE 7.8

What are the upper and lower bounds for estimating the PSD when 30 spectra are averaged and a 99% confidence interval is desired?

$$L1 = \chi^2_{60,0.005} = 91.95 \quad \text{and} \quad L2 = \chi^2_{60,0.995} = 35.53$$

The bounds are

$$\frac{60I_K(m)}{91.95} \le S(m) \le \frac{60I_K(m)}{35.53}$$

### EXAMPLE 7.9

Let us now determine the limits for the estimates of the PSD of a white-noise process by determining the boundaries as shown in Examples 7.7 and 7.8. A sample function containing 960 points was divided into 30 segments and the average periodogram calculated. The 95% confidence limits are

$$0.65I_K(m) \le S(m) \le 1.69I_K(m)$$

These bounds are plotted in Figure 7.13 along with the estimate. Again, notice the asymmetry in the upper and lower bounds. The ideal spectrum lies within these bounds with a probability of 0.95.

### EXAMPLE 7.10

The confidence limits are established for the estimate of the PSD of the random process in Example 7.6 with $\sigma_x^2 = 5$. The result is plotted in Figure 7.14. The ideal spectrum is found in Figure 7.11g by correcting the power scale for a lower value of $\sigma_x^2$ (divide by 4.) The ideal spectrum is well within the confidence limits.

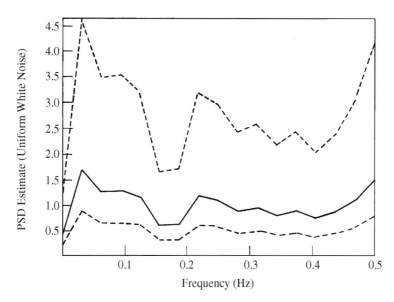

**FIGURE 7.13** The estimated spectrum (—) from a white-noise signal, $N = 960$, with $T = 1.0$ and $\sigma^2 = 1.0$, and its confidence limits (- - -) for $K = 30$.

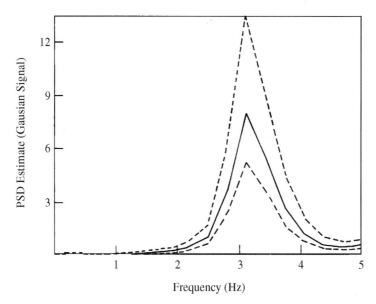

**FIGURE 7.14** The estimated spectrum (—) from a second-order process, $a_1 = a_2 = 0.75$, $\sigma_x^2 = 5$, $T = 0.1$, $K = 30$, and $N = 960$, and its confidence limits (- - -).

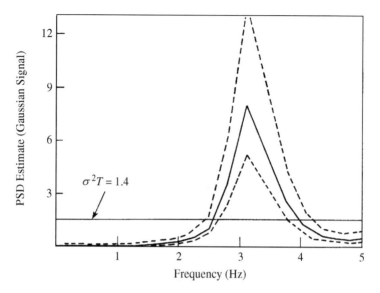

**FIGURE 7.15** The estimated spectrum (—) from a second-order process, its confidence limits (- - -), and an ideal white-noise spectrum with the same sample variance.

Now let us suppose that we do not know anything about the characteristics of the random process. We hypothesize that it is white noise with the calculated sample variance. The estimated spectrum and the limits for the random process are shown in Figure 7.15. The hypothesized spectrum is outside of the designated boundaries. Thus we would reject the hypothesis and state definitely that the process is not white noise but has some structure. The next step is to propose several alternate hypotheses and test them also.

---

**EXAMPLE 7.11**

The second-order process in Example 7.10 is divided into 60 segments. The averaged estimate and confidence limits are plotted in Figure 7.16. The confidence limits are much closer together and permit more stringent judgment about possible ideal spectra that are consistent with the data. The frequency resolution has now been reduced substantially. Notice that the peak of the spectra is broader and its magnitude is less.

---

### 7.3.3 Summary of Procedure for Spectral Averaging

The procedure for spectral averaging is sequential and is as follows:

1. Decide on the spectral resolution needed and thus choose $M$.
2. Divide the sample function into $K$ segments.
3. Detrend and apply a data window to each segment.
4. Zero pad the signals in each segment to either use an FFT algorithm or change the resolution.

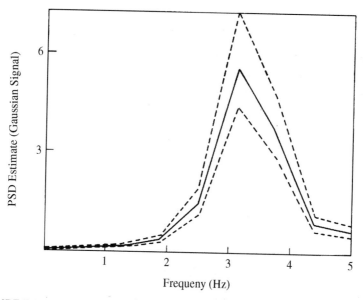

**FIGURE 7.16** Spectral estimate (—) and confidence limits (- - -) for a second-order AR signal. The estimate was obtained by averaging 60 periodogram segments.

5. Calculate the periodogram for each segment.
6. Correct for process loss.
7. Calculate the ensemble average.
8. Calculate the chi-square confidence bounds.
9. Plot $I_k(m)$ and the confidence bounds.
10. Test for consistency with any hypothesized ideal spectrum.

If one decides that the frequency spacing is smaller than necessary and that closer confidence bounds are desired, then the number of segments can be increased and the entire procedure repeated. Detrending needs to be reemphasized here. It was defined in Section 3.5.6 as fitting each segment of the signal with a low-order polynomial and subtracting the values of the fitted curve. Detrending is necessary to remove bias in the low-frequency range of the spectral estimate.

### 7.3.4 Welch Method

A variation of the segmentation and averaging (or Bartlett method) is the *Welch Method* (Welch, 1967). In this method the sample function is also divided into $K$ segments containing $M$ sample points. However, overlapping of segments is allowed and the initial data points in each segment are separated by $D$ time units. A schematic is shown in Figure 7.17: There is an overlap of $M - D$ sample points, and $K = \dfrac{(N - M)}{D} + 1$ segments are formed. The *percent overlap* is defined as $100 \cdot \dfrac{(M - D)}{M}$. Overlapping allows more segments to be created with an acquired sample function. Alternatively, this allows more points in a segment for a given number of segments, thus reducing the bias. However, the segments are not strictly independent and the variance reduction factor, VR, is not $K$ and depends on the data window. For any data window, we define the function $p(l)$ as

$$p(l) = \frac{\left( \sum\limits_{n=0}^{M-1} d(n)d(n + lD) \right)^2}{\left( \sum\limits_{n=0}^{M-1} d^2(n) \right)^2} \tag{7.62}$$

The variance reduction factor is

$$\frac{K}{1 + 2 \sum\limits_{l=1}^{K-1} \dfrac{K - 1}{K} p(l)} \tag{7.63}$$

The equivalent degrees of freedom is twice this factor. Studying PSD estimates of simulated signals indicates that using a Hanning or Parzen data window with a 50 to 65% overlap yields good results (Marple, 1987).

### 7.3.5 Spectral Smoothing

Perhaps the simplest method for reducing the variance of PSD estimates is *spectral smoothing or the Daniell method*. It is also based on the notion of averaging independent spectral values; however, in this method the magnitudes are averaged over the frequency domain. For *boxcar or rectangular smoothing*, the smoothed spectral estimate, $\tilde{I}_K(m)$, is obtained by first calculating the

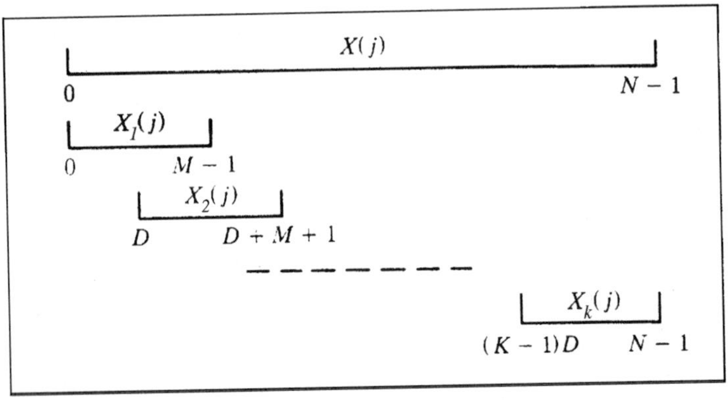

**FIGURE 7.17**   Illustration of the signal segments containing $M$ points with an overlap of $(M - D)$ points. [Adapted from Welch, fig. 1, with permission.]

periodogram or BT estimate and then averaging spectral magnitudes over a contiguous set of harmonic frequencies of $\dfrac{1}{NT}$; that is,

$$\tilde{I}_K(m) = \frac{1}{K} \sum_{j=-J}^{J} I(m-j), \quad J = \frac{(K-1)}{2} \tag{7.64}$$

The smoothing is symmetric and $K$ is an odd integer. Again, the variance of the estimate is reduced by a factor of $K$ and

$$\mathrm{Var}\left[\frac{2K\tilde{I}_K(m)}{S(m)}\right] = 2v = 2 \cdot 2K \quad \text{or} \quad \mathrm{Var}[\tilde{I}_K(m)] = \frac{S^2(m)}{K} \tag{7.65}$$

As with averaging, the bias must be reduced by applying a data window to the signal and correcting $I(m)$ for process loss. There are some important differences from the results of the averaging technique. First, the frequency spacing is still $\dfrac{1}{NT}$; there are still $\left[\dfrac{N}{2}\right]$ harmonic values in $\tilde{I}_K(m)$. However, smoothing makes them no longer independent; any value of $\tilde{I}_K(m)$ is composed of $K$ consecutive harmonics of $I(m)$. Thus the frequency resolution is decreased. The width in the frequency range over which the smoothing is done is called the *bandwidth*, $B$ where $B = \dfrac{2J}{NT}$. This is shown schematically in Figure 7.18. Another factor is that bias can be introduced. Consider the expectation of the smoothed estimate

$$E[\tilde{I}_K(m)] = \frac{1}{K} \sum_{j=-J}^{J} E[I(m-j)] \approx \frac{1}{K} \sum_{j=-J}^{J} S(m-j) \tag{7.66}$$

If consecutive values of the actual spectra are not approximately equal, then a bias can be caused. Obviously local maxima will be underestimated and local minima will be overestimated. This only becomes a matter of concern when *narrowband* signals are being analyzed.

---

### EXAMPLE 7.12

An AR(2) process is generated with $a(1) = a(2) = 0.75$, $\sigma_x^2 = 5$, $T = 0.02$, and $N = 64$. The periodogram is plotted in Figure 7.19a and is smoothed over three and seven harmonics, respectively. The resulting smoothed spectral estimates are shown in Figure 7.19b and 7.19c and tabulated in Appendix 7.5. Verify some of these values. The bandwidths of three and seven-point smoothing are $B = \dfrac{2}{NT} = 1.56\,\mathrm{Hz}$ and $B = \dfrac{6}{NT} = 4.69\,\mathrm{Hz}$, respectively. As expected, the seven-point smoothing produces estimates with smaller variance and less erratic characteristics. The penalty for this is that the resolution has decreased; that is, the bandwidth is wider or spectral values that are independent of one another are

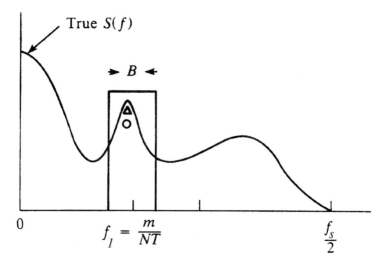

**FIGURE 7.18** Schematic of rectangular smoothing. The actual spectrum and rectangular spectral window of bandwidth $B$ are shown. The triangle and square represent peak values resulting from different smoothing bandwidths.

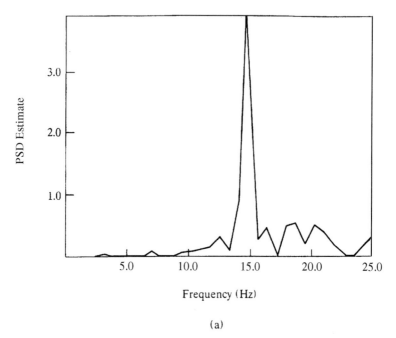

(a)

**FIGURE 7.19** Spectral estimates with three- and seven-point smoothing for an AR signal with 64 points. A Hamming data window was used. (a) Original estimate.

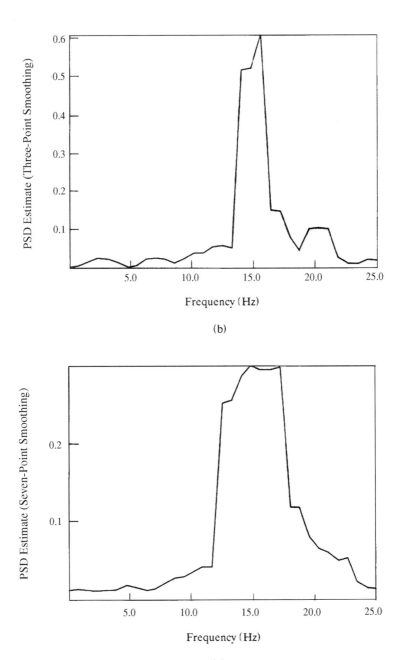

**FIGURE 7.19** Spectral estimates with three- and seven-point smoothing for an AR signal with 64 points. A Hamming data window was used. (b) Estimate with three-point smoothing. (c) Estimate with seven-point smoothing.

separated by seven harmonics. This is not a problem for the white-noise process but is for a correlated random process. Observe in Figure 7.19c that the magnitude for $\tilde{I}_7(m)$ is less than that for $\tilde{I}_3(m)$. Thus more of a negative bias is introduced because of the narrow spectral peak.

---

There are other smoothing weighting functions. Another one that is sometimes used is the triangular function (Schwartz and Shaw, 1975). For smoothing over $K$ terms,

$$\tilde{I}_K^t(m) = \frac{1}{J} \sum_{i=-J}^{J} \left( 1 - \frac{|i|}{J} \right) I(m-i), \quad J = \frac{(K-1)}{2} \tag{7.67}$$

The benefit is that less bias is introduced for smoothing over the same number of harmonics. The cost is that the variance is not reduced as much. The number of degrees of freedom is

$$v = 2 \cdot \frac{2J+1}{\dfrac{4}{3} + \dfrac{2J^2 + 2J + 1}{3J^3}} \tag{7.68}$$

Bandwidth is difficult to define for nonrectangular smoothing. Equivalent bandwidth, $B_e$, is defined as the width of the rectangular window that has the same peak magnitude and area as the triangular window. It is $B_e = \dfrac{J}{NT}$. The proof is left as an exercise.

In general, it has been shown that bias is proportional to the second derivative of the true spectrum (Jenkins and Watts, 1968). The proofs are quite complicated. A computer exercise at the end of this chapter will emphasize this concept.

### 7.3.6   Additional Applications

Several additional applications are briefly summarized to demonstrate how some of the principles that have been explained in the previous sections have been implemented.

In the development of magnetic storage media for computer disks, the size of the magnetic particles seems to influence the amount of noise generated during the reading of the stored signal. In particular, it was suspected that the amount of noise is dependent on the frequency. To test this phenomenon, periodic square wave signals were sampled and stored on a disk with the frequency being changed at repeated trials. The recorded signals were read and their PSD estimated using periodogram averaging. The fundamental and harmonic frequency components of the square wave were removed from the spectra in

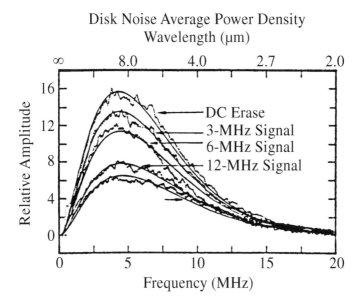

**FIGURE 7.20** Noise spectra from a recording medium as a function of signal frequency. [Adapted from Anzaloni and Barbosa, fig. 3, with permission.]

order to study the noise spectra. They are plotted in Figure 7.20. The shapes of the PSDs are the same for all the signals. However, it is evident that the higher frequency signals produce less noise.

Several biological parameters are monitored during maternal pregnancy and labor. One of these is the electrical activity of the uterus recorded from a site on the abdomen, called the electrohysterogram (EHG). The characteristics of the EHG change as parturition becomes immanent. Figure 7.21a and 7.21b show that as labor processes, the EHG has more high-frequency components. The signals are monitored over an 8-hour period and are sampled at a rate of 20 Hz. Each sample function must be at least 50 seconds long to capture the results of an entire contraction. For convenience, $N = 1024$ for each sample function. The periodogram is calculated for each sample function and is smoothed using a Hanning spectral window. Representative spectra are shown in Figures 7.21c and 7.21d. The actual changes in the spectra are monitored by calculating the energy in different frequency bands. Frequency bands of low (L), 0.2–0.45 Hz, high (H), 0.8–3.0 Hz, and total (T), 0.2–3.0 Hz are stipulated. An increase in the H/T ratio with a corresponding decrease in the L/T ratio indicated that the uterus was in its labor mode of contraction.

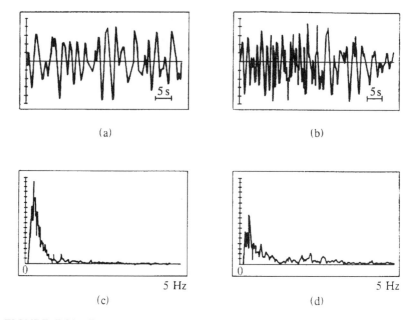

**FIGURE 7.21** The EHG during pregnancy and labor contraction: (a) EHG during pregnancy. (b) EHG during labor. (c) PSD of (a). (d) PSD of (b). [Adapted from Marque et al., fig. 2, with permission.]

## 7.4 CONSISTENT ESTIMATORS: INDIRECT METHODS

### 7.4.1 Spectral and Lag Windows

Spectral smoothing is actually one mode of a very good general method for improving spectral estimates. Since the smoothing functions are symmetric, the smoothing operation is mathematically equivalent to the convolution of the periodogram with an even function, which is now represented by $W(f)$ or $W(m)$ depending on whether the continuous or discrete frequency domain is being used. Again, because of the convolution in the frequency domain,

$$\tilde{I}_K(m) = I(m) * W(m) \tag{7.69}$$

the lag domain multiplication

$$\tilde{R}_K(k) = \hat{R}(k) \cdot w(k) \tag{7.70}$$

is implied. The functions $W(m)$ and $w(k)$ are Fourier transform pairs and are called the *spectral window* and the *lag window*, respectively. The difference with respect to smoothing is that emphasis is placed on operating in the lag domain because the spectral windows usually have a more complicated mathematical representation. The procedure follows.

a. $\hat{R}(k)$ is estimated.
b. The lag window is selected and the operation in equation 7.70 is performed.
c. The Fourier transform of $\tilde{R}_K(k)$ is calculated.

This procedure is called the *Blackman-Tukey (BT)* approach, named after the two researchers who developed this technique. To distinguish the windowing approach from the smoothing approach, the notation will be changed slightly. Equations 7.69 and 7.70 become

$$\tilde{S}_M(m) = I(m) * W(m) \tag{7.71}$$

and

$$\tilde{R}_M(k) = \hat{R}(k) \cdot w(k) \tag{7.72}$$

The use of the subscript $M$ will become apparent. Many spectral–lag window pairs have been developed, and most have a specialized usage. The ones most often used will be discussed. The article by Harris (1978) presents a comprehensive discussion of the characteristics of all the windows. Appendix 7.4 contains the mathematical formulations for the lag and spectral windows. Notice that all these windows have the same mathematical form as the data windows listed in Appendix 3.6. The only significant difference is that the lag windows range from $-MT$ to $+MT$ and the data windows range from 0 to $(N - 1)T$. The triangular or Bartlett lag window is plotted in Figure 7.5 with $M = N$. Consequently, the lag spectral windows do not have a phase component but the magnitude components are the same as those of the data spectral windows. They are plotted in Figure 3.17.

Again, the basic concept is for the windowing operation to increase the number of degrees of freedom and hence the variation reduction factor. Focusing on the representation in equation 7.2, the PSD estimate from a truncated ACF estimate is

$$\hat{S}_M(f) = T\left(\hat{R}(0) + 2 \sum_{k=1}^{M} \hat{R}(k) \cos(2\pi f k T)\right) \tag{7.73}$$

This truncation represents the function of the rectangular lag window. For a weighted window, equation 7.73 becomes

$$\tilde{S}_M(f) = T\left(\hat{R}(0) + 2 \sum_{k=1}^{M} w(k)\hat{R}(k) \cos(2\pi f k T)\right) \tag{7.74}$$

The research into effective windows was to eliminate the leakage and bias produced in $\tilde{S}_M(f)$ through the implied convolution operation. Thus the goal was to develop lag windows whose spectral windows had small side lobe magnitudes, which minimized leakage, and narrow main lobe bandwidths, which minimized bias. The design of various windows to optimize certain characteristics of the estimation procedure is called *window carpentry* in the older literature. It was discovered that optimizing one property precluded optimizing the other. These characteristics of windows were discussed in detail in Chapter 3. It can be understood from the discussion on smoothing that a wide main lobe tends to induce more bias while a small side lobe tends to minimize leakage. Under the supposition that most spectra have a small first derivative, then the concentration on small side lobes is necessary. Thus the rectangular and Bartlett windows will not be good and the Hanning/Tukey, Hamming, and Parzen windows will be effective. Before continuing the discussion on additional properties of windows needed for spectral analysis, let us examine a few examples of spectral estimation using the BT approach.

---

### EXAMPLE 7.13

The time series of Example 7.3 is used and the estimate of its ACF with $N = 256$ and $M = 25$ is shown in Figure 7.22a. Simply calculating the Fourier transform after applying the rectangular window with $M = 25$ produces the estimate in Figure 7.22b. This is calculating equation 7.74 with $f = m/2MT$ and $0 \leq m \leq 25$. Compare this estimate with the original periodogram estimate and the true spectrum. This is a more stable estimate than the periodogram because some smoothing was applied, but it still does not resemble the true spectrum very much.

Actually, equation 7.74 is calculated indirectly using the FFT algorithm. $\tilde{R}_M(k)$ is shifted 25 time units and zeros are added to give a total of 128 points, as shown in Figure 7.22c. Then the FFT algorithm is used to calculate the DFT. The magnitude component of the DFT is the spectral estimate $\tilde{S}_M(m)$.

If a Parzen window is applied with $M = 25$, it results in $\tilde{R}_M(k)$ as shown in Figure 7.22d after shifting and zero padding. The resulting estimate, $\tilde{S}_M(m)$, is much better and is shown in Figure 7.22e. A discussion of the variance reduction will make this result expected.

---

## 7.4.2   Important Details for Using FFT Algorithms

Before proceeding, we must briefly discuss the procedures for calculating the BT estimate. If equation 7.74 is calculated directly, then the results are straightforward. However, one finds that this requires a large computation time, so instead the FFT algorithm is used for calculating the DFT. A statement of caution is needed. The FFT algorithms usually assume that the time function begins at $n = 0$. However, $\tilde{R}_M(k)$ begins at $k = -M$ and the total number of points is $2M + 1$. This means that to use the FFT algorithm $\tilde{R}_M(k)$ must be shifted $M$ time units and it must be zero padded. This is shown in Figure 7.22c. What does the

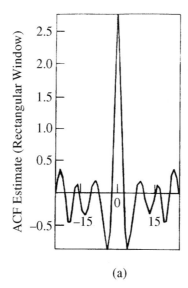

(a)

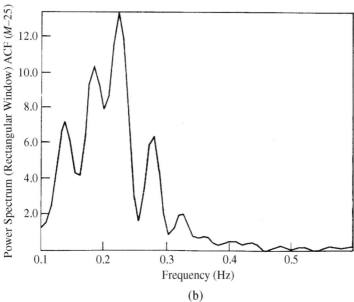

(b)

**FIGURE 7.22**  From the time series in Example 7.3 the following estimates are plotted.
a. Autocorrelation function.
b. Spectrum from using rectangular lag window, $M = 25$.

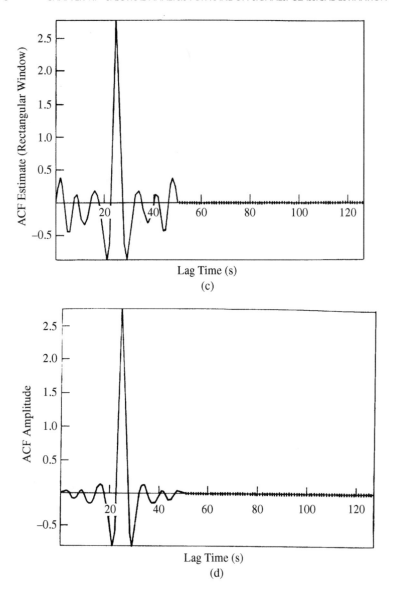

**FIGURE 7.22** From the time series in Example 7.3 the following estimates are plotted.
c. Autocorrelation function shifted and zero padded.
d. Product of ACF in (a) and Parzen lag window, $M = 25$.

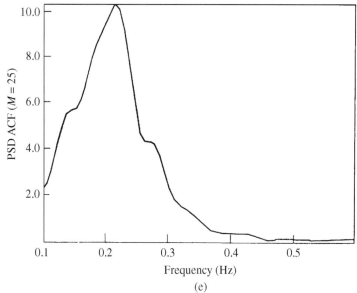

**FIGURE 7.22** From the time series in Example 7.3 the following estimates are plotted. e. Spectrum using Parzen lag window, $M = 25$.

time shift imply? Will the resulting $\tilde{S}_M(m)$ be real or complex? It is known that the estimate must be a real function.

### 7.4.3 Statistical Characteristics of BT Approach

Examination of equation 7.71 shows that the initial estimated spectrum is convolved with the spectral window. How the bias, consistency, and confidence limits are evaluated in this situation must be examined.

#### 7.4.3.1 Bias

Fortunately, for bias considerations the width of the spectral window dominates and the effect of the Fejer kernel can be neglected. Let us examine this briefly. The effect of applying a lag window produces a spectral estimate such that

$$\tilde{S}(f) = W(f) * \hat{S}(f)$$

$$= W(f) * \int_{-1/2T}^{1/2T} S(g)D(f - g)\, dg$$

$$= \int_{-1/2T}^{1/2T} W(f - h) \int_{-1T}^{1/2T} S(g)D(h - g)\, dg\, dh \qquad (7.75)$$

Interchanging the order of the independent variables for integrating produces

$$\tilde{S}(f) = \int_{-1/2T}^{1/2T} S(g) \left( \int_{-1/2T}^{1/2T} W(f-h)D(h-g)\, dh \right) dg \qquad (7.76)$$

Concentrate on the integration within the brackets. The two windows are sketched in Figure 7.23. Since $M \leq 0.3N$, $D(f)$ is much narrower than $W(f)$ and can be considered as a delta function. Therefore,

$$\tilde{S}(f) = \int_{-1/2T}^{1/2T} S(g) \left( \int_{-1/2T}^{1/2T} W(f-h)\delta(h-g)\, dh \right) dg$$

and

$$\tilde{S}(f) \approx \int_{-1/2T}^{1/2T} S(g)W(f-g)\, dg \qquad (7.77)$$

and the lag spectral window dominates the smoothing operation. The bias that can be induced is expressed as an expectation of equation 7.77 and is

$$E[\tilde{S}(f)] = E\left( \int_{-1/2T}^{1/2T} S(g)W(f-g)\, dg \right)$$

If the spectrum varies slightly over the width of the window, then

$$E[\tilde{S}(f)] = S(f) \int_{-1/2T}^{1/2T} W(g)\, dg \qquad (7.78)$$

since $S(f)$ and $W(f)$ are deterministic. To maintain an unbiased estimate, the spectral window must have an area of 1. If the spectrum does not vary smoothly relative to the window, then there is a bias proportional to the second derivative of the spectrum. Formulas for approximation can be found in Jenkins and Watts (1968).

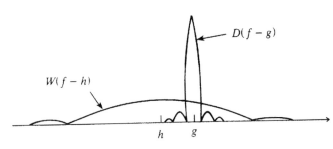

**FIGURE 7.23** Sketch of spectral windows $D(f)$ and $W(f)$. [Adapted from Fante, fig. 9.23, with permission.]

### 7.4.3.2 Variance

The variance of the Blackman-Tukey procedure can be shown to be

$$\text{Var}[\tilde{S}_M] = \frac{S^2(f)}{NT} \int_{-1/2T}^{1/2T} W^2(f)\, df = \frac{S^2(f)}{N} \sum_{k=-M}^{M} w^2(k) \qquad (7.79)$$

The proof is given in Appendix 7.2. To obtain a measure of the amount of smoothing that occurs when using the various windows, the notion of equivalent bandwidth is used again. $B_e$ is the bandwidth of the rectangular spectral window $W_e(f)$, which produces the same variance as given in equation 7.79. Using the previous definition and knowing that the area must be 1, we have

$$W_e(f) = \begin{cases} \dfrac{1}{B_e}, & -\dfrac{B_e}{2} \leq f \leq \dfrac{B_e}{2} \\ 0 & \text{elsewhere} \end{cases} \qquad (7.80)$$

then

$$\int_{-1/2T}^{1/2T} W_e^2(f)\, df = \frac{1}{B_e} \qquad (7.81)$$

The spectral variance in terms of the equivalent bandwidth is

$$\text{Var}[\tilde{S}_M(f)] = \frac{S^2(f)}{NTB_e} \qquad (7.82)$$

The point of reference for the BT approach is the rectangular lag window. That is, $\hat{R}(k)$ is simply truncated at lags $\pm M$. The PSD is estimated with equation 7.73. The frequency spacing is $\dfrac{1}{2MT}$. The variance reduction, VR, is simply calculated with equation 7.79. Then

$$\text{Var}[\tilde{S}_M(f)] = \frac{S^2(f)}{N} \sum_{k=-M}^{M} w^2(k) \approx S^2(f)\frac{2M}{N} \qquad (7.83)$$

and $v = \dfrac{N}{M}$. The equivalent bandwidth is simply found by equating equation 7.83 to equation 7.82. Thus,

$$\frac{2M}{N} = \frac{1}{NTB_e} \quad \text{or} \quad B_e = \frac{1}{2MT}$$

and $B_e$ is equal to the frequency spacing. All of the other lag windows have some shaping and have greater variance reduction at the cost of a greater equivalent bandwidth for a given value of $M$. That is, there is an effective smoothing of the initial estimate caused by shaping the lag window. All of the lag windows have the following properties,

    a. $w(0) = 1$, preserves variance.
    b. $w(k) = w(-k)$, even symmetry.
    c. $w(k) = 0$, $|k| > M$ and $M < N$, truncation.

Because of these properties, the spectral windows have the following corresponding general properties,

a. $\int_{-1/2T}^{1/2T} W(f)\, df = 1$, no additional bias is induced.

b. $W(f) = W(-f)$, even symmetry.

c. Frequency spacing is $\dfrac{1}{2MT}$.

---

### EXAMPLE 7.14

For the Bartlett window, find the variance reduction factor, $v$, and $B_e$. Since all these terms are interrelated, $B_e$ will be found first using equation 7.81. Now

$$B_e = \left( \int_{-1/2T}^{1/2T} W_M^2(f)\, df \right)^{-1} = \left( T \sum_{k=-M}^{M} w^2(k) \right)^{-1}$$

$B_e$ will be calculated from the lag window since the mathematical forms are simpler. To be more general, the integral approximation of the summation will be implemented with $\tau = kT$. Hence the total energy is

$$\sum_{k=-M}^{M} w^2(k) \approx \frac{1}{T} \int_{-MT}^{MT} w^2(\tau)\, d\tau$$

$$= \frac{1}{T} \int_{-MT}^{MT} \left( 1 - \frac{|\tau|}{MT} \right)^2 d\tau$$

$$= \frac{2}{T} \int_{0}^{MT} \left( 1 - \frac{2\tau}{MT} + \frac{\tau^2}{M^2 T^2} \right) d\tau$$

$$= \frac{2}{T} \left( \tau - \frac{\tau^2}{MT} + \frac{\tau^3}{3M^2 T^2} \right) \Big|_{0}^{MT}$$

$$= \frac{2M}{3}$$

and

$$B_e = \frac{3}{2MT}$$

The equivalent bandwidth is wider than that for the rectangular window. The variance is easily found from equation 7.82:

$$\text{Var}[\tilde{S}(f)] = \frac{S^2(f)}{NTB_e} = \frac{S^2(f)2MT}{NT3} = \frac{S^2(f)M2}{N3}$$

The VR is $\dfrac{3N}{2M}$ and $v = \dfrac{3N}{M}$. Thus with respect to the rectangular window the variance has been reduced by a factor of $\frac{1}{3}$ and the degrees of freedom increased by a factor of 3.

### 7.4.3.3 Confidence Limits

Other windows that are often used are the Tukey and Parzen windows. The choice of the window and the truncation point are important factors. It is a difficult choice because there are no obvious criteria, only general guidelines. The first important factor is the number of degrees of freedom. The general formula for degrees of freedom for any window is

$$v = \frac{2N}{\sum\limits_{k=-M}^{M} w^2(k)} \tag{7.84}$$

The degrees of freedom (along with other characteristics) for various windows are listed in Appendix 7.3. For the Tukey and Parzen windows in particular they are $2.67 \dfrac{N}{M}$ and $3.71 \dfrac{N}{M}$, respectively. For a given truncation lag on the same sample function, the Parzen window will yield a lower variance. Now comes the trade-off. The measure of the main lobe width that quantitates the amount of smoothing accomplished over $I(m)$ is the effective bandwidth. It indicates the narrowest separation between frequency bands that can be detected and is a measure of resolution. These are also listed in Appendix 7.3 and are $\dfrac{2.67}{MT}$ and $\dfrac{3.71}{MT}$, respectively, for these two windows. As can be seen, these measures are consistent with our qualitative notions that the Tukey window would have better resolution. The other advantage for the Parzen function is that its spectral window is never negative and therefore a negative component is not possible in $\tilde{S}_M(m)$. However, in general it has been found that these two window functions produce comparable estimates.

The establishment of confidence limits for any spectral estimate is the same procedure as for periodogram estimates. The only difference is that equations 7.59 to 7.61 are written slightly differently to reflect the different functions used. They become

$$\chi_v^2 = \frac{v\tilde{S}_M(m)}{S(m)} \le L1 \quad \text{or} \quad S(m) \ge \frac{v\tilde{S}_M(m)}{L1} \tag{7.85}$$

$$\chi_v^2 = \frac{v\tilde{S}_M(m)}{S(m)} \ge L2 \quad \text{or} \quad S(m) \le \frac{v\tilde{S}_M(m)}{L2} \tag{7.86}$$

$$L1 = \chi_{v,\alpha/2}^2 \quad \text{and} \quad L2 = \chi_{v,1-\alpha/2}^2 \tag{7.87}$$

---

#### EXAMPLE 7.15

For Example 7.13 and Figure 7.22, the confidence limits are calculated for the BT estimates using the Hamming and rectangular spectral windows. For the

Hamming window with 95% confidence limits, $v = 2.52 \dfrac{N}{M} = 2.51 * \dfrac{256}{25} = 25.68 \approx 26$. Therefore,

$$L1 = \chi^2_{26, 0.025} = 41.92 \quad \text{and} \quad L2 = \chi^2_{26, 0.975} = 13.84$$

$$\frac{26\tilde{S}_M(m)}{41.92} \leq S(m) \leq \frac{26\tilde{S}_M(m)}{13.84}$$

The PSD estimate and the confidence limits are plotted in Figure 7.24a. The limits bound the actual PSD very well. The bounds for the rectangular lag window are plotted in Figure 7.24b. Note that they create a much larger confidence interval and reflect the erratic nature of the estimate. For a PSD that does not contain narrowband peaks, the Hamming lag window yields a much better PSD estimate than the rectangular lag window.

---

Several general principles hold for all window functions. $B_e$ and $v$ are both inversely proportional to $M$. Thus when $M$ is made smaller to increase the degrees of freedom (reduce variance), the bandwidth will also increase (reduce resolution). Hence when trying to discover whether any narrowband peaks occur, a larger variance must be tolerated. The tactic for approaching this dilemma is called *window closing*. Several estimates of a PSD are made with different values of $M$. These spectral estimates are assessed to judge whether the measured signal has a narrowband or broadband spectrum. An example will help illustrate this point.

---

### EXAMPLE 7.16

Figure 7.25a shows a segment of a radar return signal with sampling interval of 1 second and $N = 448$. We want to determine its frequency components. The BT method of estimation using the Bartlett window will be implemented. The NACF is estimated for a maximum lag of 60 and is plotted in Figure 7.25b. Several lags are used: $M = 16$, 48, and 60. With $T = 1$, the frequency spacings are 0.03125, 0.01042, and 0.00833 and the equivalent bandwidths are 0.09375, 0.03126, and 0.02499, respectively. It is seen in Figure 7.26 that the spectral estimate with $M = 16$ is smooth and does not reveal the peaks at frequencies $f = 0.07$ and 0.25 Hz that are produced with the other two estimates. This is because $B_e$ is wider than the local valleys in the PSD. Since the estimates with $M = 48$ and 60 lags are very similar, these can be considered as reasonable estimates. Naturally, the one to choose is the one with the greater degrees of freedom, that is, the estimate with lag = 48. The number of degrees of freedom is 28.

---

This example in general illustrates the fact that larger lags are necessary to reveal the possibility of narrow spectral bands. Smaller lags are adequate if the true spectrum is very smooth. The only way to determine a reasonable lag value is to perform window closing and essentially search for ranges of lag values that

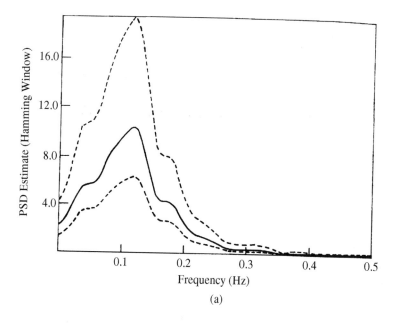

(a)

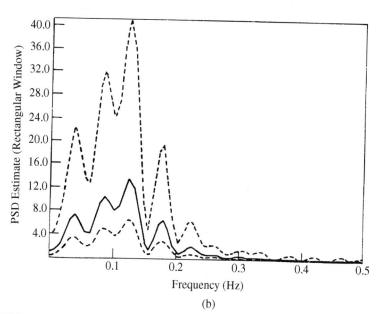

(b)

**FIGURE 7.24** BT spectral estimates (——) and 95% confidence limits (- - -) using Hamming (a) and rectangular (b) lag windows.

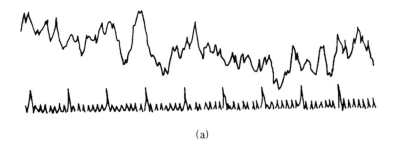

(a)

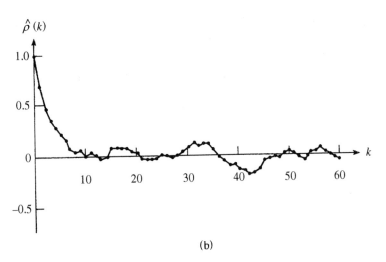

(b)

**FIGURE 7.25**   (a) A radar return signal. (b) Its sample correlation function. [Adapted from Jenkins and Watts, figs. 5.1 and 7.16, with permission.]

produce similar estimates. Once a range is determined, the smallest maximum lag is used since it will give the largest number of degrees of freedom.

A slight variation in steps can be made with the Tukey window. Its spectral functions is

$$W(m) = 0.25\delta(m-1) + 0.5\delta(m) + 0.25\delta(m+1) \tag{7.88}$$

and a convolution can easily be performed in the frequency domain. The alternative approach is as follows.

a. $\hat{R}(k)$ is estimated.

b. The Fourier transform of $\hat{R}(k)$ is calculated with selected value of maximum lag.

c. $\hat{S}(m)$ is smoothed with equation 7.88.

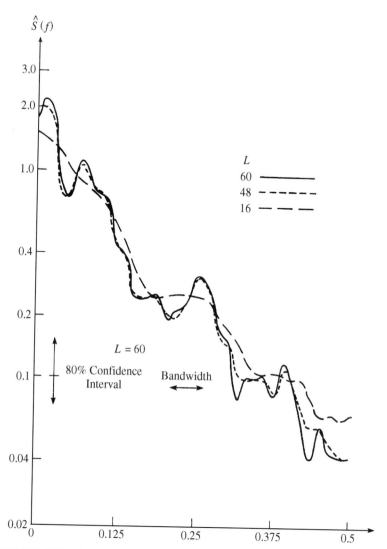

**FIGURE 7.26** Estimates of the power spectral density function of the radar return signal in Figure 7.25. The BT method is used and $M$ = maximum lag. [Adapted from Jenkins and Watts, fig. 7.17, with permission.]

## 7.5 AUTOCORRELATION ESTIMATION

The sum of the lag products is the traditional and direct method for estimating the autocorrelation function. With the advent of the FFT an indirect but much faster method is commonly used. It is based on the periodogram and the Fourier transform relationship between the PSD and the ACF. Now the estimate of the

ACF is

$$\hat{R}(k) = \text{IDFT}[I(m)] \tag{7.89}$$

The difficulty arises because discretization in the frequency domain induces periodicity in the time domain. (Refer to Chapter 3.) Thus equation 7.89 actually forms the circular ACF. To avoid this error, the original data must be padded with at least $N$ zeros to form the ordinary ACF with equation 7.89. The steps in this alternate method are as follows.

    a. Zero pad the measured data with at least $N$ zeros.
    b. Calculate the periodogram with an FFT algorithm.
    c. Estimate $\hat{R}(k)$ through an inverse transform of $I(m)$.
    d. Multiply $\hat{R}(k)$ by a selected lag window.
    e. Calculate the Fourier transform of $\tilde{R}_M(k)$ to obtain $\tilde{S}_M(m)$.

The estimate produced using this method is not different from the direct BT approach. The only difference that will occur is that the frequency spacing will differ.

## REFERENCES

A. Anzaloni and L. Barbosa, "The Average Power Density Spectrum of the Readback Voltage from Particulate Media," IBM Research Report RJ 4308 (47136); IBM Research Laboratory, San Jose, CA, 1984.

J. Bendat and A. Piersol, *Random Data, Analysis and Measurement Procedures*, Wiley, New York, 1986.

P. Bloomfield, *Fourier Analysis of Time Series—An Introduction*, Wiley, New York, 1976.

R. Fante, *Signal Analysis and Estimation*, Wiley, New York, 1988.

N. Geckinli and D. Yavuz, *Discrete Fourier Transformation and Its Applications to Power Spectra Estimation*, Elsevier Scientific Publishing, Amsterdam, 1983.

V. Haggan and O. Oyetunji, "On the Selection of Subset Autoregressive Time Series Models," *J. Time Series Analysis* **5**:103–114 (1984).

F. Harris, "On the Use of Windows for Harmonic Analysis with the Discrete Fourier Transform," *Proc. IEEE* **66**:51–83 (1978).

G. Jenkins and D. Watts, *Spectral Analysis and Its Applications*, Holden-Day, San Francisco, 1968.

S. Kay, *Modern Spectral Estimation, Theory and Applications*, Prentice-Hall, Englewood Cliffs, NJ, 1988.

T. Landers and R. Lacoss, "Geophysical Applications of Spectral Estimates," *IEEE Trans. Geos. Electr.* **15**:26–32 (1977).

S. Marple, *Digital Spectral Analysis with Applications*, Prentice-Hall, Englewood Cliffs, NJ, 1987.

C. Marque, J. Duchene, S. LeClercq, G. Panczer, and J. Chaumont, "Uterine EHG Processing for Obstetrical Monitoring," *IEEE Trans. Biomed. Eng.* **33**:1182–1187 (1986).

M. Noori and H. Hakimmashhadi, "Vibration Analysis," in *Signal Processing Handbook*, C. Chen, Ed., Dekker, New York, 1988.

M. Priestley, *Spectral Analysis and Time Series: Volume 1—Univariate Series*, Academic Press, New York, 1981.

S. Reddy, S. Collins, and E. Daniel, "Frequency Analysis of Gut EMG," *Crit. Rev. in Biomed. Eng.* **15**(2):95–116 (1987).

M. Schwartz and L. Shaw, *Signal Processing: Discrete Spectral Analysis, Detection, and Estimation*, McGraw-Hill, New York, 1975.

P. Welch, "The Use of Fast Fourier Transform for the Estimation of Power Spectra: A Method Based on Time Averaging over Short, Modified Periodograms," *IEEE Trans. Audio and Electroacoustics* **15**:70–73 (1967).

## EXERCISES

**7.1** Prove that the correlation of the rectangular data window with itself yields the triangular lag window, equation 7.13.

**7.2** Derive Fejer's kernel from the properties of the rectangular window.

**7.3** A signal has the triangular PSD in Figure 7.6. Assume that the data acquisition and signal conditioning produces a rectangular spectral window of unity magnitude over the frequency range $-0.05 \leq f \leq 0.05$ Hz.
a. Sketch the spectral window.
b. Convolve the spectral window with the PSD and show the bias that will be produced in the estimated spectrum.

**7.4** In the derivation of the variance for the variance of the periodogram in Appendix 7.1, some summations are simplified. Derive equation A7.11 from equation A7.10.

**7.5** Prove that the variance of the real part of the estimate of the PSD for white noise is

$$
E[A^2(m)] = \begin{cases} T\sigma^2/2, & m \neq 0 \quad \text{and} \quad \left[\dfrac{N}{2}\right] \\[2em] T\sigma^2, & m = 0 \quad \text{and} \quad \left[\dfrac{N}{2}\right] \end{cases}
$$

Start with the defining equation 7.24,

$$E[A^2(m)] = \frac{T\sigma^2}{N} \sum_{n=0}^{N-1} \cos^2(2\pi mn/N)$$

expand it using Euler's formula, and simplify using the geometric sum formulas.

**7.6**  In Section 7.2.1 the real part of the DFT of a white-noise process is $A(m)$. Derive the expression for the variance of $A(m)$ that is found in equation 7.25. [A possible approach to the derivation is to expand the squared cosine term found in equation 7.24 by using a double angle formula and then to express the cosine term as a summation of complex exponentials using Euler's formula. Find the resultant sum using the geometric sum formula.] If you have any difficulties, consult Bloomfield (1976), page 15.

**7.7**  Prove that the harmonic components in a white-noise process are uncorrelated as stated in equation 7.28; that is,

$$E[X(m)X(p)] = \begin{cases} NT^2\sigma^2, & m = p = 0 \quad \text{and} \quad m = p = \left[\dfrac{N}{2}\right] \\ 0, & \text{otherwise} \end{cases}$$

**7.8**  Prove that the real and imaginary parts of the harmonics of the DFT are uncorrelated. Start with equation 7.29.

**7.9**  Using the covariance expression in Appendix 7.1, prove that the PSD values evaluated at integer multiples of $\frac{1}{NT}$ are uncorrelated; that is, for $f_1 = \frac{m_1}{NT}$, $f_2 = \frac{m_2}{NT}$, and $m_1 \neq m_2$, prove that $\text{Cov}[I_y(f_1), I_y(f_2)] = 0$.

**7.10**  Use the sample spectrum plotted in Figure 7.8a to do the following.
  a.  Sketch the spectrum if $T = 1$ ms and $T\sigma^2 = 1$.
  b.  Sketch the spectrum if $T = 1$ ms and $\sigma^2 = 1$.
  c.  Sketch the spectrum if $T = 50$ ms and $\sigma^2 = 1$.

**7.11**  A short time series, $x(n)$, has the values 6, 4, 2, −1, −5, −2, 7, 5.
  a.  What is the sample variance?
  b.  Create a Hamming data window, $d(n)$, for $x(n)$.
  c.  Multiply $x(n)$ by $d(n)$. What are the resulting values?
  d.  What is the new sample variance? Is it consistent with the process loss?

**7.12**  Derive the process loss for the Hanning data window.

**7.13**  For Figure 7.11g, verify the magnitudes of the averaged spectral estimate at the frequencies, 2, 3 and 4 Hz.

**7.14**  Verify the confidence bounds on the spectral averaging used in Example 7.8.

**7.15** What are the 95% confidence bounds on the procedure performed in Example 7.6?

**7.16** In the derivation of the variance of the spectral estimates found in Appendices 7.1 and 7.2, the following form of the triangular spectral window is encountered.

$$G(f) = \left(\frac{\sin(\pi f\, TN)}{N \sin(\pi f\, T)}\right)^2$$

a. Show that $G(0) = 1$.

b. For $G(f)$ to approximate a delta function as $N \rightarrow \infty$, its area must be 1. Show that $NTG(f)$ could be a good approximation to $\delta(f)$ for a large $N$.

**7.17** For the smoothed spectral estimate in Figure 7.19b, verify the magnitudes at the frequencies nearest 13, 15, and 20 Hz in the unsmoothed spectrum. The values are tabulated in Appendix 7.5. Repeat this for Figure 7.19c.

**7.18** What are the confidence limits on the seven-point spectral smoothing used in Example 7.12 if

a. a 95% confidence level is desired?

b. a 99% confidence level is desired?

**7.19** What are the bandwidths and degrees of freedom for nine-point rectangular and triangular spectral smoothing? How do they compare?

**7.20** Derive the variance reduction factor, $v/2$, for triangular smoothing as given in equation 7.68. Assume that the PSD is constant over the interval of smoothing. The following summation formula may be useful.

$$\sum_{i=1}^{N} i^2 = \frac{N(N+1)(2N+1)}{6}$$

**7.21** Prove that $B_e = J$ for triangular smoothing.

**7.22** What type of bias will be produced in the variance reduction procedure if, with respect to a smoothing window.

a. the actual PSD has a peak,

b. the actual PSD has a trough.

**7.23** Verify the frequency spacings, equivalent bandwidths, and degrees of freedom in Example 7.14.

**7.24** For the Tukey window, calculate the degrees of freedom and the equivalent bandwidth from equation 7.84.

**7.25** For the BT estimate of the PSD using the rectangular lag window and plotted in Figure 7.24b, verify the confidence limits also shown in the figure.

## Computer Exercises

**7.26** Generate a white-noise process with a variance of 2 and a sampling frequency of 50 Hz.

   a. Generate at least 1000 signal points.

   b. Divide the signal into at least four segments and calculate the individual spectra and plot them.

   c. What are the range of magnitudes of the individual periodograms?

   d. Produce a spectral estimate through averaging.

   e. What is the range of magnitudes? Did it increase from those found in part (c)?

   f. Find the 95% confidence limits and plot them with the spectral estimate.

   g. Does the spectrum of the ideal process lie between the confidence limits?

**7.27** The annual trapping of Canadian lynx from 1821 to 1924 has been modeled extensively (Haggan and Oyetunji, 1984). Two models are

$$y(n) - 1.48y(n-2) + 0.54y(n-4) = x(n), \quad \sigma_x^2 = 1.083$$
$$y(n) - 1.492y(n-1) + 1.324y(n-2) = x(n), \quad \sigma_x^2 = 0.0505$$

after the data has been logarithmically transformed and detrended and $x(n)$ is white noise. It is claimed that the trappings have a period of 9.5 years.

   a. Generate a 200-point random process for each model.

   b. Plot the true spectra for each model.

   c. Using the same procedure estimate the spectra for each model.

   d. Which model more accurately reflects the period claimed?

**7.28** Consider an MA process with an autocorrelation function estimate

$$\hat{R}(0) = 25, \quad \hat{R}(\pm 1) = 15, \quad \hat{R}(k) = 0 \quad \text{for} \quad |k| \geq 2$$

   a. Calculate $\hat{S}_1(m)$ directly using equation 7.73.

   b. Calculate $\hat{S}_2(m)$ using an FFT algorithm with $N = 8$.

   1. Is $\hat{S}_2(m)$ real or complex?

   2. What part of $\hat{S}_2(m)$ is equal to $\hat{S}_1(m)$?

   3. How do the frequency spacings differ?

**7.29** A daily census for hospital inpatients is tabulated in Appendix 7.6.

   a. Plot the time series.

   b. Estimate the ACF with $M = 20$.

   c. What periodicities are present as ascertained from (a) and (b).

   d. Estimate the PSD using the BT technique with a suitable lag window.

   e. What information is present in this particular $\tilde{S}_M(m)$?

   f. Repeat steps (b) through (e) with $M = 10$ and $M = 30$.

   g. How do these last two PSD estimates differ from the first one?

## APPENDIX 7.I VARIANCE OF PERIODOGRAM

The bias of the periodogram was discussed in Section 7.1.2. The variance was also stated, and it is important to have its derivation available. The system model will be used so that the variation is imbedded in the white-noise input, which has a zero mean and a variance $\sigma_x^2$. Direct derivation of the variance under different conditions can be found in several references, such as Jenkins and Watts (1968), Kay (1988), or Fante (1988). The periodogram estimate is

$$I_y(f) = \frac{1}{NT} \hat{Y}(-f)\hat{Y}(f) = \frac{1}{NT} \hat{Y}*(f)\hat{Y}(f) \qquad (A7.1)$$

where

$$Y(f) = T \sum_{n=0}^{N-1} y(n)e^{-j2\pi fnT} \quad \text{and} \quad Y(f) = H(f)X(f) \qquad (A7.2)$$

When $N$ is large, the covariance at different frequencies is written

$$\text{Cov}[I_y(f_1), I_y(f_2)] = \text{Cov}[H(f_1)H^*(f_1)I_x(f_1), H(f_2)H^*(f_2)I_x(f_2)] \qquad (A7.3)$$

Since the variability is restricted to $x(n)$, equation A7.3 becomes

$$\text{Cov}[I_y(f_1), I_y(f_2)] = H(f_1)H^*(f_1) \cdot H(f_2)H^*(f_2) \cdot \text{Cov}[I_x(f_1), I_x(f_2)] \qquad (A7.4)$$

and

$$\text{Cov}[I_x(f_1), I_x(f_2)] = E[I_x(f_1)I_x(f_2)] - E[I_x(f_1)] \cdot E[I_x(f_2)] \qquad (A7.5)$$

Examine the second term on the right side of equation A7.5. Each factor is the mean of the periodogram of a white-noise process and was shown in Section 7.2.1.2 to be unbiased and to be

$$E[I_x(f_1)] = E[I_x(f_2)] = T\sigma_x^2 \qquad (A7.6)$$

The mean product is much more complicated and becomes the mean product of four terms. Let the letters $s$, $t$, $k$, and $l$ represent the summing indices. Now

$$I_x(f_1) = \frac{T}{N} \sum_{s=0}^{N-1} \sum_{t=0}^{N-1} x(s)x(t) \exp(-j(s-t)2\pi f_1 T) \qquad (A7.7)$$

$I_x(f_2)$ is similarly expressed and

$$E[I_x(f_1)I_x(f_2)] = \frac{T^2}{N^2} \sum_{s=0}^{N-1} \sum_{t=0}^{N-1} \sum_{k=0}^{N-1} \sum_{l=0}^{N-1} E[x(s)x(t)x(k)x(l)]$$
$$\cdot \exp(-j(s-t)2\pi f_1 T - j(k-l)2\pi f_2 T) \qquad (A7.8)$$

As with the proofs of stationarity, it will be assumed that the fourth moment behaves approximately as that of a Gaussian process. Therefore,

$$E[x(s)x(t)x(k)x(l)] = R_x(s-t)R_x(k-l) + R_x(s-k)R_x(t-l) + R_x(s-l)R_x(t-k) \qquad (A7.9)$$

Equation A7.8 will be evaluated on a term-by-term basis using equation A7.9. Examine the sums produced from individual terms on the right of equation A7.9. Remember that $R_x(s - t) = \sigma_x^2 \delta(s - t)$ and $R_x(k - l) = \sigma_x^2 \delta(k - 1)$. The first term has exponents of zero when the autocorrelation values are nonzero, and summation produces the term $\Sigma_1 = N^2 \sigma_x^4$. The second term has nonzero values only when $s = k$ and $t = l$, and its summation reduces to

$$\Sigma_2 = \sum_{s=0}^{N-1} \sum_{t=0}^{N-1} \sigma_x^4 \exp(-js2\pi(f_1 + f_2)T + jt2\pi(f_1 + f_2)T)$$

$$= \sigma_x^4 \sum_{s=0}^{N-1} \exp(-js2\pi(f_1 + f_2)T) \cdot \sum_{t=0}^{N-1} \exp(+jt2\pi(f_1 + f_2)T) \quad \text{(A7.10)}$$

Using the geometric sum formula and then Euler's formula,

$$\Sigma_2 = \sigma_x^4 \left( \frac{1 - \exp(-j2\pi(f_1 + f_2)TN)}{1 - \exp(-j2\pi(f_1 + f_2)T)} \cdot \frac{1 - \exp(+j2\pi(f_1 + f_2)TN)}{1 - \exp(+j2\pi(f_1 + f_2)T)} \right)$$

$$= \sigma_x^4 \left( \frac{\sin(\pi(f_1 + f_2)TN)}{\sin(\pi(f_1 + f_2)T)} \right)^2 \quad \text{(A7.11)}$$

The third term has similar properties to $\Sigma_2$ when $s = l$ and $t = k$, and its summation is

$$\Sigma_3 = \sigma_x^4 \left( \frac{\sin(\pi(f_1 - f_2)TN)}{\sin(\pi(f_1 - f_2)T)} \right)^2 \quad \text{(A7.12)}$$

Summing $\Sigma_1$, $\Sigma_2$, and $\Sigma_3$ produces

$$E[I_x(f_1)I_x(f_2)] = \frac{T^2 \sigma_x^4}{N^2} \left( N^2 + \left( \frac{\sin(\pi(f_1 + f_2)TN)}{\sin(\pi(f_1 + f_2)T)} \right)^2 + \left( \frac{\sin(\pi(f_1 - f_2)TN)}{\sin(\pi(f_1 - f_2)T)} \right)^2 \right) \quad \text{(A7.13)}$$

Subtracting the product of the means of the periodograms from equation A7.13 yields the covariance between periodogram values of white noise, which is

$$\text{Cov}[I_x(f_1), I_x(f_2)] = \sigma_x^4 T^2 \left( \left( \frac{\sin(\pi(f_1 + f_2)TN)}{N \sin(\pi(f_1 + f_2)T)} \right)^2 + \left( \frac{\sin(\pi(f_1 - f_2)TN)}{N \sin(\pi(f_1 - f_2)T)} \right)^2 \right) \quad \text{(A7.14)}$$

From system theory it is known that

$$S_y(f) = H(f)H^*(f)S_x(f) \quad \text{(A7.15)}$$

Substituting equation A7.15 and equation A7.14 into equation A7.4 produces

$$\text{Cov}[I_y(f_1), I_y(f_2)] = S_y(f_1)S_y(f_2)\left(\left(\frac{\sin(\pi(f_1+f_2)TN)}{N\sin(\pi(f_1+f_2)T)}\right)^2 + \left(\frac{\sin(\pi(f_1-f_2)TN)}{N\sin(\pi(f_1-f_2)T)}\right)^2\right) \tag{A7.16}$$

This is a complicated expression that has a very important result. Since in discrete time processing the spectral values are evaluated at integer multiples of the resolution $\frac{1}{NT}$, the periodogram magnitude values are uncorrelated. This proof is left as an exercise. The variance is found when $f_1 = f_2$ and

$$\text{Var}[I_y(f)] = S_y^2(f)\left(1 + \left(\frac{\sin(2\pi fTN)}{N\sin(2\pi fT)}\right)^2\right) \tag{A7.17}$$

Thus even when $N$ approaches infinity,

$$\text{Var}[I_y(f)] \rightarrow S_y^2(f) \tag{A7.18}$$

and the periodogram is an inconsistent estimator.

## APPENDIX 7.2   PROOF OF VARIANCE OF BT SPECTRAL SMOOTHING

From Section 7.4.3 the BT estimator is

$$\tilde{S}(f) = T\sum_{k=-M}^{M} w(k)\hat{R}(k)e^{-j2\pi fkT} = \int_{-1/2T}^{1/2T} \hat{S}(g)W(f-g)\,dg \tag{A7.19}$$

Its variance is then

$$\text{Var}[\tilde{S}(f)] = E\left[\left(\int_{-1/2T}^{1/2T} \hat{S}(g)W(f-g)\,dg\right)^2\right] - E[\tilde{S}(f)]^2$$
$$= E\left[\left(\int_{-1/2T}^{1/2T} \hat{S}(g)W(f-g)\,dg\right)\left(\int_{-1/2T}^{1/2T} \hat{S}(h)W(f-h)\,dh\right)\right]$$
$$- E[\tilde{S}(f)]^2 \tag{A7.20}$$

Expanding the squared mean and collecting terms, equation A7.20 becomes

$$\text{Var}[\tilde{S}(f)] = \int_{-1/2T}^{1/2T}\int_{-1/2T}^{1/2T} W(f-g)W(f-h)\,\text{Cov}[\hat{S}(g),\hat{S}(h)]\,dg\,dh \tag{A7.21}$$

Since $\hat{S}(f)$ is a periodogram, its covariance is given by equation A7.16 in Appendix 7.1. For large $N$, each squared term approaches $\dfrac{\delta(\cdot)}{(NT)}$ and equation A7.21 becomes

$$\mathrm{Var}[\tilde{S}(f)] \approx \int_{-1/2T}^{1/2T} \int_{-1/2T}^{1/2T} W(f-g)W(f-h)\,\frac{S(g)S(h)}{NT}$$
$$\times\,(\delta(g+h) + \delta(g-h))\,dg\,dh \qquad (A7.22)$$

$$\mathrm{Var}[\tilde{S}(f)] \approx \frac{1}{NT}\int_{-1/2T}^{1/2T} (W(f+h)W(f-h) + W(f-h)^2)S(h)^2\,dh \quad (A7.23)$$

Equation A7.23 is again simplified by assuming $N$ is large and consequently the spectral window is narrow. Then there will not be appreciable overlap of windows located at frequencies $(f+h)$ and $(f-h)$. Thus the term in the brackets of the integrand can be considered zero. Also the PSD can be considered constant over the window width. Thus

$$\mathrm{Var}[\tilde{S}(f)] \approx \frac{S(f)^2}{NT}\int_{-1/2T}^{1/2T} W(h)^2\,dh \qquad (A7.24)$$

## APPENDIX 7.3   WINDOW CHARACTERISTICS

| Window | Equivalent Bandwidth | Ratio: Highest Side Lobe Level to Peak | Degrees of Freedom | Process Loss |
|---|---|---|---|---|
| Rectangular | $\dfrac{0.5}{MT}$ | 0.220 | $\dfrac{N}{M}$ | 1.00 |
| Bartlett | $\dfrac{1.5}{MT}$ | 0.056 | $3\,\dfrac{N}{M}$ | 0.33 |
| Hanning-Tukey | $\dfrac{1.33}{MT}$ | 0.026 | $2.67\,\dfrac{N}{M}$ | 0.36 |
| Hamming | $\dfrac{1.25}{MT}$ | 0.0089 | $2.51\,\dfrac{N}{M}$ | 0.40 |
| Parzen | $\dfrac{1.86}{MT}$ | 0.0024 | $3.71\,\dfrac{N}{M}$ | 0.28 |

## APPENDIX 7.4 LAG WINDOW FUNCTIONS

| Lag Window | Spectral Window |
|---|---|

### Rectangular

$$w_R(k) = 1, \quad |k| \le M \qquad\qquad W_R(f) = T\,\frac{\sin(2\pi f\,MT)}{\sin(\pi f\,T)}$$

### Triangular–Bartlett

$$w_B(k) = 1 - \frac{|k|}{M}, \quad |k| \le M \qquad\qquad W_B(f) = \frac{T}{M}\left(\frac{\sin(\pi f\,MT)}{\sin(\pi f\,T)}\right)^2$$

### Hanning-Tukey

$$w_T(k) = \frac{1}{2}\left(1 + \cos\left(\pi\frac{k}{M}\right)\right)$$

$$W_T(f) = 0.5 W_R(f)$$
$$+ 0.25 W_R\left(f + \frac{1}{2MT}\right)$$
$$+ 0.25 W_R\left(f - \frac{1}{2MT}\right)$$

### Hamming

$$w_H(k) = 0.54 + 0.45\cos\left(\pi\frac{k}{M}\right)$$

$$W_H(f) = 0.54 W_R(f)$$
$$+ 0.23 W_R\left(f + \frac{1}{2MT}\right)$$
$$+ 0.23 W_R\left(f - \frac{1}{2MT}\right)$$

### Parzen

$$W_P(k) = \begin{cases} 1 - 6\left(\dfrac{k}{M}\right)^2 + 6\left(\dfrac{|k|}{M}\right)^3, & |k| \le \dfrac{M}{2} \\[2mm] 2\left(1 - \dfrac{|k|}{M}\right)^3, & \dfrac{M}{2} < |k| \le M \end{cases}$$

$$W_P(f) = \frac{8T}{M^3}\left(\frac{3}{2}\,\frac{\sin^4(\pi fMT/2)}{\sin^4(\pi fT)} - \frac{\sin^4(\pi fMT/2)}{\sin^2(\pi fT)}\right)$$

All lag windows have $w(k) = 0$ for $|k| > M$.

## APPENDIX 7.5    SPECTRAL ESTIMATES FROM SMOOTHING

Table of Spectral Estimates from Three- and Seven-Point Smoothing in Figures 7.19b and 7.19c

| Frequency | Original Periodogram | Periodogram with Three-Point Smoothing | Periodogram with Seven-Point Smoothing |
|---|---|---|---|
| 0.00000 | 0.00003 | 0.00454 | 0.01344 |
| 0.78125 | 0.00680 | 0.00667 | 0.01491 |
| 1.56250 | 0.01318 | 0.01568 | 0.01384 |
| 2.34375 | 0.02705 | 0.02586 | 0.01309 |
| 3.12500 | 0.03735 | 0.02336 | 0.01369 |
| 3.90625 | 0.00567 | 0.01487 | 0.01406 |
| 4.68750 | 0.00158 | 0.00382 | 0.01992 |
| 5.46875 | 0.00420 | 0.00507 | 0.01762 |
| 6.25000 | 0.00941 | 0.02260 | 0.01321 |
| 7.03125 | 0.05418 | 0.02484 | 0.01545 |
| 7.81250 | 0.01094 | 0.02386 | 0.02194 |
| 8.59375 | 0.00646 | 0.01293 | 0.02909 |
| 9.37500 | 0.02139 | 0.02495 | 0.03059 |
| 10.15625 | 0.04699 | 0.04088 | 0.03673 |
| 10.93750 | 0.05427 | 0.04039 | 0.04386 |
| 11.71875 | 0.01991 | 0.05711 | 0.04353 |
| 12.50000 | 0.09714 | 0.05930 | 0.25210 |
| 13.28125 | 0.06086 | 0.05404 | 0.25618 |
| 14.06250 | 0.00413 | 0.51547 | 0.28618 |
| 14.84375 | 1.48142 | 0.52037 | 0.29994 |
| 15.62500 | 0.07555 | 0.60708 | 0.29561 |
| 16.40625 | 0.26426 | 0.15202 | 0.29617 |
| 17.18750 | 0.11624 | 0.14909 | 0.29847 |
| 17.96875 | 0.06678 | 0.08262 | 0.11871 |
| 18.75000 | 0.06482 | 0.05061 | 0.11793 |
| 19.53125 | 0.02024 | 0.10270 | 0.08260 |
| 20.31250 | 0.22305 | 0.10446 | 0.06660 |
| 21.09375 | 0.07008 | 0.10338 | 0.06000 |
| 21.87500 | 0.01701 | 0.03044 | 0.05273 |
| 22.65625 | 0.00423 | 0.01394 | 0.05513 |
| 23.43750 | 0.02058 | 0.01291 | 0.02525 |
| 24.21875 | 0.01392 | 0.02384 | 0.01818 |
| 25.00000 | 0.03702 | 0.02162 | 0.01635 |

## APPENDIX 7.6   HOSPITAL CENSUS DATA

*Hospital Inpatient Daily Census (April 21 to July 19)*

| | | | | | | | | | |
|---|---|---|---|---|---|---|---|---|---|
| 397 | 462 | 486 | 483 | 477 | 438 | 407 | 421 | 480 | 484 |
| 486 | 479 | 415 | 400 | 419 | 477 | 510 | 503 | 500 | 435 |
| 408 | 417 | 478 | 497 | 500 | 512 | 450 | 421 | 423 | 471 |
| 496 | 478 | 463 | 413 | 396 | 366 | 375 | 444 | 469 | 480 |
| 439 | 402 | 442 | 492 | 507 | 518 | 493 | 439 | 399 | 428 |
| 476 | 499 | 488 | 460 | 419 | 380 | 406 | 472 | 502 | 495 |
| 490 | 443 | 398 | 417 | 490 | 505 | 499 | 484 | 430 | 384 |
| 392 | 452 | 455 | 426 | 414 | 405 | 379 | 410 | 485 | 514 |
| 525 | 511 | 461 | 436 | 444 | 488 | 494 | 510 | 493 | 429 |
| 392 | 420 | 466 | 476 | 494 | 484 | 423 | 388 | 411 | 472 |

Total 100 observations (read row-wise)

# VIII

## RANDOM SIGNAL MODELING AND MODERN SPECTRAL ESTIMATION

### 8.1 INTRODUCTION

The preceding analyses of properties of random signals are focused on estimating the first- and second-order stochastic characteristics of the time series and their spectral density composition. More detailed information about a time series can be obtained if the time series itself can be modeled. In addition, its power spectrum can be obtained from the model. This approach has been given increased emphasis in engineering during the last several decades with new algorithms being developed and many useful applications appearing in books and journal publications.

Basically, the concept is to model the signal using the principles of causal and stable discrete time systems as studied in Chapter 6. Recall that in the time domain the input, $x(n)$, and output, $y(n)$, of a system are related by

$$y(n) = -\sum_{i=1}^{p} a(i)y(n-i) + \sum_{l=0}^{q} b(l)x(n-l) \qquad (8.1)$$

In signal modeling the output function is known. From what principle does the input function arise? As was learned in Chapter 6, the system can, when the input is white noise, create an output with properties of other random signals. Thus the properties of $y(n)$ depend on the variance of the white noise, and the parameters $a(i)$, $b(l)$, $p$, and $q$. This approach is called *parametric signal modeling*. Since equation 8.1 also defines an autoregressive moving-average model, the approach is synonymously called *ARMA modeling*. Also recall that if the system's parameters and input are known, the power spectral density function of the output

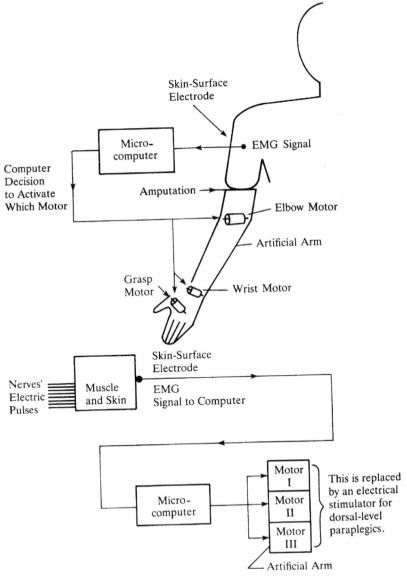

**FIGURE 8.1** Schematic diagram of arm prosthesis (a) and signal identification and control process (b); $f_s = 2\,\text{KHz}$. [Adapted from Graupe, fig. B2, with permission.]

can be determined through the power transfer function. With $x(n)$ being white noise, then

$$S_y(f) = |H(f)|^2 S_x(f) = |H(f)|^2 \sigma_x^2 T \tag{8.2}$$

Thus an alternative approach to determining the PSD of a signal can be defined on a parametric or ARMA model. Collectively this approach to signal processing is still known as modern signal processing.

An application in the time domain will help illustrate the utility of this approach. If a signal containing 1000 points can be modeled by equation 8.1 with the orders $p$ and $q$ being 20 or less, then the information in this signal can be represented with relatively few parameters. This condition is true for most signals (Chen, 1988; Jansen, 1985). Thus further analyses and interpretation are facilitated. One such application is the use of the electromyogram (EMG) from an leg or arm muscle to control a device for a paralyzed person or an artificial limb for an amputee. A schematic diagram for controlling a limb prosthesis is shown in Figure 8.1. The concept is that different intensities of contraction will produce a signal with different structure and thus different parameter sets. These sets will make the prosthesis perform a different function. Their change in structure is

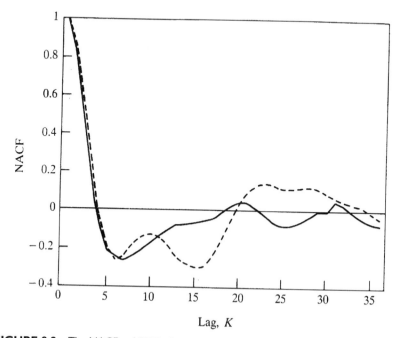

**FIGURE 8.2**   The NACFs of EMGs from a contracting muscle, 500 signal points: NACF at 25% MVC (–); NACF at 50% MVC (– –). [Adapted from Triolo et al., figs. 6 and 8, with permission.]

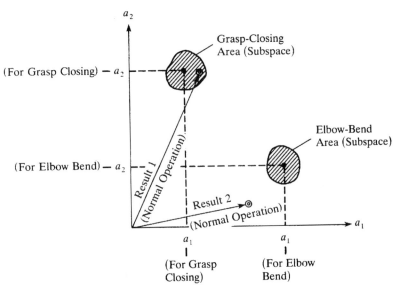

**FIGURE 8.3** Parameter spaces for two movements based on the first two AR parameters. [Adapted from Graupe, fig. B1, with permission.]

indicated by the change in NACF, as shown in Figure 8.2. When the parameter sets caused by different levels of contraction are plotted (Figure 8.3) one can see that the parameter sets are disjoint. It is then straightforward to have a multiple task controller with this scheme, AR(4) models have been found to be successful (Triolo et al., 1988; Hefftner et al., 1988).

## 8.2 MODEL DEVELOPMENT

Model development is based on the fact that signal points are interrelated across time, as expressed by equation 8.1. This interrelatedness can be expressed by the autocorrelation function. Thus there is information contained in several signal values that can be used to estimate future values. For instance, what is the estimate of the present value of $y(n)$ based on two previously measured values? Mathematically, this is expressed as

$$\hat{y}(n) = h(1)y(n-1) + h(2)y(n-2) \tag{8.3}$$

This weighted sum is called a *linear prediction*.

Naturally, the desire is to estimate the value of $y(n)$ with the smallest error possible. We define the error as

$$\epsilon(n) = y(n) - \hat{y}(n) \tag{8.4}$$

Notice that the error is also a time series, which is stationary since $y(n)$ is stationary. Again the mean squared error, MSE, will be minimized. Notice that the MSE is also the variance, $\sigma_\epsilon^2$, of the error time series. By definition,

$$\text{MSE} = E[\epsilon^2(n)] = E[(y(n) - \hat{y}(n))^2]$$
$$= E[(y(n) - h(1)y(n - 1) - h(2)y(n - 2))^2] \tag{8.5}$$

The coefficients will be those that minimize the MSE. To find $h(1)$,

$$\frac{\partial}{\partial h(1)} \text{MSE} = \frac{\partial}{\partial h(1)} E[(y(n) - h(1)y(n - 1) - h(2)y(n - 2))^2]$$
$$= E\left[\frac{\partial}{\partial h(1)} (y(n) - h(1)y(n - 1) - h(2)y(n - 2))^2\right]$$
$$= -2E[(y(n) - h(1)y(n - 1) - h(2)y(n - 2))y(n - 1)] = 0 \tag{8.6}$$

The factor, $-2$, can be disregarded and the terms within the square brackets must be expanded to simplify the expectations:

$$(y(n) - h(1)y(n - 1) - h(2)y(n - 2))y(n - 1)$$
$$= y(n)y(n - 1) - h(1)y(n - 1)y(n - 1) - h(2)y(n - 2)y(n - 1) \tag{8.7}$$

The terms in equation 8.7 are cross products, and its expectation will yield an equation containing some terms of the autocorrelation function of $y(n)$. The expectation is

$$E[y(n)y(n - 1) - h(1)y(n - 1)y(n - 1) - h(2)y(n - 2)y(n - 1)]$$
$$= E[y(n)y(n - 1)] - E[h(1)y(n - 1)y(n - 1)] - E[h(2)y(n - 2)y(n - 1)]$$
$$= R_y(-1) - h(1)R_y(0) - h(2)R_y(1) = 0$$

Since all ACFs are even functions,

$$R_y(1) - h(1)R_y(0) - h(2)R_y(1) = 0 \tag{8.8}$$

Another equation is obtained similarly. Minimizing equation 8.5 with respect to $h(2)$, produces the equation

$$R_y(2) - h(1)R_y(1) - h(2)R_y(0) = 0 \tag{8.9}$$

As can be appreciated, if some values of the ACF of $y(n)$ are known, then the weighting coefficients can be found. The value of one more quantity must be found. Since the MSE was minimized, its magnitude is important in order to judge the accuracy of the model. It can be directly obtained from equations 8.5, 8.8, and 8.9 and is derived in Appendix 8.1. It is

$$\sigma_\epsilon^2 = R_y(0) - h(1)R_y(1) - h(2)R_y(2) \tag{8.10}$$

The last three equations are usually represented in matrix form as

$$\begin{bmatrix} R(0) & R(1) & R(2) \\ R(1) & R(0) & R(1) \\ R(2) & R(1) & R(0) \end{bmatrix} \begin{bmatrix} 1 \\ -h(1) \\ -h(2) \end{bmatrix} = \begin{bmatrix} \sigma_\epsilon^2 \\ 0 \\ 0 \end{bmatrix} \qquad (8.11)$$

This solution form is very prevalent for parametric modeling and is known as the *Yule-Walker* (*YW*) equations. The $3 \times 3$ matrix also has a special form, called a *Toeplitz* matrix. Notice that all the terms on the diagonal, subdiagonal, and supradiagonal are the same.

Now combine the prediction equations 8.3 and 8.4 and write them as

$$y(n) - h(1)y(n-1) - h(2)y(n-2) = \epsilon(n) \qquad (8.12)$$

This is exactly the form of a second-order AR system with input $\epsilon(n)$. The nature of $\epsilon(n)$ determines the adequacy of the model. We stipulated that not only must the variance of $\epsilon(n)$ be minimum but it also must be a white-noise sequence. We know from Chapter 6 that an AR system can produce an output with structure when its input is uncorrelated. Thus, knowing the autocorrelation function of a signal permits the development of models for it.

The AR model is used most often because the solution equations for its parameters are simpler and more developed than those for either MA or ARMA models. Fortunately, this approach is valid as well as convenient because of the *Wold Decomposition Theorem* (Fante, 1988; Priestley, 1981). The theorem states that the random component of any stationary process can be modeled by a stable causal linear system with a white-noise input. The impulse response is essentially an MA process that can have infinite order. This was demonstrated in Chapter 5. The AR and ARMA process models also produce such impulse responses. Thus ARMA or MA processes can be adequately represented by an equivalent AR process of suitable, possibly infinite, order. Typically, one uses different criteria to determine the model order, criteria that are functions of estimates of $\sigma_\epsilon^2$. However, in most signal processing situations the values of $R(k)$ are not known either. Thus the modeling must proceed from a different viewpoint. Within the next several sections, methods will be derived to develop signal models. The derivations are very similar to the preceding procedure and parallel the philosophy of other estimating procedures. That is, the first attempt for implementation is to use the theoretical procedure with the theoretical functions being replaced by their estimates.

## 8.3   RANDOM DATA MODELING APPROACH

### 8.3.1   Basic Concepts

The *random data model* is the name given to the approach for modeling signals based on the notion of linear prediction when the autocorrelation function is unknown. Again, we want to select model parameters that minimize the squared error. We start with the general equation for linear prediction using the previous $p$ terms; that is,

$$\hat{y}(n) = h(1)y(n-1) + h(2)y(n-2) + \cdots + h(p)y(n-p) \tag{8.13}$$

with $\hat{y}(n)$ representing the predicted value of the time series at time $n$. The prediction error is

$$\epsilon(n) = y(n) - \hat{y}(n) \tag{8.14}$$

and the error is a random time sequence itself. Combining equations 8.14 and 8.13 to express the signal and error sequence in a system form produces

$$\epsilon(n) = y(n) - h(1)y(n-1) - h(2)y(n-2) - \cdots - h(p)y(n-p) \tag{8.15}$$

The error sequence is the output of a $p$th-order MA system with the signal as the input and weighting coefficients $b(i) = -h(i)$, $0 \leq i \leq p$. If the sequence $\epsilon(n)$ is a white-noise sequence, the process defined by equation 8.15 is called *prewhitening* the signal $y(n)$.

Another aspect to this situation is that usually the mean, $m$, of the process is also unknown. Thus, to be all inclusive,

$$y(n) = z(n) - m \tag{8.16}$$

where $z(n)$ is the signal to be modeled. The error is then explicitly,

$$\epsilon(n) = (z(n) - m) + a(1)(z(n-1) - m) + \cdots + a(p)(z(n-p) - m) \tag{8.17}$$

Now the error is based not only on the signal points available but also the mean value. So a slightly different approach must be considered. Let the total squared error, TSE, be the error criterion. Since $p + 1$ points are used in equation 8.17, the error sequence begins at time point $p$ and

$$\text{TSE} = \sum_{n=p}^{N-1} \epsilon^2(n) = \sum_{n=p}^{N-1} ((z(n) - m) + a(1)(z(n-1) - m) + \cdots$$
$$+ a(p)(z(n-p) - m))^2 \tag{8.18}$$

Examination of equation 8.18 reveals that because of the different time indices in each term, the resulting summations are over different time ranges of the process $z(n)$. This can have a significant effect on the estimation of the parameters when $N$ is small. In the following example, we develop the procedure for estimating $m$ and $a(1)$ in a first-order model to appreciate the potential complexities.

**EXAMPLE 8.1**

For any point in time, the error sequence and the TSE for a first-order model are

$$\epsilon(n) = (z(n) - m) + a(1)(z(n-1) - m)$$

$$\text{TSE} = \sum_{n=1}^{N-1} ((z(n) - m) + a(1)(z(n-1) - m))^2$$

The parameters $m$ and $a(1)$ are found through the standard minimization procedure. Thus

$$\frac{\partial}{\partial a(1)} \text{TSE} = \sum_{n=1}^{N-1} ((z(n) - m) + a(1)(z(n-1) - m))(z(n-1) - m) = 0$$

and

$$\frac{\partial}{\partial m} \text{TSE} = \sum_{n=1}^{N-1} ((z(n) - m) + a(1)(z(n-1) - m))(1 + a(1)) = 0$$

Rearranging the last equation and indicating the estimates of $m$ and $a(1)$ with circumflexes yields

$$\sum_{n=1}^{N-1} (z(n) - \hat{m}) + \hat{a}(1) \sum_{n=1}^{N-1} (z(n-1) - \hat{m}) = 0$$

The subtle point to notice is that the summation in the above equation is over different time spans of $z(n)$. The first summation involves the last $N - 1$ points and the second summation involves the first $N - 1$ points. Dividing the equation by $N - 1$ and designating the sums at different estimates of the mean as $\hat{m}_2$ and $\hat{m}_1$, respectively, gives

$$(\hat{m}_2 - \hat{m}) + \hat{a}(1)(\hat{m}_1 - \hat{m}) = 0$$

One can easily ascertain that if $N$ is large then

$$\hat{m}_2 \approx \hat{m}_1 \approx \hat{m}$$

and the conventional estimate of the mean is valid. Notice very clearly that this is not true if $N$ is small.

The solution for the other equation is more complicated and contains cross products. Expanding produces

$$\sum_{n=1}^{N-1} (z(n) - \hat{m})(z(n-1) - \hat{m}) + \hat{a}(1)(z(n-1) - \hat{m})^2 = 0$$

or

$$\hat{a}(1) = \frac{-\sum_{n=1}^{N-1}(z(n) - \hat{m})(z(n-1) - \hat{m})}{\sum_{n=1}^{N-1}(z(n-1) - \hat{m})^2}$$

Again, if the numerator and denominator are divided by $(N-1)$, they are estimates of the covariance of lag 1 and the variance, respectively. Thus

$$\hat{a}(1) = -\frac{\hat{C}(1)}{\hat{C}(0)} = -\hat{\rho}(1),$$

the value of the negative of the sample NACF at lag 1. Notice that the estimator for the variance did not utilize all of the available sample points of $z(n)$. The total squared error is

$$\text{TSE} = (N-1)(\hat{C}(0) + \hat{a}(1)\hat{C}(1))$$

The development of the models for signals will be based on the knowledge gained from Example 8.1. The signals will be considered to be long enough so that summations that contain the same number of points but differ by their ranges of summations can be considered as approximately equal. This means that *the mean value can be estimated immediately and subtracted from the signal*. That is, before being analyzed, the signal will be detrended as is necessary in conventional spectral analysis. The models will be developed on the zero average sample signal. The criterion used to derive the solutions for the models' parameters will be the minimization of the total squared error for linear prediction. Also remember that the estimate for the parameter of the first-order model was the negative value of the sample NACF at lag 1. It shall be seen that the NACF values are essential to this model development. The models derived in this manner are called the *linear prediction coefficient* (LPC) models.

The development of signal modeling techniques for small sample signals requires detailed study of the effects of summing over different time ranges of the signal. These estimating techniques are a current topic of research interest and can be studied in more advanced textbooks and in the current literature. See, for example, relevant material in Jenkins and Watts (1968), Kay (1988), and Wei (1990).

## EXAMPLE 8.2

One signal that seems to be modeled well by a first-order AR process is the surface of a grinding wheel. The roughness of the surface dictates the texture of the surface of the object being ground. The surface is quantized by measuring the height of the wheel above a designated reference point. The sampling interval is

0.002 inches. A sample measurement is shown in Figure 8.4a and tabulated in Appendix 8.3. The units are in milli-inches. The measurements have a nonzero mean of 9.419, which had to be subtracted from the signal before the parameter $a(1)$ could be calculated. The model is

$$y(n) = 0.627y(n-1) + \epsilon(n) \quad \text{with } \sigma_\epsilon^2 = 6.55$$

The sample standard deviation of the signal is 3.293. If one simulates the process with a generated $\epsilon(n)$, a resulting signal $s(n)$, such as that in Figure 8.4b, is produced. The mean of the measured signal was added to the simulated signal for this figure. Notice that $y(n)$ and $s(n)$ are not identical but that the qualitative appearances are similar. Since the pdf for $y(n)$ is approximately Gaussian, the pdf for $\epsilon(n)$ was chosen to be Gaussian also, with a mean of zero and a variance equal to the error variance.

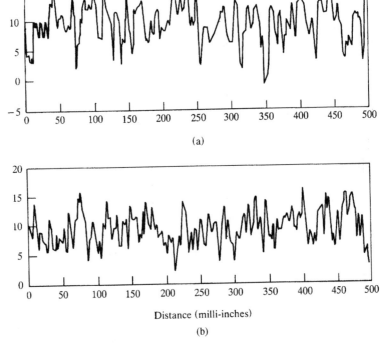

**FIGURE 8.4** Grinding wheel profile: (a) Surface height in milli-inches measured every 2 milli-inches. (b) Simulated profile with first-order model.

## 8.3.2 Solution of the General Model

After detrending, the general signal model is equation 8.15 with $a(i) = -h(i)$. The parameters are found through the minimization of the TSE, where

$$\text{TSE} = \sum_{n=p}^{N-1} e^2(n) = \sum_{n=p}^{N-1} (y(n) + a(1)y(n-1) + a(2)y(n-2)$$
$$+ \cdots + a(p)y(n-p))^2 \tag{8.19}$$

The general solution becomes

$$\frac{\partial}{\partial a(i)} \text{TSE} = 0 = \sum_{n=p}^{N-1} (y(n) + a(1)y(n-1) + a(2)y(n-2)$$
$$+ \cdots + a(p)y(n-p))y(n-i) \tag{8.20}$$

for $1 \leq i \leq p$. Thus there are $p$ equations of the form

$$- \sum_{n=p}^{N-1} y(n)y(n-i) = \sum_{n=p}^{N-1} a(1)y(n-1)y(n-i) + a(2)y(n-2)y(n-i)$$
$$+ \cdots + a(p)y(n-p)y(n-i) \tag{8.21}$$

If each term is divided by $N$, the summations become the biased estimates of the autocovariance and autocorrelation function or,

$$-\hat{R}(i) = a(1)\hat{R}(i-1) + a(2)\hat{R}(i-2) + \cdots + a(p)\hat{R}(i-p) \tag{8.22}$$

The unbiased estimates could also have been used by dividing the summations by $(N - p)$. The lowercase symbol $r(k)$ is usually used as the symbol for the sample autocorrelation function, $\hat{R}(k)$. This will be used to make the writing simpler. Equation 8.22 then becomes

$$a(1)r(i-1) + a(2)r(i-2) + \cdots + a(p)r(i-p) = -r(i) \tag{8.23}$$

Examine the arguments of $r(k)$, and notice that some are negative. In rewriting these $p$ equations, their even property will be utilized; thus in matrix form they can be written as

$$
\begin{bmatrix}
r(0) & r(1) & \cdots & r(p-1) \\
r(1) & r(0) & \cdots & r(p-2) \\
\vdots & \vdots & \cdots & \vdots \\
r(p-1) & r(p-2) & \cdots & r(0)
\end{bmatrix}
\begin{bmatrix}
a(1) \\
a(2) \\
\vdots \\
a(p)
\end{bmatrix}
=
\begin{bmatrix}
-r(1) \\
-r(2) \\
\vdots \\
-r(p)
\end{bmatrix}
\tag{8.24}
$$

Notice that this matrix equation has the same form as equation 8.11 and contain the Yule-Walker equations for modeling measured signals. The square matrix of dimension $p \times p$ is called the *autocorrelation matrix*. The solution can be found with any technique that solves simultaneous linear equations, such as the Gaussian elimination method. Because the solution of equation 8.24 estimates the model's parameters, a circumflex is put over their symbols. The total squared error then becomes

$$\text{TSE} = N\left(r(0) + \sum_{i=1}^{p} \hat{a}(i)r(i)\right) \tag{8.25}$$

The TSE is also called the *sum of squared residuals*. Because of its correspondence to the theoretical model, the variance of the error sequence is estimated to be

$$s_p^2 = \frac{\text{TSE}}{N} = r(0) + \sum_{i=1}^{p} \hat{a}(i)r(i) \tag{8.26}$$

Now that we can model the signal, it would be good to be assured that the system model is a stable one. Analytical derivations and research using simulated random signals have shown that the use of the biased autocorrelation estimates in the Yule-Walker equations always produces an ACF matrix that is positive definite. This in turn ensures that the signal models will be stable. This will not be the situation if the unbiased estimates for $\hat{R}(k)$ are used (Kay, 1988; Marple, 1987). As a reminder, a matrix is *positive definite* if its determinant and those of all of its principal minors are positive. For a formal definition, consult any introductory textbook on matrices (such as Ayres, 1962).

### EXAMPLE 8.3

Develop a second-order model for a temperature signal. Twelve measurements of the signal and the estimates of its autocovariance function for five lags are tabulated in Table 8.1. The sample mean and variance are 9.467 and 21.9, respectively. The equations for a second-order model of a signal are

$$\hat{a}(1)r(0) + \hat{a}(2)r(1) = -r(1)$$

and

$$\hat{a}(1)r(1) + \hat{a}(2)r(0) = -r(2)$$

From Table 8.1 and the variance, the sample NACF values are [1, 0.687, 0.322, −0.16, −0.447, −0.547]. The solution for the model's parameters are

$$\hat{a}(1) = \frac{r(1)(r(2) - r(0))}{r^2(0) - r^2(1)} = \frac{\hat{\rho}(1)(\hat{\rho}(2) - 1)}{1 - \hat{\rho}^2(1)} = -0.883$$

$$\hat{a}(2) = \frac{(r^2(1) - r(0)r(2))}{r^2(0) - r^2(1)} = \frac{(\hat{\rho}^2(1) - \hat{\rho}(2))}{1 - \hat{\rho}^2(1)} = 0.285$$

The variance of the residual is found from equation 8.26;

$$s_p^2 = r(0)[1 + \hat{a}(1)\hat{\rho}(1) + \hat{a}(2)\hat{\rho}(2)] = 21.9 \cdot 0.48 = 10.6$$

At this point it is very important to discuss some subtleties that occur when applying the AR modeling approach. In general, what has been discussed is a *least squares* technique for finding the model coefficients. There are several

**TABLE 8.1**

| Month | Temp, °C | $\hat{C}(k)$ |
|-------|----------|--------------|
| 1 | 3.4 | 15.1 |
| 2 | 4.5 | 7.05 |
| 3 | 4.3 | −3.5 |
| 4 | 8.7 | −9.79 |
| 5 | 13.3 | −12.0 |
| 6 | 13.8 | |
| 7 | 16.1 | |
| 8 | 15.5 | |
| 9 | 14.1 | |
| 10 | 8.9 | |
| 11 | 7.4 | |
| 12 | 3.6 | |

versions of this technique, and their differences occur in the range of summation when estimating the prediction error and the autocorrelation function. Notice that in the beginning of this section the summations occurred over the range of $n$ from $p$ to $N − 1$. Thus different estimates of $R(k)$ for a specific lag will result when $N$ is relatively small. This is different when considering the estimates of the autocorrelation function defined in Section 5.5.1, where the range of summation was over all possible cross products. When implementing the Yule-Walker approach, these definitions are used when solving equations 8.24 and 8.26 as in the previous examples. This is the usual practice when $N$ is not small.

### 8.3.3 Model Order

As one can surmise, the model for a signal can be any order that is desired. However, it should be as accurate as possible. After the parameters have been estimated, the actual error process can be generated from the signal using the MA system equation

$$\epsilon(n) = y(n) + \hat{a}(1)y(n − 1) + \hat{a}(2)y(n − 2) + \cdots + \hat{a}(p)y(n − p) \qquad (8.27)$$

Our knowledge from fitting regression equations to curves provides us with some general intuition. A model with too low an order will not represent the properties of the signals; one with too high an order will also represent any measurement noise or inaccuracies and not be a reliable representation of the true signal. Thus methods that will determine model order must be used. Some methods can be ascertained from the nature of the modeling process, some of which depend on the same concepts that are used in regression. The squared error is being minimized so it seems that a plot of TSE or $s_p^2$ against model order, $p$, should reveal the optimal value of $p$. The plot should be monotonically decreasing and asymptotically approach zero. The lowest value of $p$ associated with no appreciable decrease in $s_p^2$ should indicate the model order.

Consider the error plot in Figure 8.5. The residual error does decrease as expected but their is no definitive slope change to indicate clearly any asymptotic behavior. There are large decrements from orders 2 to 3 and from 4 to 6. Otherwise the error does not seem to decrease much. Thus one could judge that the order for the signal is approximately either 3 or 6. This is typical of the characteristic of residual error curves. As can be seen, there is no definitive curve characteristic and, in general, the residual variance is only an approximate indicator of model order.

Another approach based on the principles of prediction is that one simply increases the model order until the residual process, $\epsilon(n)$, in equation 8.27 becomes white noise. This is the approach used by Triolo et al. (1988) and Hefftner et al. (1988) for modeling the electromyogram from skeletal muscle. The NACF of the residuals was calculated from fitting models of various orders. Representative results are shown in Figure 8.6. The majority of signals studied in this manner produced white-noise residuals with fourth-order models and hence this order was used. A typical model found is

$$y(n) - 0.190y(n-1) + 0.184y(n-2) + 0.192y(n-3) + 0.125y(n-4) = \epsilon(n)$$
(8.28)

with $\hat{\sigma}_y^2 = 0.016$ (units are in volts).

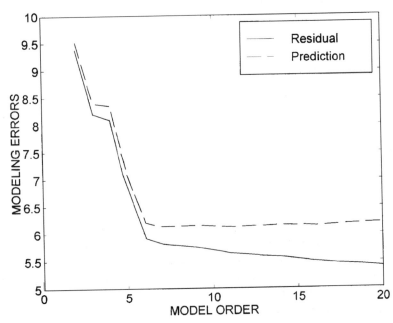

**FIGURE 8.5**   Plot of residual error (solid) and FPE (dashed) vs model order for a signal to be analyzed in Example 8.4.

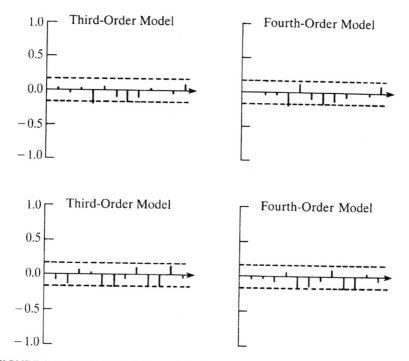

**FIGURE 8.6** The NACFs of the residuals of some third- and fourth-order models fitted to EMG signals; $T = 2$ ms. The dashed lines are the $\pm 1.96/\sqrt{N}$ confidence limits. [Adapted from Hefftner et al., fig. 7, with permission.]

Other criteria have been developed that are functions of the residual variance and tend to have a more defined minimum. These criteria are based on concepts in mathematical statistics but are only approximate solutions; their validity is tested with simulations. These developments have been pioneered by Akaike, who has developed two criteria whose properties are summarized in recent advanced treatments (Kay and Marple, 1981). Both criteria presume that the signal has been detrended. The first of these is the *final prediction error* (*FPE*), where

$$\text{FPE} = s_p^2 \frac{N+p+1}{N-p-1} \tag{8.29}$$

The FPE is based on the following concept: given all the signals with the same statistical properties as the measured signal, the FPE is the variance of all possible prediction errors for a given order. Thus the minimum FPE indicates the best model order. The fractional portion of FPE increases with $p$ and accounts for the inaccuracies in estimating $a(i)$. A study of the characteristics of this criterion shows that it tends to have a minimum at values of $p$ that are less than the model order when using the Yule-Walker method.

The other criterion, called the *Akaike's information criterion* (*AIC*), is

$$\text{AIC} = N \ln s_p^2 + 2p \qquad (8.30)$$

The term $2p$ is a penalty for higher orders. This criteria also has its shortcomings and tends to overestimate model order. In spite of these shortcomings, both criteria are used quite frequently in practical applications. An example will help illustrate the use of these criteria.

---

### EXAMPLE 8.4

Consider the signal generated by the fourth-order AR system

$$y(n) - 2.7607y(n-1) + 3.816y(n-2)$$
$$- 2.6535y(n-3) + 0.9238y(n-4) = x(n)$$

where $x(n)$ is white noise with unit variance and $T = 2\,\text{ms}$. A realization is plotted in Figure 8.7. The signal was modeled with orders up to 20 and the FPE is plotted in Figure 8.5. Notice that the first local minima appear for $p = 4$. As the model order increases, a global minimum appears for $p = 6$. Since ordinarily one does not know the model order, one would choose either a fourth- or sixth-order model for the signal, depending on the complexity or simplicity desired. One could also argue that the percentage decrease in criterion values does not warrant the increased model complexity. In any event, a judgment is necessary.

The parameter sets for the two models are:

[1    − 1.8046    1.6128    − 0.5079    0.0897]
[1    − 1.6292    1.1881    0.1896    − 0.1482    − 0.2105    0.3464]

This signal model is used in the signal processing literature and one can find extensive analyses of it. For instance, refer to Robinson and Treital (1980) and Ulrych and Bishop (1975).

---

### EXAMPLE 8.5

Meteorological data is always of great interest to modelers and scientists, especially with the real-life problems of droughts and hunger. It is then of interest to model data such as rainfall accurately in order to predict the future trends. Figure 8.8 is the plot of the deviation of the averge annual rainfall in the eastern United States for the years 1817 through 1922, $\sigma^2 = 89.9$. (The values are listed in Appendix 8.4.) It is our task to model this time series. AR models with orders from 1 to 20 were estimated for it. The FPE and AIC criteria are plotted in Figure 8.9. There are local minima for order 4 and global minima for order 8. For the sake of simplicity, order 4 shall be chosen, and the model coefficients are [1    − 0.065    0.122    − 0.325    − 0.128].

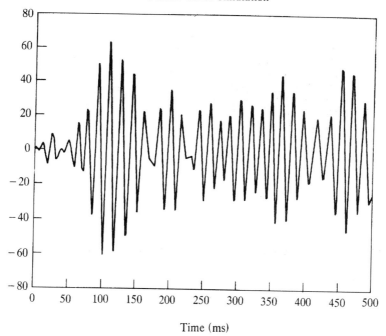

Fourth-Order Simulation

**FIGURE 8.7** A sample function of a realization of the fourth-order process described in Example 8.4; $T = 2$ ms.

Paralleling this concept is the study of the value of the parameters, $\hat{a}(i)$. Intuitively, one would guess that when the values of the parameters become very small, they are insignificant and should be considered zero. The largest value of $p$ associated with a nonzero value of $a(i)$ is probably the model order. Testing whether specific values of $a(i)$ are nonzero, although based on a complicated aspect of correlation, can be easily implemented. Reconsider now the first-order model of the grinding wheel profile in Example 8.2. The estimate of $a(1)$ is actually the value of the sample NACF at lag 1. One can test whether a correlation coefficient is nonzero using the Gaussian approximation studied in Section 5.5.5. Thus if $\hat{a}(1)$ lies within the range $\pm 1.96/\sqrt{N}$, then it is actually zero at the 95% confidence level and the signal can be considered as white noise. However, if $a(1) \neq 0$, then the signal is at least a first-order process. For the profile, $N = 250$, and the confidence limits are $\pm 0.124$. Since $\hat{a}(1) = -0.627$, the profile is at least a first-order process.

It would be easy if all the other coefficients could be tested as simply. However, notice from the Yule-Walker equations that $a(i)$ depends on all the other parameters. Thus it is not a simple correlation coefficient. Another concept, *partial correlation*, must be introduced. When fitting the signal to a model of

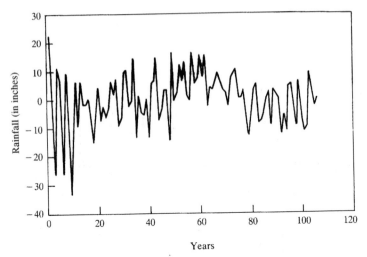

**FIGURE 8.8**   The average annual rainfall in the eastern United States for the years 1817 through 1922.

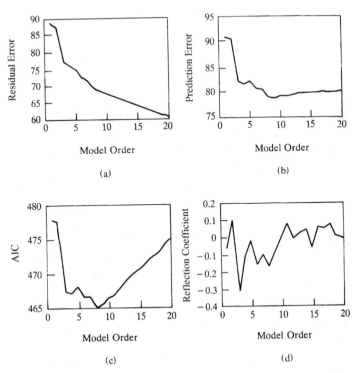

**FIGURE 8.9**   The model selection criteria for the rainfall time series vs model order: (a) Residual error. (b) FPE. (c) AIC. (d) Partial correlation coefficient. The Yule-Walker method was used.

order $p$, designate the last coefficient as $\hat{\pi}_p = \hat{a}(p)$. The parameter $\hat{\pi}_p$ is the $p$th partial correlation coefficient, a measure of the amount of correlation at lag $p$ *not accounted for by a $(p - 1)$-order model*. Thus it can be tested as an ordinary correlation coefficient. Its derivation will be examined in the next section. A plot of $\hat{\pi}_p$ vs $p$ with the correlation confidence limits superimposed presents another order testing modality. This plot is shown in Figure 8.9d for the process used in the previous example. Notice the value of $\hat{\pi}_p$ is always within the confidence limits for $p \geq 9$. Thus the criteria indicates that an eighth-order model is appropriate.

### 8.3.4 Levinson-Durbin Algorithm

Autoregressive modeling has become a very powerful approach for signal modeling and spectral estimation. One simply solves the Yule-Walker matrix equation with increasing dimensionality until the best model order is determined. The major shortcoming of this approach is that the solution of this matrix equation fifteen or twenty times in a batch mode requires too much computational time. Thus a recursive algorithm has been developed. The recursive scheme is based on estimating the parameters of a model of order $p$ from the parameters of a model of order $p - 1$. The scheme is also based on matrix equation 8.24 and is dependent on the Toeplitz form of the correlation matrix.

At this point it is helpful to rewrite the basic equations of the AR system that were presented in Chapter 6. The equation for a system of order $p$ is

$$y(n) + a(1)y(n - 1) + a(2)y(n - 2) + \cdots + a(p)y(n - p) = x(n) \qquad (8.31)$$

When $x(n)$ is a white-noise process, all types of correlation functions have a recursive relationship among magnitudes at different lag times. The relationship for the autocovariance function is

$$C(k) + a(1)C(k - 1) + a(2)C(k - 2) + \cdots + a(p)C(k - p) = 0 \qquad (8.32)$$

for $k \neq 0$. For $k = 0$,

$$C(0) + a(1)C(1) + a(2)C(2) + \cdots + a(p)C(p) = \sigma_x^2 \qquad (8.33)$$

Remember that in general the mean values of $x(n)$ and $y(n)$ are zero so that $R(k) = C(k)$. The development of the recursive scheme will begin by deriving the recursive algorithm for a second-order model. The matrix equation is

$$\begin{bmatrix} r(0) & r(1) \\ r(1) & r(0) \end{bmatrix} \begin{bmatrix} \hat{a}_2(1) \\ \hat{a}_2(2) \end{bmatrix} = \begin{bmatrix} -r(1) \\ -r(2) \end{bmatrix} \qquad (8.34)$$

Notice that the symbols for the estimates of the parameters now have a subscript. The value of the subscript indicates the model order. This is because in general the value of $a(2)$ for a second-order model is different than that for a third-order model. From Section 8.3.1 it was found that for a measured signal, a first-order model has the solution

$$\hat{a}_1(1) = -\frac{r(1)}{r(0)} \tag{8.35}$$

with MSE

$$\sigma_{\epsilon,1}^2 = r(0)(1 - \hat{a}_1(1)^2) \tag{8.36}$$

Now rewrite the first equation of matrix equation 8.34 solving for $\hat{a}_2(1)$, or

$$r(0)\hat{a}_2(1) = -r(1) - r(1)\hat{a}_2(2) \tag{8.37}$$

This can be expressed in terms of parameters as

$$\hat{a}_2(1) = -\frac{r(1)}{r(0)} - \hat{a}_2(2)\frac{r(1)}{r(0)} = \hat{a}_1(1) + \hat{a}_2(2)\hat{a}_1(1) \tag{8.38}$$

The first parameter of the second-order model is expressed in terms of the parameter of the first-order model and $\hat{a}_2(2)$. To find the latter parameter the augmented Yule-Walker equations must be used. The augmented equations use the equation for the noise variance, equation 8.26. The augmented matrix equation containing sample autocovariances is

$$\begin{bmatrix} r(0) & r(1) & r(2) \\ r(1) & r(0) & r(1) \\ r(2) & r(1) & r(0) \end{bmatrix} \begin{bmatrix} 1 \\ \hat{a}_2(1) \\ \hat{a}_2(2) \end{bmatrix} = \begin{bmatrix} \sigma_{\epsilon,2}^2 \\ 0 \\ 0 \end{bmatrix} \tag{8.39}$$

The critical point is expressing the vector on the left side of equation 8.39 in terms of $\hat{a}_2(2)$. Using equation 8.38,

$$\begin{bmatrix} 1 \\ \hat{a}_2(1) \\ \hat{a}_2(2) \end{bmatrix} = \begin{bmatrix} 1 \\ \hat{a}_1(1) + \hat{a}_1(1)\hat{a}_2(2) \\ \hat{a}_2(2) \end{bmatrix} = \begin{bmatrix} 1 \\ \hat{a}_1(1) \\ 0 \end{bmatrix} + \hat{a}_2(2)\begin{bmatrix} 0 \\ \hat{a}_1(1) \\ 1 \end{bmatrix} \tag{8.40}$$

Expanding the left side of equation 8.39 using equation 8.40 gives a sum of two vectors. Each will be found in succession. For the first one,

$$T_1 = \begin{bmatrix} r(0) & r(1) & r(2) \\ r(1) & r(0) & r(1) \\ r(2) & r(1) & r(0) \end{bmatrix} \begin{bmatrix} 1 \\ \hat{a}_1(1) \\ 0 \end{bmatrix} = \begin{bmatrix} r(0) + r(1)\hat{a}_1(1) \\ r(1) + r(0)\hat{a}_1(1) \\ r(2) + r(1)\hat{a}_1(1) \end{bmatrix}$$

Examine the elements of the vector $T_1$ with respect to the equations for a first-order model. The first element is equal to the error variance. The second element is given the symbol $\Delta_2$. If order 1 was sufficient, this term would equal zero. The third element is given the symbol $\Delta_3$, and now

$$T_1 = \begin{bmatrix} \sigma_{\epsilon,1}^2 \\ \Delta_2 \\ \Delta_3 \end{bmatrix} \tag{8.41}$$

The second term of equation 8.39 is

$$T_2 = \hat{a}_2(2) \begin{bmatrix} r(0) & r(1) & r(2) \\ r(1) & r(0) & r(1) \\ r(2) & r(1) & r(0) \end{bmatrix} \begin{bmatrix} 0 \\ \hat{a}_1(1) \\ 1 \end{bmatrix} = \hat{a}_2(2) \begin{bmatrix} r(1)\hat{a}_1(1) + r(2) \\ r(0)\hat{a}_1(1) + r(1) \\ r(1)\hat{a}_1(1) + r(0) \end{bmatrix}$$

Examining the elements of the vector on a term-by-term basis shows that the elements are the same as in $T_1$ except that they are in reverse order. Now,

$$T_2 = \hat{a}_2(2) \begin{bmatrix} \Delta_3 \\ \Delta_2 \\ \sigma^2_{\epsilon,1} \end{bmatrix} \tag{8.42}$$

Summing the two terms together and equating them to equation 8.39 gives

$$T_1 + T_2 = \begin{bmatrix} \sigma^2_{\epsilon,1} \\ \Delta_2 \\ \Delta_3 \end{bmatrix} + \hat{a}_2(2) \begin{bmatrix} \Delta_3 \\ \Delta_2 \\ \sigma^2_{\epsilon,1} \end{bmatrix} = \begin{bmatrix} \sigma^2_{\epsilon,2} \\ 0 \\ 0 \end{bmatrix} \tag{8.43}$$

The second parameter is found using the third element of equation 8.43 as

$$\hat{a}_2(2) = -\frac{\Delta_3}{\sigma^2_{\epsilon,1}} = -\frac{r(1)\hat{a}_1(1) + r(2)}{\sigma^2_{\epsilon,1}} \tag{8.44}$$

Similarly, the error variance is found using the first element of equation 8.43 and equation 8.44 as

$$\sigma^2_{\epsilon,2} = \sigma^2_{\epsilon,1} + \hat{a}_2(2)\Delta_3 = (1 - \hat{a}_2(2)^2)\sigma^2_{\epsilon,1} \tag{8.45}$$

---

## EXAMPLE 8.6

Consider again the temperature signal. What is the residual error for the first-order model? From equation 8.36 it is known that

$$\sigma^2_{\epsilon,1} = r(0)(1 - \hat{a}_1(1)^2) = 21.9(1 - 0.472) = 11.56$$

Calculate the parameters of a second-order model. From equations 8.44, 8.45, and 8.38,

$$\hat{a}_2(2) = -\frac{r(1)\hat{a}_1(1) + r(2)}{\sigma^2_{\epsilon,1}} = -\frac{15.1 \cdot (-0.687) + 7.05}{11.56} = 0.285$$

$$\hat{a}_2(1) = \hat{a}_1(1) + \hat{a}_2(2)\hat{a}_1(1) = -0.687 + 0.285 \cdot (-0.687) = -0.882$$

$$\sigma^2_{\epsilon,2} = (1 - \hat{a}_2(2)^2)\sigma^2_{\epsilon,1} = (1 - 0.8267)11.56 = 10.6$$

---

This process can be continued and general equations can be written to summarize this approach. The algorithm is called *Levinson's algorithm*. The steps in this algorithm follow.

1. Start with a zero-order model; $p = 0$, $\hat{a}_0(0) = 1$, $\sigma_{\epsilon,0}^2 = r(0)$,

2. Generate the last parameter of the next-higher-order model and designate it $\hat{\pi}_{p+1}$. The equation is

$$\hat{\pi}_{p+1} = \frac{-\sum_{i=0}^{p} r(p + 1 - i)\hat{a}_p(i)}{\sigma_{\epsilon,p}^2}$$

3. Find the error variance,

$$\sigma_{\epsilon,p+1}^2 = (1 - \hat{\pi}_{p+1}^2)\sigma_{\epsilon,p}^2$$

4. Find the remaining parameters, $\hat{a}_{p+1}(i)$,

$$\hat{a}_{p+1}(0) = 1, \qquad \hat{a}_{p+1}(p + 1) = \hat{\pi}_{p+1},$$

$$\hat{a}_{p+1}(i) = \hat{a}_p(i) + \hat{\pi}_{p+1}\hat{a}_p(p + 1 - i) \quad \text{for } 1 \le i \le p$$

5. Set $p = p + 1$ and return to step 2 until parameters for the maximum-order model are calculated.

The additional parameter needed for the next-higher-order model is called the *reflection coefficient*. Examining the error equation in step 3, it is desired that

$$0 \le |\hat{\pi}_p| \le 1 \tag{8.46}$$

so that

$$\sigma_{\epsilon,p+1}^2 \le \sigma_{\epsilon,p}^2 \tag{8.47}$$

This means that the successive error variances associated with higher-order models decreases. In Section 8.3.3 it is shown that $\hat{\pi}_p$ is a correlation coefficient; therefore equation 8.46 is true. A close examination of the numerator of the equation for $\hat{\pi}_{p+1}$ shows that it equals zero if the model order is correct and $\hat{a}_p(i) = a(i)$ and $r(k) = R(k)$ for the signal.

All of the equations can be derived in an elegant manner using matrices. One derivation is included in Appendix 8.2. An analysis of the steps in this algorithm shows that not only are all lower-order models estimated when estimating a model of order $p$ but the number of mathematical operations is less. For a $p$th-order model, a conventional linear equation solution of the Yule-Walker equations—such as Gaussian elimination—requires approximately $p^3$ operations. For Levinson's algorithm, each iteration for an $m$th-order model requires $2m$ operations. Thus a $p$th-order model requires $\sum_{m=1}^{p} 2m = p(p - 1)$ operations, which is much fewer (Kay, 1988).

### 8.3.5 Burg Method

The Burg method was developed because of the inaccuracies of the parameter estimation that sometimes occurred when the Yule-Walker equations were used directly. This became evident when models were used to simulate data that was then used to estimate model parameters. This is evident in Example 8.4. One of the hypothesized sources of error was the bias in the autocorrelation function estimate. A possible solution was to change the solution equations so that the data could be used directly. Also perhaps more signal points could be utilized simultaneously.

The Burg method was developed using the Levinson-Durbin algorithm. Notice that the only parameter that is directly a function of $r(k)$ is the reflection coefficient in step 2. Thus an error criterion was needed that was not only a function of $\hat{\pi}_p$ but also of more signal points. Thus to utilize more points a *backward prediction error*, $\epsilon^b(n)$, is defined. The error that has been used is called the *forward prediction error*, $\epsilon^f(n)$. The new error criterion is the average of the mean squared value of both errors.

The backward predictor uses future points to predict values in the past; for a one-step predictor,

$$\hat{y}(n-p) = -\hat{a}_p(1)y(n-p+1) - \hat{a}_p(2)y(n-p+2) - \cdots - \hat{a}_p(p)y(n) \quad (8.48)$$

and

$$\epsilon_p^b(n) = \hat{y}(n-p) + \hat{a}_p(1)y(n-p+1) + \hat{a}_p(2)y(n-p+2) + \cdots + \hat{a}_p(p)y(n)$$
$$(8.49)$$

The forward prediction error is equation 8.27. It is rewritten explicitly as a function of model order and

$$\epsilon_p^f(n) = y(n) + \hat{a}_p(1)y(n-1) + \hat{a}_p(2)y(n-2) + \cdots + \hat{a}_p(p)y(n-p) \quad (8.50)$$

These errors can now be made a function of the reflection coefficient. Substituting the recursive relationship for the AR parameters in step 4 of Levinson's algorithm into equations 8.49 and 8.50 yields

$$\epsilon_p^b(n) = \epsilon_{p-1}^b(n-1) + \hat{\pi}_p \epsilon_{p-1}^f(n) \quad (8.51)$$

$$\epsilon_p^f(n) = \epsilon_{p-1}^f(n) + \hat{\pi}_p \epsilon_{p-1}^b(n-1) \quad (8.52)$$

where
$$\epsilon_0^b(n) = \epsilon_0^f(n) = y(n) \quad (8.53)$$

The average prediction error is now

$$\sigma_{\epsilon,p}^2 = \frac{1}{2}\left(\frac{1}{N-p}\sum_{n=p}^{N-1}|\epsilon_p^f(n)|^2 + \frac{1}{N-p}\sum_{n=p}^{N-1}|\epsilon_p^b(n)|^2\right) \quad (8.54)$$

Substituting equations 8.51 and 8.52 into equation 8.54 produces

$$\sigma_{\epsilon,p}^2 = \frac{1}{2(N-p)}\sum_{n=p}^{N-1}(|\epsilon_{p-1}^f(n) + \hat{\pi}_p\epsilon_{p-1}^b(n-1)|^2 + |\epsilon_{p-1}^b(n-1) + \hat{\pi}_p\epsilon_{p-1}^f(n)|^2)$$
$$(8.55)$$

Differentiating this equation with respect to $\hat{\pi}_p$ and setting the result equal to zero will yield the solution for the reflection coefficient:

$$\frac{\partial \sigma^2_{\epsilon,p}}{\partial \hat{\pi}_p} = 0 = \frac{1}{N-p} * \tag{8.56}$$

$$\sum_{n=p}^{N-1} ((\epsilon^f_{p-1}(n) + \hat{\pi}_p \epsilon^b_{p-1}(n-1))\epsilon^b_{p-1}(n-1) + (\epsilon^b_{p-1}(n-1) + \hat{\pi}_p \epsilon^f_{p-1}(n))\epsilon^f_{p-1}(n))$$

Rearranging terms produces

$$\hat{\pi}_p = \frac{-2 \sum_{n=p}^{N-1} \epsilon^f_{p-1}(n)\epsilon^b_{p-1}(n-1)}{\sum_{n=p}^{N-1} ((\epsilon^f_{p-1}(n))^2 + (\epsilon^b_{p-1}(n-1))^2)} \tag{8.57}$$

Study in detail the terms in equation 8.57. The numerator is a sum of cross products and the denominator contains two sums of squares. This is the definition of the correlation coefficient between the forward and backward prediction errors using the pooled variance as the variance of both error sequences. This reflection coefficient is a partial correlation coefficient and thus

$$-1 \le \hat{\pi}_p \le 1 \tag{8.58}$$

The steps for implementing the Burg method follow.

1. *Initial conditions*

$$s_0^2 = r(0)$$

$$\epsilon^f_0(n) = y(n); \qquad n = 0, 1, \ldots, N-1$$
$$\epsilon^b_0(n) = y(n); \qquad n = 0, 1, \ldots, N-1$$

2. *Reflection coefficients*
   For $p = 1, \ldots, P$,

$$\hat{\pi}_p = \frac{-2 \sum_{n=p}^{N-1} \epsilon^f_{p-1}(n)\epsilon^b_{p-1}(n-1)}{\sum_{n=p}^{N-1} ((\epsilon^f_{p-1}(n))^2 + (\epsilon^b_{p-1}(n-1))^2)}$$

$$s_p^2 = (1 - |\hat{\pi}_p|^2)s_{p-1}^2$$

   For $p = 1$,

$$\hat{a}_1(1) = \hat{\pi}_1$$

   For $p > 1$,

$$a_p(i) = \begin{cases} a_{p-1}(i) + \hat{\pi}_p a_{p-1}(p - i), & i = 1, 2, \ldots, p - 1 \\ \hat{\pi}_p & i = p \end{cases}$$

3. *Prediction errors for next order*

$$\epsilon_p^f(n) = \epsilon_{p-1}^f(n) + \hat{\pi}_p \epsilon_{p-1}^b(n-1), \qquad n = p, p+1, \ldots, N-1$$

$$\epsilon_p^b(n) = \epsilon_{p-1}^b(n-1) + \hat{\pi}_p \epsilon_{p-1}^f(n), \qquad n = p, p+1, \ldots, N-1$$

---

**EXAMPLE 8.7**

The parameters of the fourth-order signal model were estimated with the Burg method. The estimates of $a_4(1)$ through $a_4(4)$ are

$$[-2.7222 \quad 3.7086 \quad -2.5423 \quad 0.8756]$$

These are almost exactly the theoretical values.

---

**EXAMPLE 8.8**

Since the order criteria were contradictory for the rainfall data (Example 8.5 and Appendix 8.4), the criteria and model parameters were recalculated using the Burg method. The residual error and criteria are plotted in Figure 8.10. If the residual error in Figure 8.10a is compared with that in Figure 8.9a, it is seen that the Burg method produces a slightly larger error at all orders. Examining Figure 8.10b, 8.10c, and 8.10d reveals that all criteria suggest the same model order, that is, 3. The estimates of $a_3(1)$ through $a_3(3)$ are

$$[-0.1088 \quad 0.14 \quad -0.3392] \quad \text{with } s_3^2 = 78.47$$

---

Simulations have shown that in general the model order criteria are fairly accurate when the Burg method is used. However, when a local minimum occurs at a lower order than the global minimum, the order associated with the local minimum is usually more accurate (Ulrych and Bishop, 1975). The Burg method is also used when $N$ is small, a situation in which the constraint $p \leq N/2$ must be used.

There are several other methods for determining AR parameters using the least squares criterion. Some involve finding all the parameters directly. Notice that in the Burg method only the reflection coefficient is found directly. The other parameters are found using the Levinson-Durbin algorithm. For further study on this topic refer to Marple (1987).

## 8.3.6   Summary of Signal Modeling

The minimization of the error of linear prediction is the basic modern approach to signal modeling. Many books treat this topic; for instance, refer to Cadzow (1987), Kay (1988), or Marple (1987). The direct application of this principle leads to the Yule-Walker equations. The use of estimates of the autocovariance, autocorrelation function in the YW matrix equation comprises the autocorrelation

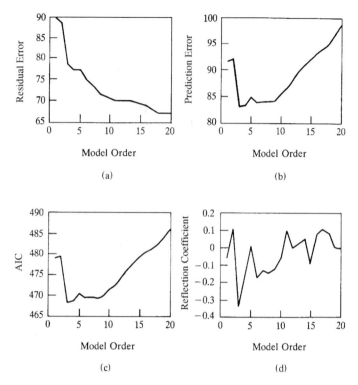

**FIGURE 8.10**  The model selection criteria for the rainfall time series vs model order: (a) Residual error. (b) FPE. (c) AIC. (d) Partial correlation coefficient. The Burg method was used.

method for estimating the parameters of the signal model. It is a direct approach for which there are well-known techniques for solving the matrix equations. The need for faster solution techniques led to the development of the Levinson algorithm for solving the YW equations recursively. This has the benefit of being computationally faster as well as producing all the lower-order models.

Since it is easy to produce models of any order, it is necessary to have rational criteria for estimating model order. We have described several criteria: the final prediction error, Akaike's information criteria, and the partial correlation coefficient. It seems that all have their shortcomings, but if they are used with these shortcomings in mind, they are good criteria.

In examining the models of simulated signals, we discovered that the autocorrelation method is not always accurate. It was hypothesized that this is because the estimates of the autocorrelation function are used. Thus another method was developed by Burg. Using the same signal values but minimizing the forward and backward prediction errors simultaneously, it has proven in general to be a much more accurate method.

Notice that because of the random nature of signals not all of the variations can be modeled. That is, some signals have an inherent large white-noise component. This is why the rainfall signal could not be modeled very well compared with the other signals. This is ascertained by the fact that the residual error for the appropriate model only decreased by 13% compared to decreases on the order of 35% for the temperature and grinding wheel signals.

## 8.4  SPECTRAL DENSITY FUNCTION ESTIMATION

### 8.4.1  Definition and Properties

The power spectral density (PSD) function of a signal can be estimated after a model has been found. Once the appropriate order has been decided then it is assumed that the optimal prediction error sequence is a white-noise sequence and $s_p^2 = \hat{\sigma}_x^2$. Using the system input–output relationship, it is known that

$$S_y(f) = |H(f)|^2 S_x(f) = |H(f)|^2 \sigma_x^2 T \tag{8.59}$$

Also it is known that for an AR system the transfer function is

$$H(f) = \frac{1}{\displaystyle\sum_{i=0}^{p} a(i)e^{-j2\pi f i T}} \tag{8.60}$$

The PSD for a signal is estimated using equations 8.59 and 8.60 and substituting estimates for the noise variance and system parameters; that is,

$$\hat{S}(f) = \frac{s_p^2 T}{\left| \displaystyle\sum_{i=0}^{p} \hat{a}(i)e^{-j2\pi f i T} \right|^2} \tag{8.61}$$

This estimator is the foundation for what is still called *modern spectral estimation* but is now more appropriately called *parametric spectral estimation*. There are some advantages of this approach over the classical approach. One of them is in the frequency spacing. Examination of equation 8.61 shows that frequency is still a continuous parameter. Thus the picket fence effect is eliminated, and one is not concerned with zero padding and the number of signal points for this purpose. Another advantage is that only $p$ parameters are being estimated directly instead of all the spectral values. This enables a reliable estimate of a spectrum with small values of $N$, naturally $p < N$.

Several differences exist between the two techniques because of some underlying assumptions. In the periodogram approach, the Fourier transform induces a periodic repetition on the signal outside the time span of its measurement.

In the modern approach, one predicts nonmeasured future values using the linear predictor. However, the prediction is not of concern to us now. Another difference comes about in the assumptions about the ACF. In the BT approach, it is assumed that the ACF is zero for lags greater than a maximum lag $M$, whereas the AR model has a recursion relationship for the ACF and values for lags greater than $M$. As with all techniques, there are also some inaccuracies in parametric spectral estimation. These will be explained in subsequent paragraphs. Many of the numerical inaccuracies have been reduced by using the Burg method for the parameter estimation.

### EXAMPLE 8.9

The signal described by the fourth-order AR system in Example 8.4 has the PSD

$$S(f) = \frac{T}{\left| \sum_{i=0}^{4} a(i)e^{-j2\pi f i T} \right|^2}$$

with $T = 2\,\text{ms}$ and $a(1) = -2.76$, $a(2) = 3.82$, $a(3) = -2.65$, and $a(4) = 0.92$. As we know, the range for frequency is $0 \leq f \leq 250\,\text{Hz}$. For a fine resolution for plotting, let $f$ equal an integer. A plot of the PSD is shown in Figure 8.11.

### EXAMPLE 8.10

A random signal was generated using the model in Example 8.9. Its time series and estimated parameters using the Burg method are described in Examples 8.4 and 8.7. The spectral estimate from this model is plotted in Figure 8.12. Notice that it agrees quite well with the theoretical spectrum.

At this point it is advantageous to demonstrate the need for a second methodology for spectral estimation. Because the AR model is the result of an all-pole system, narrowband spectral peaks can be estimated with more accuracy and closely spaced spectral peaks can be distinguished. Consider target reflections in radar systems. The Doppler frequencies are used to estimate target velocities. Figure 8.13 shows the output spectrum estimated with both the AR method and the BT method. Two spectral peaks are evident at approximately 225 and 245 Hz. The stronger energy peak is from a discrete target. The AR spectrum possesses a more narrow peak at 245 Hz, which is consistent with a narrow frequency band of power, whereas the BT spectra tends to blend the two peaks together (Kaveh and Cooper, 1976).

This same good characteristic can also produce artifactual peaks in the estimated spectra. If the model order is too high then false peaks appear; this has been proved using simulations. Figure 8.14 is an example. The theoretical spectrum comes from the second-order AR system

$$y(n) = 0.75y(n-1) - 0.5y(n-2) + x(n); \qquad \sigma_x^2 = 1, \quad T = 1 \qquad (8.62)$$

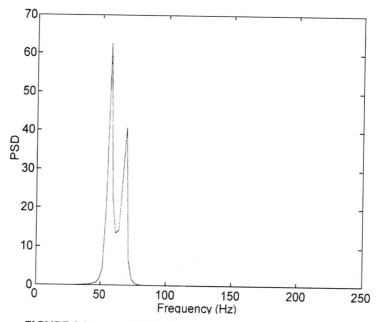

**FIGURE 8.11**  The PSD for the theoretical fourth-order AR model.

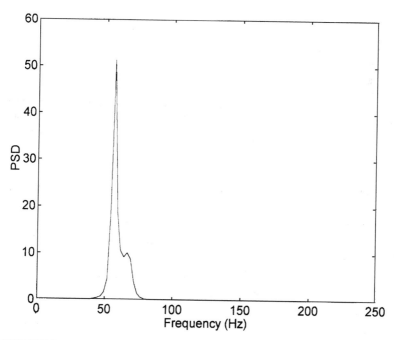

**FIGURE 8.12**  Estimate of a PSD from a simulated random signal; $p = 4$, $N = 250$.

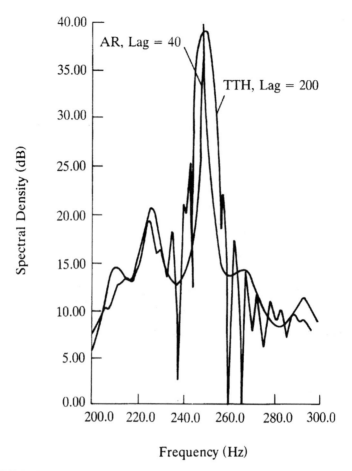

**FIGURE 8.13**   Radar output spectrum for a 2000-point Doppler signal with a sampling interval of I ms. TTH indicates the BT estimate with a Hanning window and maximum ACF lag of 200; AR indicates the AR estimate using 40 ACF lags. [Adapted from Kaveh and Cooper, fig. 10, with permission.]

A realization containing 50 points was generated and its PSD estimated with both second- and eleventh-order models. It can be seen quite clearly in the figure that the spectrum of the second-order model estimates the actual spectrum very closely. However, in the other model, the broad spectral peak has been split into two peaks. This phenomena is called *line splitting* (Ulrych and Bishop, 1975). This emphasizes even more the importance of model order determination.

An application that illustrates the possible occurrence of line splitting is the study of geomagnetic micropulsations. These micropulsations are disturbances on Earth's geomagnetic field and have magnitudes on the order of several hundred gamma (Earth's field strength is on the order of 50,000 gammas). The

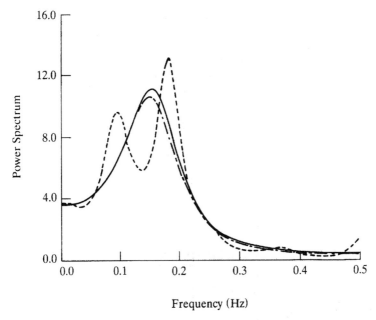

Frequency (Hz)

**FIGURE 8.14** Illustration of line splitting caused by overestimating the model order: actual PSD of second-order signal (solid); PSD estimates using second-order (dot-dash) and eleventh-order (dash-dash) models. [Adapted from Ulrych and Bishop, fig. 1b, with permission.]

disturbances arise from the magnetosphere, and their spectra are calculated so that models of this disturbance can be developed. Figure 8.15a shows the $y$ component of a 15-minute record that was sampled every 3.5 seconds ($f_s = 0.283$ Hz), giving $N = 257$. The AR spectrum was estimated using 80 parameters and is shown in Figure 8.15b. Notice that this is quite high compared to modeling that emphasizes the time domain properties. Two narrowband spectral peaks occur at 7 mHz (millihertz) and 17 mHz. To determine if any of these peaks are composed of multiple peaks closely spaced in frequency, the spectrum was reestimated using 140 parameters (Figure 8.15c). All the peaks are sharper, and the one at 17 mHz appears as two peaks. Now one must be careful that this is not an artifact of line splitting. One must resort to theory of the phenomena or other spectral estimation methods (such as Pisarenko harmonic decomposition) to verify these closely spaced peaks (Kay, 1988).

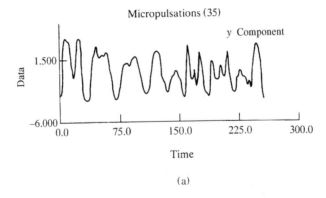

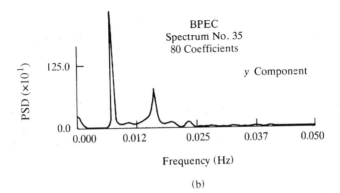

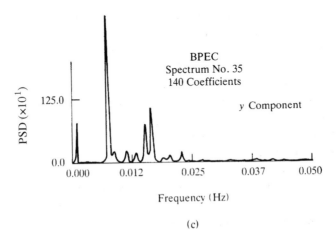

**FIGURE 8.15** Micropulsations and spectra. (a) 15-minute record of micropulsations. (b) AR PSD using 80 parameters. (c) AR PSD using 140 parameters. [Adapted from Radowski et al., figs. 2, 5, and 7, with permission.]

## 8.4.2   Statistical Properties

The exact results for the statistics of the AR spectral estimator are not available. Approximations based on large samples show that for stationary processes $\hat{S}(f)$ is distributed according to a Gaussian pdf and is an asymptotically unbiased and consistent estimator of the PSD. Its variance depends on the model order. In particular,

$$E[\hat{S}(f)] = S(f)$$

and

$$\text{Var}[\hat{S}(f)] = \begin{cases} \dfrac{4p}{N}\,S^2(f), & f = 0 \quad \text{and} \quad \pm\dfrac{1}{2T} \\[2ex] \dfrac{2p}{N}\,S^2(f), & f = \text{otherwise} \end{cases} \tag{8.63}$$

As with the periodogram method, the magnitudes are uncorrelated; that is,

$$\text{Cov}[\hat{S}(f_1), \hat{S}(f_2)] = 0 \quad \text{for } f_1 \neq f_2 \tag{8.64}$$

The upper and lower limits for the $100(1 - \alpha)\%$ confidence interval are

$$\hat{S}(f)\left[1 \pm \sqrt{\frac{2p}{N}}\,z\left(1 - \frac{\alpha}{2}\right)\right] \tag{8.65}$$

where $z(\beta)$ represents the $\beta$ percentage point of a zero mean, unit variance, Gaussian pdf (Kay, 1988). Using the symbols in Chapter 4, $\Phi(z) = \beta$.

### EXAMPLE 8.11

The PSD estimate for the rainfall data using the AR(4) model in Example 8.5 is

$$\hat{S}(f) = \frac{s_p^2 T}{\left|\sum\limits_{i=0}^{p} \hat{a}(i)e^{-j2\pi f i T}\right|^2} = \frac{77.19}{\left|\sum\limits_{i=0}^{4} \hat{a}(i)e^{-j2\pi f i}\right|^2}$$

with the coefficients being $[-0.065 \quad 0.122 \quad -0.325 \quad -0.128]$. The 95% confidence limits for this estimate are

$$\hat{S}(f)\left[1 \pm \sqrt{\frac{2p}{N}}\,z\left(1 - \frac{\alpha}{2}\right)\right] = \hat{S}(f)\left[1 \pm \sqrt{\frac{2 \cdot 4}{106}} \cdot 1.96\right] = \hat{S}(f)[1 \pm 0.54]$$

The estimate and its limits are plotted in Figure 8.16.

### EXAMPLE 8.12

Speech is a short-term stationary signal. Stationary epochs are approximately 120-ms long. Speech signals are sampled at the rate of 10 kHz, and 30-ms epochs

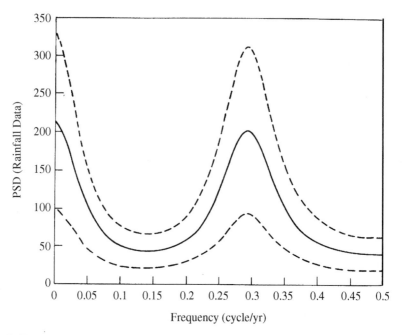

**FIGURE 8.16** The LPC PSD estimate of the rainfall data (−) and the 95% confidence limits (−−−).

are studied for their frequency content. Each epoch overlaps 20 ms with the preceding epoch. The LPC method is the preferred approach because of the small values of $N$ (approximately 300) available and the speed and low variance required. AR(14) models are accurate. Figure 8.17 shows the spectrogram produced for performing a spectral analysis of the phrase "oak is strong".

### 8.4.3   Other Spectral Estimation Methods

There are many other spectral estimation methods. Most of them are designed for special purposes, such as accurately estimating the frequencies of sinusoids whose frequencies are close together or estimating the PSD for signals with a small number of points. The study of these topics is undertaken in advanced treatments. The one method that must be mentioned explicitly is the *maximum entropy method* (*MEM*). It is also an AR modeling method that places a constraint on unavailable data. The principle is that given the known or estimated autocorrelation function, one can predict future or unknown signal values such that they are not only consistent with the known values but also are the most random set of signal points possible. The MEM produces the Yule-Walker equations and hence is the same as the LPC method.

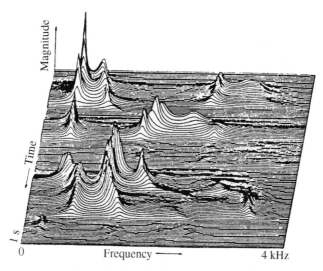

**FIGURE 8.17** Digital spectrogram of the utterance "oak is strong" using AR(14) models. [Adapted from Veeneman, fig. 17.23, with permission.]

There are alternative algorithms and methods for solving Yule-Walker equations, and most of them depend on the Levinson-Durbin recursion. They are more accurate and necessary for short-duration signals. One of these is the Burg method, which uses not only the forward but also the backward prediction error. It minimizes the sum of both of them. For long-duration signals, it yields the same results as the autocorrelation method studied.

### 8.4.4 Comparison of Parametric and Classical Methods

At this point it is worthwhile to compare the applicability of classical and parametric AR methods for estimating power spectra.

1. The AR method is definitely superior when the signal being analyzed is appropriately modeled by an AR system. In addition, one is not concerned about frequency resolution.
2. The AR approach is superior for estimating narrowband spectra. If the spectra are smooth, the classical approach is adequate and more reasonable.
3. When additive white-noise components dominate the signal, the classical approach is more accurate. AR spectral estimates are very sensitive to high noise levels.
4. The variance of both methods are comparable when the ratio of maximum autocorrelation lag in the BT method or the segment length in the

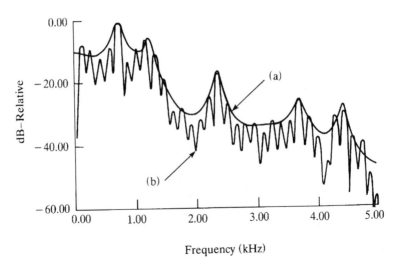

**FIGURE 8.18** The periodogram and AR spectra ($p = 16$) of a speech segment; $N = 256$, $T = 0.1$ ms. [Adapted from Childers, fig. 7, with permission.]

periodogram averaging method are approximately equal to the model order.

There is a major qualitative difference between the two spectral estimation methods that arises because of the difference in the number of parameters being estimated. In the classical approach, the magnitude of the PSD is being calculated at every harmonic frequency. In the parametric approach, only several parameters are being calculated and the spectrum from the model is used. This results in the estimates from the latter approach being much smoother. Figure 8.18 illustrates this quite readily.

There are other methods of spectral estimation based on MA and ARMA signal models, but their usage is not as widespread and the theory and implementation are more complex. Hence these methods are left as topics of advanced study. One important generality is that MA-based methods are good for estimating broadband and narrowband reject spectra.

## REFERENCES

F. Ayres, *Theory and Problems of Matrices*, Schaum Publishing, New York, 1962.

*J. Burg, "A New Analysis Technique for Time Series Data," NATO Advanced Study Institute on Signal Processing with Emphasis on Underwater Acoustics, August, 1968.

J. Cadzow, *Foundations of Digital Signal Processing and Data Analysis*, Macmillan, New York, 1987.

*Reprinted in *Modern Spectral Analysis*, D. Childers, Ed., IEEE Press, New York, 1978.

C. Chen, *One Dimensional Signal Processing*, Marcel Dekker, New York, 1988.

D. Childers, "Digital Spectral Analysis," in *Digital Waveform Processing and Recognition*, C. Chen, Ed., CRC Press, Boca Raton, FL, 1982.

R. Fante, *Signal Analysis and Estimation—An Introduction*, Wiley, New York, 1988.

D. Graupe, *Time Series Analysis, Identification and Adaptive Filtering*, Robert E. Krieger Publishing, Malabar, FL, 1984.

G. Hefftner, W. Zucchini, and G. Jaros, "The Electromyogram (EMG) as a Control Signal for Functional Neuromuscular Stimulation—Part 1: Autoregressive Modeling as a Means of EMG Signature Discrimination," *IEEE Trans. Biomed. Eng.* **35**:230–237 (1988).

B. Jansen, "Analysis of Biomedical Signals by Means of Linear Modeling," *Crit. Rev. in Biomed. Eng.* **12**(4): (1985).

G. Jenkins and D. Watts, *Spectral Analysis and Its Applications*, Holden-Day, San Francisco, 1968.

*M. Kaveh and G. Cooper, "An Empirical Investigation of the Properties of the Autoregressive Spectral Estimator," *IEEE Trans. Inform. Theory* **22**:313–323 (1976).

S. Kay, *Modern Spectral Estimation, Theory and Applications*, Prentice-Hall, Englewood Cliffs, NJ, 1988.

S. Kay and S. Marple, "Spectrum Analysis—A Modern Perspective," *Proc. IEEE* **69**:1380–1419 (1981).

J. Lim and A. Oppenheim, *Advanced Topics in Signal Processing*, Prentice-Hall, Englewood Cliffs, NJ, 1988.

S. Marple, *Digital Spectral Analysis with Applications*, Prentice-Hall, Englewood Cliffs, NJ, 1987.

*H. Newton, "TIMESLAB: A Time Series Analysis Laboratory," Wadsworth, Pacific Grove, CA, 1988.

S. Pandit and S. Wu, *Time Series Analysis and Applications*, Wiley, New York, 1983.

M. Priestley, *Spectral Analysis and Time Series: Volume 1—Univariate Series*, Academic Press, New York, 1981.

*H. Radowski, E. Zawalick, and P. Fougere, "The Superiority of Maximum Entropy Power Spectrum Techniques Applied to Geomagnetic Disturbances," *Phys. Earth Planetary Interiors* **12**:208–216 (1976).

E. Robinson and S. Treital, *Geophysical Analysis*, Prentice-Hall, New York, 1980.

R. Triolo, D. Nash, and G. Moskowitz, "The identification of Time Series Models of Lower Extremity EMG for the Control of Prostheses Using Box-Jenkins Criteria," *IEEE Trans. Biomed. Eng.* **35**:584–594 (1988).

*T. Ulrych and T. Bishop, "Maximum Entropy Spectral Analysis and Autoregressive Decomposition," *Rev. Geophysics and Space Phys.* **13**:183–200 (1975).

D. Veeneman, "Speech Signal Analysis," in *Signal Processing Handbook*, C. Chen, Ed., Dekker, New York, 1988.

W. Wei, *Time Series Analysis, Univariate and Multivariate Methods*, Addison-Wesley, Reading, MA, 1990.

## EXERCISES

**8.1** Prove that minimizing the MSE in equation 8.5 with respect to $h(2)$ yields equation 8.9.

**8.2** Prove the expression for the total squared error of a first-order AR signal model is that given in Example 8.1.

**8.3** Starting with equation 8.19, show that the estimate of the minimum error variance is equation 8.25.

**8.4** Write the Yule-Walker equations for a fourth-order signal model.

**8.5** Modeling the temperature signal in Example 8.3 produced parameter estimates of $a(1) = -0.883$ and $a(2) = 0.285$. Use the unbiased covariance estimator and show that the ACF and parameters estimates change.

**8.6** Derive the solutions for the parameters of a second-order signal model. Prove that they are

$$\hat{a}_1 = \frac{r(1)(r(2) - r(0))}{r^2(0) - r^2(1)}$$

$$\hat{a}_2 = \frac{(r^2(1) - r(0)r(2))}{r^2(0) - r^2(1)}$$

**8.7** Find the second-order model for the first 10 points of the rainfall data in Appendix 8.4.

**8.8** The NACF of an EMG signal is plotted in Figure 8.2.

    a. Find the AR parameters for model orders of 1, 2, and 3; $\hat{\rho}(1) = 0.84$, $\hat{\rho}(2) = 0.5$, $\hat{\rho}(3) = 0.15$.

    b. Assuming that $\hat{R}(0) = 0.04$ volts$^2$, find the MSE for each model. Does the third-order model seem better than the first- or second-order ones?

**8.9** Write the augmented Yule-Walker equations for a first-order AR signal model. Solve for the parameter and squared error in the form of the recursive equations.

**8.10** The Yule-Walker equations can be derived directly from the definition of an all-pole system with a white-noise input. Show this by starting with the system definition and making the autocorrelations

$$E[y(n)y(n-k)] \quad \text{for } 0 \le k \le p$$

**8.11** Consider the second-order model of the temperature signal in Example 8.3.
a. Calculate the residual error sequence.
b. Estimate $\hat{\rho}_\epsilon(1)$ and $\hat{\rho}_\epsilon(2)$.
c. Do these correlation values test as zero?

**8.12** For the grinding wheel signal in Example 8.2, generate the error sequence. Is it white noise?

**8.13** Write the Levinson recursion equations for a fourth-order signal model. Expand the summation in step 2 and write all the equations in step 4. Refer to Section 8.3.4.

**8.14** Redo Exercise 8.8a using the Levinson-Durbin algorithm.

**8.15** In Section A8.2.2 (Appendix 8.2), the recursion relationships for the reflection coefficient and squared error are derived. Perform this derivation for $p = 4$ without partitioning the matrices. That is, keep all of the elements explicit in the manipulations.

**8.16** Derive the recursive relationship for the forward and backward prediction errors in equations 8.51 and 8.52.

**8.17** Prove that the denominator, $D(p-1)$, in the reflection coefficient as defined by step 2 in Burg's algorithm (Section 8.3.5) can be derived in a recursive manner by

$$D(p) = (1 - |\hat{\pi}_{p-1}|^2)D(p-1) - \epsilon_{p-1}^f(p-1)^2 - \epsilon_{p-1}^b(N-1)^2$$

**8.18** What is the PSD of the temperature signal in Example 8.3? Plot it.

**8.19** Generate a first-order AR signal with $a(1) = 0.7$, unit variance white-noise input, a sampling interval of 1, and $N = 100$.
a. What is the theoretical power spectrum?
b. What are $\hat{a}(1)$, the error variance, and the estimated PSD?
c. What are any differences between the theoretical and estimated spectra?

**8.20** For the signal in Exercise 8.19 the model order is estimated to be 3. Estimate the PSD using the Burg algorithm. How does it differ from what is expected?

**8.21** Consider the spectral analysis of the speech signal in Example 8.12.
a. What are the degrees of freedom, number of unknown parameters, in the LPC analysis?

   b. Consider using a BT approach with $M = 10$ and 20 for each epoch. What are the resulting degrees of freedom? What are the frequency spacings?

   c. State which method is preferable and give reasons.

**8.22** A discrete time signal with $T = 1$ has an ACF

$$R(k) = \delta(k) + 5.3\cos(0.3\pi k) + 10.66\cos(0.4\pi k)$$

   a. Plot $R(k)$ for $0 \le k \le 20$.

   b. Find the coefficients and white-noise variance for an AR(3) model. Use the Levinson-Durbin algorithm and perform the calculations manually or write your own software to perform them.

**8.23** An AR(4) model of a biological signal was found to have the parameters $[-0.361 \quad 0.349 \quad 0.212 \quad -0.005]$, $\sigma_y^2 = 0.059$, and $T = 2$ ms. What is its PSD?

## Computer Exercises

**8.24** Generate two white-noise sequences with zero mean and variance 6.55. Use them to drive the first-order model of the grinding wheel signal in Example 8.2. How does each compare visually to the original signal?

**8.25** Use the vibration data from an operating mechanical system listed in Appendix 8.5.

   a. Model this data for $1 \le p \le 20$.

   b. Plot the FPE and $\sigma_\epsilon^2$ vs model order.

   c. What is the best model for this signal?

**8.26** The hospital census data in Appendix 7.6 is modeled well with an ARMA(7,6) system. The data has a mean of 461.5 and $\sigma_\epsilon^2 = 119.4$. The parameters are

$$a(i) = [1 \quad 0.27 \quad 0.24 \quad 0.25 \quad 0.28 \quad 0.28 \quad 0.20 \quad -0.72]$$
$$b(i) = [1 \quad 0.91 \quad 0.98 \quad 1.24 \quad 1.30 \quad 1.53 \quad 1.26]$$

Find a good AR model for this data. Use any model order selection criteria studied.

**8.27** A discrete time signal with $T = 1$ has an ACF

$$R(k) = \delta(k) + 5.33\cos(0.3\pi k) + 10.66\cos(0.4\pi k)$$

   a. Calculate its PSD using an AR(10) model.

   b. Calculate its PSD using a BT approach having a triangular lag window and maximum lag of 10.

   c. What are the differences between these PSDs?

**8.28** Estimate the PSD of the hospital census data in Appendix 7.6 using the YW and Burg algorithms.

   a. What are the orders and residual errors of the chosen AR spectra?

   b. Do either of them make more sense when compared to what is expected from the time series?

## APPENDIX 8.1  MEAN SQUARED ERROR FOR TWO-TERM LINEAR PREDICTION

For the two-term prediction, the MSE or error variance is

$$\text{MSE} = \sigma_\epsilon^2 = E[(y(n) - h(1)y(n-1) - h(2)y(n-2))^2] \qquad (A8.1)$$

This expression is expanded and simplified using the recursive forms for the ACF as

$$R_y(1) - h(1)R_y(0) - h(2)R_y(1) = 0 \qquad (A8.2)$$
$$R_y(2) - h(1)R_y(1) - h(2)R_y(0) = 0 \qquad (A8.3)$$

The term within the square brackets of equation A8.1 is expanded as

$$y(n)y(n) - h(1)y(n-1)y(n) - h(2)y(n-2)y(n)$$
$$- h(1)(y(n)y(n-1) - h(1)y(n-1)y(n-1) - h(2)y(n-2)y(n-1))$$
$$- h(2)(y(n)y(n-2) - h(1)y(n-1)y(n-2) - h(2)y(n-2)y(n-2)) \qquad (A8.4)$$

Examine the sum of terms within the brackets of lines 2 and 3 in equation A8.4. If expectations are made, they become identical to equations A8.2 and A8.3 and equal zero. Thus the error variance is the expectation of line 1 and is

$$\sigma_\epsilon^2 = R_y(0) - h(1)R_y(1) - h(2)R_y(2) \qquad (A8.5)$$

## APPENDIX 8.2  MATRIX FORM OF LEVINSON-DURBIN RECURSION

### A8.2.1  General Coefficients

The recursion form can be developed from the matrix representation of the Yule-Walker equations. The general form is

$$
\begin{bmatrix}
c(0) & c(1) & \cdots & c(p-1) \\
c(1) & c(0) & \cdots & c(p-2) \\
\vdots & \vdots & \vdots & \vdots \\
c(p-1) & c(p-2) & \cdots & c(0)
\end{bmatrix}
\begin{bmatrix}
\hat{a}(1) \\
\hat{a}(2) \\
\vdots \\
\hat{a}(p)
\end{bmatrix}
=
\begin{bmatrix}
-c(1) \\
-c(2) \\
\vdots \\
-c(p)
\end{bmatrix}
\qquad (A8.6)
$$

where $c(k)$ is the sample autocovariance function. Using bold letters to represent matrices and vectors,

$$\mathbf{C_p a_p}(p) = -\mathbf{c_p} \tag{A8.7}$$

where the subscript denotes the order of the square matrix and vector,

$$\mathbf{a_p}(m) = [\hat{a}_p(1), \hat{a}_p(2), \dots, \hat{a}_p(m)]'$$

and

$$\boldsymbol{\alpha_p}(m) = [\hat{a}_p(m), \hat{a}_p(m-1), \dots, \hat{a}_p(1)]' \tag{A8.8}$$

Begin with the solution to the first-order model from Section 8.3.1. That is,

$$\hat{a}_1(1) = -\frac{c(1)}{c(0)}, \qquad \sigma_{\epsilon,1}^2 = c(0)(1 - |\hat{a}_1(1)|^2) \tag{A8.9}$$

For a second-order system,

$$\begin{bmatrix} c(0) & c(1) \\ c(1) & c(0) \end{bmatrix} \begin{bmatrix} \hat{a}_2(1) \\ \hat{a}_2(2) \end{bmatrix} = \begin{bmatrix} -c(1) \\ -c(2) \end{bmatrix} \tag{A8.10}$$

Using the first equation of equation A8.10 and equation A8.9 yields

$$\hat{a}_2(1) = \hat{a}_1(1) - \hat{a}_2(2)\frac{c(1)}{c(0)} = \hat{a}_1(1) + \hat{a}_2(2)\hat{a}_1(1) \tag{A8.11}$$

For a third-order and larger model, the same general procedure is utilized and implemented through matrix partitioning. The goal of the partitioning is to isolate the $(p, p)$ element of $\mathbf{C_p}$. Thus for a third-order model,

$$\begin{bmatrix} c(0) & c(1) & c(2) \\ c(1) & c(0) & c(1) \\ c(2) & c(1) & c(0) \end{bmatrix} \begin{bmatrix} \hat{a}_3(1) \\ \hat{a}_3(2) \\ \hat{a}_3(3) \end{bmatrix} = \begin{bmatrix} -c(1) \\ -c(2) \\ -c(3) \end{bmatrix} \tag{A8.12}$$

becomes

$$\begin{bmatrix} \mathbf{C_2} & \boldsymbol{\psi_2} \\ \boldsymbol{\psi_2'} & c(0) \end{bmatrix} \begin{bmatrix} \mathbf{a}_3(2) \\ \hat{a}_3(3) \end{bmatrix} = \begin{bmatrix} -\mathbf{c_2} \\ -c(3) \end{bmatrix} \tag{A8.13}$$

where

$$\boldsymbol{\psi_p} = [c(p), c(p-1), \dots, c(1)]' \tag{A8.14}$$

Now, solving for $\mathbf{a}_3(2)$ yields

$$\mathbf{a}_3(2) = -\mathbf{C_2^{-1}}\boldsymbol{\psi_2}\hat{a}_3(3) - \mathbf{C_2^{-1}}\mathbf{c_2} \tag{A8.15}$$

From the second-order model it is known that $\mathbf{C_2 a_2}(2) = -\mathbf{c_2}$. Thus

$$\mathbf{a}_2(2) = -\mathbf{C_2^{-1}}\mathbf{c_2} \tag{A8.16}$$

Because of the property of Toeplitz matrices, $\mathbf{C_p \alpha_p}(p) = -\mathbf{\psi_p}$ and

$$\alpha_2(2) = -\mathbf{C_2^{-1} \psi_2} \tag{A8.17}$$

Now equation A8.15 can be written in recursive form as

$$\mathbf{a_3}(2) = \mathbf{a_2}(2) + \hat{a}_3(3)\alpha_2(2) \tag{A8.18}$$

Notice that this procedure yields all but the reflection coefficient. The recursive relationship for this parameter and the error variance will be developed in the next section.

The general recursive rule for the reflection coefficient is also found by first partitioning the $p \times p$ covariance matrix to isolate the $\hat{a}_p(p)$ parameter. Equation A8.6 is written as

$$\begin{bmatrix} \mathbf{C_{p-1}} & \mathbf{\psi_{p-1}} \\ \mathbf{\psi'_{p-1}} & c(0) \end{bmatrix} \begin{bmatrix} \mathbf{a_p(p-1)} \\ \hat{a}_p(p) \end{bmatrix} = \begin{bmatrix} -\mathbf{c_{p-1}} \\ -c(p) \end{bmatrix} \tag{A8.19}$$

The first equation is solved for $\mathbf{a_p}(p-1)$ and is

$$\mathbf{a_p}(p-1) = -\mathbf{C_{p-1}^{-1}\psi_{p-1}}\hat{a}_p(p) - \mathbf{C_{p-1}^{-1}c_{p-1}} \tag{A8.20}$$

Since

$$\alpha_{p-1}(p-1) = -\mathbf{C_{p-1}^{-1}\psi_{p-1}} \quad \text{and} \quad \mathbf{a_{p-1}}(p-1) = -\mathbf{C_{p-1}^{-1}c_{p-1}} \tag{A8.21}$$

substituting the equations A8.21 into equation A8.20 yields

$$\mathbf{a_p}(p-1) = \mathbf{a_{p-1}}(p-1) + \hat{a}_p(p)\alpha_{p-1}(p-1) \tag{A8.22}$$

Thus the first $p-1$ parameters of the model for order $p$ are found.

## A8.2.2 Reflection Coefficient and Variance

The reflection coefficient and the squared error are found using the augmented Yule-Walker equations. These are

$$\begin{bmatrix} c(0) & c(1) & \cdots & c(p) \\ c(1) & c(0) & \cdots & c(p-1) \\ \vdots & \vdots & \vdots & \vdots \\ c(p) & c(p-1) & \cdots & c(0) \end{bmatrix} \begin{bmatrix} 1 \\ \hat{a}_p(1) \\ \vdots \\ \hat{a}_p(p) \end{bmatrix} = \begin{bmatrix} \sigma_{\epsilon,p}^2 \\ 0 \\ \vdots \\ 0 \end{bmatrix} \tag{A8.23}$$

The $\mathbf{a_p}(p+1)$ vector is expanded using equation A8.22 and becomes

$$\mathbf{a_p}(p+1) = \begin{bmatrix} 1 \\ \hat{a}_p(1) \\ \vdots \\ \hat{a}_p(p) \end{bmatrix} = \begin{bmatrix} 1 \\ \hat{a}_{p-1}(1) \\ \vdots \\ 0 \end{bmatrix} + \hat{a}_p(p) \begin{bmatrix} 0 \\ \hat{a}_{p-1}(p-1) \\ \vdots \\ 1 \end{bmatrix}$$

or

$$\mathbf{a_p}(p+1) = \begin{bmatrix} 1 \\ \mathbf{a_{p-1}}(p-1) \\ 0 \end{bmatrix} + \hat{a}_p(p) \begin{bmatrix} 0 \\ \boldsymbol{\alpha_{p-1}}(p-1) \\ 1 \end{bmatrix} \qquad (A8.24)$$

The $\mathbf{C_{p+1}}$ covariance matrix is partitioned such that multiplication by equation A8.24 can be accomplished. For the first vector,

$$T_1 = \begin{bmatrix} c(0) & \mathbf{c'_{p-1}} & c(p) \\ \mathbf{c_{p-1}} & \mathbf{C_{p-1}} & \boldsymbol{\psi_{p-1}} \\ c(p) & \boldsymbol{\psi'_{p-1}} & c(0) \end{bmatrix} \begin{bmatrix} 1 \\ \mathbf{a_{p-1}}(p-1) \\ 0 \end{bmatrix} = \begin{bmatrix} \sigma^2_{\epsilon,p-1} \\ \boldsymbol{\Delta_2} \\ \Delta_3 \end{bmatrix} \qquad (A8.25)$$

For the second vector,

$$T_2 = \begin{bmatrix} c(0) & \mathbf{c'_{p-1}} & c(p) \\ \mathbf{c_{p-1}} & \mathbf{C_{p-1}} & \boldsymbol{\psi_{p-1}} \\ c(p) & \boldsymbol{\psi'_{p-1}} & c(0) \end{bmatrix} \begin{bmatrix} 0 \\ \boldsymbol{\alpha_{p-1}}(p-1) \\ 1 \end{bmatrix} = \begin{bmatrix} \Delta_3 \\ \boldsymbol{\Delta_2} \\ \sigma^2_{\epsilon,p-1} \end{bmatrix} \qquad (A8.26)$$

Combining the last three equations with equation A8.23 produces

$$\begin{bmatrix} \sigma^2_{\epsilon,p} \\ \mathbf{0_{p-1}} \\ 0 \end{bmatrix} = \begin{bmatrix} \sigma^2_{\epsilon,p-1} \\ \boldsymbol{\Delta_2} \\ \Delta_3 \end{bmatrix} + \hat{a}_p(p) \begin{bmatrix} \Delta_3 \\ \boldsymbol{\Delta_2} \\ \sigma^2_{\epsilon,p-1} \end{bmatrix} \qquad (A8.27)$$

The reflection coefficient is found using the $p + 1$ equation from matrix equation A8.27:

$$\begin{aligned} \hat{a}_p(p) &= -\frac{\Delta_3}{\sigma^2_{\epsilon,p-1}} \\ &= -\frac{c(p) + \mathbf{c_{p-1}} \cdot \boldsymbol{\alpha_{p-1}}(p-1)}{\sigma^2_{\epsilon,p-1}} \\ &= -\frac{1}{\sigma^2_{\epsilon,p-1}} \sum_{i=0}^{p-1} c(p-i-1)\hat{a}_{p-1}(i) \end{aligned} \qquad (A8.28)$$

The recursion equation for the squared error is found using the first equation of matrix equation A8.27 and equation A8.28.

$$\sigma^2_{\epsilon,p} = \sigma^2_{\epsilon,p-1} + \hat{a}_p(p)\Delta_3 = \sigma^2_{\epsilon-1}(1 - |\hat{a}_p|^2) \qquad (A8.29)$$

## APPENDIX 8.3   SURFACE ROUGHNESS FOR A GRINDING WHEEL

**Height $\times$ $10^3$ inches, sampling interval $= 0.002$ inch, read across rows**

| | | | | | | | | | |
|---|---|---|---|---|---|---|---|---|---|
| 13.5 | 4.0 | 4.0 | 4.5 | 3.0 | 3.0 | 10.0 | 10.2 | 9.0 | 10.0 |
| 8.5 | 7.0 | 10.5 | 7.5 | 7.0 | 10.5 | 9.5 | 7.0 | 12.0 | 13.5 |
| 12.5 | 15.0 | 13.0 | 11.0 | 9.0 | 10.5 | 10.5 | 11.0 | 10.5 | 9.0 |
| 8.2 | 8.5 | 9.2 | 8.5 | 10.0 | 14.5 | 13.0 | 2.0 | 6.0 | 6.0 |
| 11.0 | 9.5 | 12.5 | 13.8 | 12.0 | 12.0 | 12.0 | 13.0 | 12.0 | 14.0 |
| 14.5 | 13.5 | 12.3 | 7.0 | 7.0 | 7.0 | 6.5 | 12.5 | 15.0 | 12.5 |
| 11.6 | 11.0 | 10.0 | 8.5 | 3.0 | 11.5 | 11.5 | 11.5 | 11.0 | 9.0 |
| 2.5 | 7.0 | 6.0 | 6.6 | 14.0 | 11.0 | 9.0 | 6.5 | 4.0 | 6.0 |
| 12.0 | 11.0 | 12.0 | 12.5 | 12.5 | 13.6 | 13.0 | 8.0 | 6.5 | 6.8 |
| 6.0 | 7.2 | 10.2 | 8.0 | 7.5 | 11.0 | 11.8 | 11.8 | 6.5 | 8.0 |
| 9.0 | 8.0 | 8.0 | 9.0 | 9.5 | 10.0 | 9.0 | 12.0 | 13.5 | 13.8 |
| 15.0 | 12.5 | 11.0 | 11.5 | 14.5 | 11.5 | 11.8 | 13.0 | 15.0 | 14.5 |
| 13.0 | 9.0 | 11.0 | 9.0 | 10.0 | 14.0 | 13.5 | 3.0 | 2.2 | 6.0 |
| 8.0 | 9.0 | 9.0 | 9.0 | 7.0 | 6.0 | 6.5 | 7.0 | 7.5 | 8.5 |
| 9.0 | 9.5 | 10.0 | 11.5 | 11.2 | 12.5 | 11.6 | 8.0 | 7.0 | 6.0 |
| 6.0 | 6.0 | 9.0 | 12.0 | 13.5 | 13.0 | 3.5 | 1.8 | 1.6 | 7.5 |
| 8.0 | 7.9 | 11.6 | 12.5 | 10.5 | 8.0 | 9.0 | 11.6 | 11.8 | 12.6 |
| 10.2 | 10.0 | 5.0 | 7.0 | −1.0 | 0.0 | 0.0 | 3.0 | 11.0 | 12.0 |
| 12.2 | 11.0 | 8.0 | 7.0 | 5.5 | 10.0 | 11.5 | 7.0 | 4.0 | 7.0 |
| 7.0 | 10.0 | 9.0 | 8.0 | 10.0 | 13.0 | 10.0 | 6.5 | 11.0 | 13.0 |
| 13.0 | 14.0 | 13.0 | 12.5 | 12.0 | 9.0 | 8.5 | 7.0 | 8.5 | 10.0 |
| 8.0 | 4.0 | 3.0 | 10.0 | 13.0 | 13.0 | 13.0 | 12.5 | 11.0 | 11.0 |
| 11.0 | 14.5 | 14.0 | 14.0 | 13.5 | 10.0 | 9.5 | 10.0 | 12.5 | 10.0 |
| 9.0 | 9.0 | 4.0 | 3.0 | 6.0 | 5.0 | 7.0 | 6.0 | 5.0 | 8.5 |
| 10.5 | 11.1 | 11.0 | 10.0 | 11.2 | 8.0 | 2.5 | 5.0 | 13.2 | 14.0 |

*Source*: Pandit and Wu, 1983.

## APPENDIX 8.4  AVERAGE ANNUAL RAINFALL IN EASTERN UNITED STATES

Time across rows

| | | | | | | | | | |
|---|---|---|---|---|---|---|---|---|---|
| 22.54 | −17.46 | −26.46 | 11.54 | 6.54 | −26.46 | 9.54 | −1.46 | −33.46 | −14.46 |
| 6.54 | −9.46 | 6.54 | −1.46 | −1.46 | .54 | −6.46 | −15.46 | −5.46 | 4.54 |
| −7.46 | −2.46 | −6.46 | −2.46 | 6.54 | 1.54 | 7.54 | −9.46 | −6.46 | 9.54 |
| 10.54 | −2.46 | −.46 | 14.54 | −13.46 | 1.54 | −4.46 | −5.46 | .54 | −13.46 |
| 5.54 | 6.54 | 14.54 | −7.46 | −3.46 | 3.54 | 3.54 | −14.46 | 16.54 | −.46 |
| 2.54 | 12.54 | 6.54 | 13.54 | 1.54 | −.46 | 16.54 | 5.54 | 7.54 | 15.54 |
| 7.54 | 15.54 | −2.46 | 4.54 | 3.54 | 6.54 | 9.54 | 6.54 | 3.54 | 2.54 |
| −2.46 | 7.54 | 8.54 | 10.54 | .54 | .54 | 3.54 | −6.46 | −12.46 | −5.46 |
| 3.54 | 5.54 | −8.46 | −7.46 | −4.46 | .54 | 3.54 | −9.46 | 3.54 | 1.54 |
| −0.46 | −12.46 | −5.46 | −11.46 | 4.54 | 5.54 | −3.46 | −9.46 | 6.54 | −6.46 |
| −11.46 | −9.46 | 9.54 | 3.54 | −2.46 | .54 | | | | |

*Source*: Newton, 1988.

## APPENDIX 8.5  VIBRATION IN A MECHANICAL SYSTEM

sampling interval = 0.02 seconds, read across rows

| | | | | | | | | | |
|---|---|---|---|---|---|---|---|---|---|
| 30.0 | 28.0 | 25.0 | 24.0 | 23.0 | 21.0 | 20.0 | 22.0 | 24.0 | 27.0 |
| 30.0 | 31.0 | 34.0 | 37.0 | 33.0 | 28.0 | 25.0 | 23.0 | 21.0 | 19.0 |
| 18.0 | 17.0 | 16.0 | 17.0 | 18.5 | 22.0 | 29.0 | 32.0 | 32.0 | 30.0 |
| 25.0 | 20.0 | 17.0 | 14.0 | 13.0 | 17.0 | 22.0 | 27.0 | 33.0 | 30.0 |
| 21.0 | 15.0 | 12.0 | 10.0 | 9.0 | 6.0 | 6.0 | 8.0 | 10.0 | 12.0 |
| 15.0 | 16.0 | 16.5 | 18.0 | 21.0 | 15.0 | 7.0 | 4.0 | 3.0 | 7.0 |
| 15.5 | 22.0 | 30.0 | 40.0 | 40.0 | 39.0 | 38.0 | 35.0 | 30.0 | 25.0 |
| 20.0 | 18.0 | 20.0 | 22.5 | 27.0 | 32.0 | 32.5 | 33.0 | 32.0 | 30.0 |
| 25.5 | 23.3 | 23.3 | 24.0 | 27.0 | 31.5 | 35.0 | 36.0 | 34.0 | 30.5 |
| 29.0 | 25.0 | 20.0 | 19.0 | 21.0 | 23.5 | 28.0 | 33.5 | 36.0 | 37.5 |
| 38.0 | 36.0 | 33.0 | 29.5 | 28.0 | 28.0 | 30.0 | 30.5 | 30.0 | 30.0 |
| 28.0 | 25.0 | 23.0 | 24.5 | 27.0 | 31.0 | 34.0 | 33.0 | 25.0 | 16.0 |
| 13.0 | 14.0 | 17.0 | 22.0 | 29.0 | 32.0 | 30.0 | 26.0 | 24.0 | 24.0 |

*Source:* Data from H.J. Stedudel. "A Time Series Approach to Modeling Second Order Mechanical Systems,"M.S. Thesis, Univ. Of Wisconsin-Milwaukee, 1971.

# IX

## THEORY AND APPLICATION OF CROSS CORRELATION AND COHERENCE

### 9.1 INTRODUCTION

The concept of cross correlation was introduced in Chapter 6. It was defined in the context of linear systems and described a relationship between the input and output of a system. In general, a relationship can exist between any two signals whether or not they are intrinsically related through a system. For instance, we can understand that two signals such as the temperatures in two cities or the noise in electromagnetic transmissions and atmospheric disturbances can be related; however, their relationship is not well defined. An example is demonstrated in Figure 9.1, in which the time sequence of speed measurements made at different locations on a highway are plotted. It is apparent that these signals have similar shapes and that some, such as $V_1$ and $V_2$, are almost identical except for a time shift. How identical these signals are and which time shift produces the greatest similarity can be ascertained quantitatively with cross correlation functions. This same information can be ascertained in the frequency domain with cross spectral density and coherence functions. In addition, the latter functions can be used to ascertain synchronism in frequency components between two signals. Such functions will be described in the second half of this chapter.

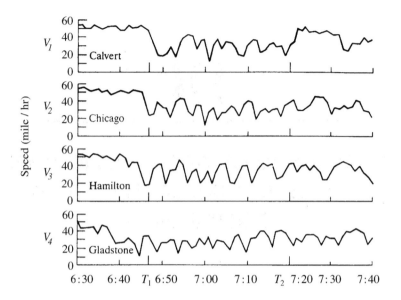

**FIGURE 9.1**   Records of average speeds at four different locations on a highway. Averages are computed over 1-minute intervals. [Adapted from Schwartz and Shaw, fig. 4.7, with permission.]

There are several comprehensive references treating cross correlation and coherence. Good treatments with applications are Bendat and Piersol (1980) and Jenkins and Watts (1968). References that focus on other developments and give detailed procedures are Carter (1988), Silvia (1987), and the special issue of *IEEE Transactions on Acoustics, Speech, and Signal Processing* in 1981.

The cross correlation function (CCF) is defined as

$$R_{yx}(k) = E[y(n)x(n+k)] \tag{9.1}$$

where $\tau_d = kT$ is the amount of time that the signal $x(n)$ is delayed with respect to $y(n)$. As with the definition of the autocorrelation functions, there are several forms. The cross covariance function (CCVF) is defined as

$$C_{yx}(k) = E[(y(n) - m_y)(x(n+k) - m_x)] = R_{yx}(k) - m_y m_x \tag{9.2}$$

and the normalized cross correlation function (NCCF) is

$$\rho_{yx}(k) = \frac{C_{yx}(k)}{\sigma_y \sigma_x} \tag{9.3}$$

We will explain the information in these functions in the context of the highway speed signals. Figure 9.2 shows the NCCF between each pair of locations.

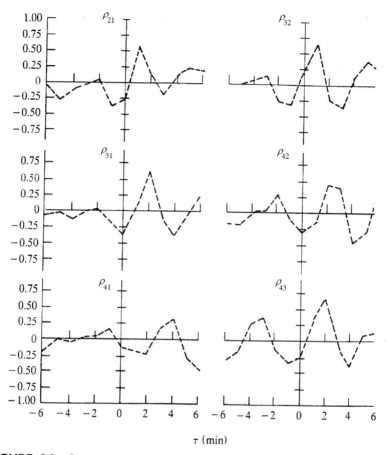

**FIGURE 9.2** Cross correlations of speed records in Figure 9.1. [Adapted from Schwartz and Shaw, fig. 4.8, with permission.]

Examination of $\rho_{21}(k)$ reveals that at a lag time of 1 minute, the NCCF achieves a maximum magnitude of 0.55. Thus the signals are most similar at this time difference and the correlation coefficient at that delay is 0.55. A realistic interpretation is that this time delay equals the time necessary to travel between locations 2 and 1. The other NCCF give the same information between the other locations.

This same type of information is needed in many other applications. Some other major applications of this measure include: the estimation of time delays to estimate ranges and bearings in radar and sonar systems, path length determination in multipath environments such as sound studios and auditoriums, and identification of systems.

## 9.2 PROPERTIES OF CROSS CORRELATION FUNCTIONS

### 9.2.1 Theoretical Function

The cross correlation function has symmetry properties that depend on the ordering of the two signals. Our intuition informs us that signals can be aligned by either shifting $y(n)$ in one direction or shifting $x(n)$ in the opposite direction. Then $R_{yx}(k) = E[y(n)x(n+k)] = E[x(n+k)y(n)]$, which by the notation is simply $R_{xy}(-k)$. Thus

$$R_{yx}(k) = R_{xy}(-k) \tag{9.4}$$

The CCF also has magnitude boundaries. Start with the inequality

$$E[(ay(n) + x(n+k))^2] \geq 0 \tag{9.5}$$

Expanding the square and taking expectations yields

$$a^2 R_{yy}(0) + 2a R_{yx}(k) + R_{xx}(0) \geq 0 \tag{9.6}$$

Solving equation 9.6 for the unknown variable, $a$, will produce complex roots since the equation is nonnegative. The discriminant then is negative and

$$4R_{yx}^2(k) - 4R_{yy}(0)R_{xx}(0) \leq 0$$

or

$$R_{yx}^2(k) \leq R_{yy}(0)R_{xx}(0) \tag{9.7}$$

Similarly, one could use the cross covariances and show that

$$C_{yx}^2(k) \leq C_{yy}(0)C_{xx}(0) \tag{9.8}$$

Another useful inequality whose proof is based on a modification of equation 9.5 is

$$|R_{yx}(k)| \leq \frac{1}{2}(R_{yy}(0) + R_{xx}(0)) \tag{9.9}$$

Its derivation is left as an exercise. Since the NCCF is itself a correlation coefficient,

$$-1 \leq \rho_{yx}(k) \leq 1 \tag{9.10}$$

## 9.2.2 Estimators

The set of cross correlation functions has estimators that are analogous to the ones for the set of autocorrelation functions. For two signals containing $N$ points and indexed over $0 \leq n \leq N - 1$, the sample CCVF is

$$\hat{C}_{yx}(k) = \begin{cases} \dfrac{1}{N} \displaystyle\sum_{n=0}^{N-k-1} (y(n) - \hat{m}_y)(x(n + k) - \hat{m}_x), & k \geq 0 \\ \dfrac{1}{N} \displaystyle\sum_{n=0}^{N-k-1} (y(n + k) - \hat{m}_y)(x(n) - \hat{m}_x), & k \leq 0 \end{cases} \qquad (9.11)$$

The sample means are defined conventionally. In fact, since the data must always be detrended first, the sample cross covariance function is equal to the sample cross correlation function, or $\hat{R}_{yx}(k) = \hat{C}_{yx}(k)$. Another symbol for $\hat{C}_{yx}(k)$ is $c_{yx}(k)$. The sample NCCF is

$$\hat{\rho}_{yx}(k) = \frac{c_{yx}(k)}{s_y s_x} \qquad (9.12)$$

Using equation 9.11 and assuming the signals have been detrended, the mean of the estimator is

$$E[c_{yx}(k)] = \frac{1}{N} \sum_{n=0}^{N-k-1} E[y(n)x(n + k)] = \frac{N - k}{N} C_{yx}(k) \quad \text{for } k \geq 0$$

The same result exists for $k \leq 0$; thus

$$E[c_{yx}(k)] = \left(1 - \frac{|k|}{N}\right) C_{yx}(k) \qquad (9.13)$$

and the estimator is biased. The derivation of the covariance for the CCVF is quite involved and similar to that for the autocorrelation function. Assuming that $x(n)$ and $y(n)$ have Gaussian distributions, the covariance is

$$\text{Cov}[c_{yx}(k)c_{yx}(l)] \approx \frac{1}{N} \sum_{r=-\infty}^{\infty} (C_{yy}(r)C_{xx}(r + l - k) + C_{yx}(r + l)C_{xy}(r - k)) \qquad (9.14)$$

and indicates that, in general, the magnitudes of CCVF estimates at different lags are correlated themselves. The exact manner is highly dependent on the inherent correlational properties of the signals themselves and thus difficult to determine without resorting to some modeling. Letting $k = l$, the variance of the estimator is

$$\text{Var}[c_{yx}(k)] \approx \frac{1}{N} \sum_{r=-\infty}^{\infty} (C_{yy}(r)C_{xx}(r) + C_{yx}(r + k)C_{xy}(r - k)) \qquad (9.15)$$

and is consistent. The variance and covariance expressions for the NCCF are much more complex (refer to Box and Jenkins (1976) for details). Another significant aspect of equation 9.15 is that the variance is dependent not only on the CCVF between both processes but also on the individual ACVFs. If both processes are uncorrelated and are white noise, then obviously

$$\mathrm{Var}[c_{yx}(k)] = \frac{\sigma_y^2 \sigma_x^2}{N} \tag{9.16}$$

Given the procedure for testing for significant correlations in a NCCF, it is tempting to conjecture that this measure could provide a basis for testing the amount of correlation between two signals. However, since the ACVFs affect this variance, caution must be used. Consider two first-order AR processes $x(n)$ and $y(n)$ with parameters $\alpha$ and $\beta$, respectively; then equation 9.15 yields

$$\mathrm{Var}[c_{yx}(k)] = \left(\frac{\sigma_y^2 \sigma_x^2}{N}\right)\left(\frac{1 + \alpha\beta}{1 - \alpha\beta}\right) \tag{9.17}$$

This means that the individual signals must be modeled first and then the estimated parameters must be used in the variance estimation.

---

**EXAMPLE 9.1**

The effect of signal structure on the variance of the estimate of CCVF is illustrated by generating two first-order AR processes with $a_y(1) = 0.9$, $a_x(1) = 0.7$, $\sigma_\epsilon^2 = T = 1$, and $N = 100$. The magnitudes of $C_{yx}(k)$ and $\rho_{yx}(k)$ are estimated and $\hat{\rho}_{yx}(k)$ is plotted in Figure 9.3a. The values should be close to zero and be within the bounds $\pm 1.96/\sqrt{N}$; however, they are not. Because of the large estimation variance, knowledge of the signal parameters is not helpful. Another approach is to remove the structure within these signals by modeling them and generating their error processes, $\epsilon_x(n)$ and $\epsilon_y(n)$. The amount of information about signal $x(n)$ within signal $y(n)$ has not changed. Now the NCCF of the error signals is estimated and plotted in Figure 9.3b. The magnitudes are much less and within the expected boundaries.

To show this effect, $y(n)$ and $\epsilon_y(n)$ are plotted in Figure 9.4. Notice that the negative correlation present in the process is absent in the error process.

---

Based on the theory of sampling, for proper testing of the significance of the cross correlation between signals, each component signal must be modeled and the variance expression in equation 9.15 must be derived. This expression cannot only be quite complex and difficult to derive but also results in a uselessly large sample variance. A simpler approach is to duplicate the general procedure used in Example 9.1. First model the signals and estimate the CCVF of the error sequences that are generated. Then test this estimate to determine the presence of any correlation between the signals. Referring to Chapter 8, this generation of the error sequence is essentially filtering the signals with an MA system and the process is called *prewhitening*.

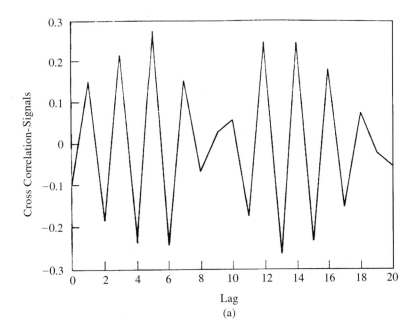

(a)

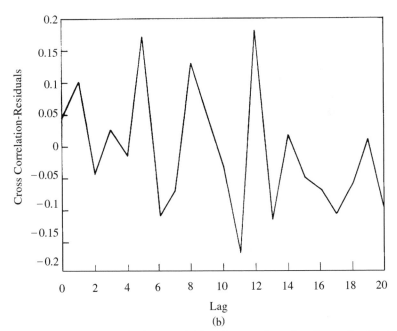

(b)

**FIGURE 9.3** The estimate of the NCCF for (a) two independent first-order AR signals and (b) their error sequences.

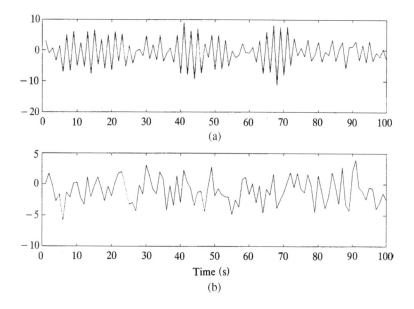

**FIGURE 9.4**    The time series for an AR(1) process with $a(1) = 0.9$. (a) Sample function, $N = 100$. (b) Error process.

## EXAMPLE 9.2

A classical correlated signal set that is analyzed in many texts is the gas furnace data that are plotted in Figure 9.5 and listed in Appendix 9.1 with $T = 9.0$ seconds. The normalized correlation function is directly estimated and plotted in Figure 9.6a. The NCCF peaks at a lag of 5 time units—45 seconds—with a magnitude near $-1$. Thus we conclude that the output strongly resembles the negative of the input and is delayed by 45 seconds. Examination of the signals would strongly suggest this. To test the significance, the prewhitening procedure must be used. The input and output are prewhitened after being modeled as sixth- and fourth-order AR signals with the parameter sets

$$\alpha_x = [1 \quad -1.93 \quad 1.20 \quad -0.19 \quad 0.13 \quad -0.27 \quad 0.11]$$
$$\alpha_y = [1 \quad -1.85 \quad 0.84 \quad 0.31 \quad -0.26]$$

respectively. The NCCF of the error sequences is plotted in Figure 9.6b along with the 95% confidence limits. It is observed that the signals are significantly correlated at lags 3 through 6, with the maximum occurring at lag 5. Thus our initial judgment is verified.

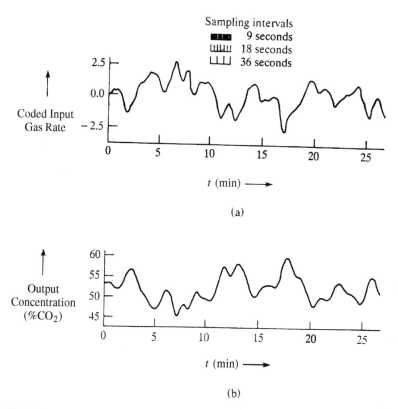

**FIGURE 9.5** Signals from a gas furnace: (a) Input gas, $ft^3/min$. (b) Output CO2, % concentration. [Adapted from Box and Jenkins, fig. 11.1, with permission.]

## 9.3 DETECTION OF TIME-LIMITED SIGNALS

A major implementation of cross correlation is for detecting the occurrence of time-limited signals. These signals can have a random or deterministic nature. One set of applications are those that seek to detect the repeated occurrences of a specific waveform. For instance, it is often necessary to know when the waveforms indicating epileptic seizures in the EEG occur. In another set of applications, it is necessary to estimate the time delay between a transmitted and received waveform. These are usually in man-made systems, so the shape of the transmitted waveform is designed to be least affected by the transmitting medium. From the time delay and knowledge of the speed of transmission in the medium, the distance of the propagation path is then calculated. In radar systems, the waveforms are transmitted electromagnetically through the atmosphere; in sonar and ultrasonic imaging systems, the waveforms are transmitted with sonic energy

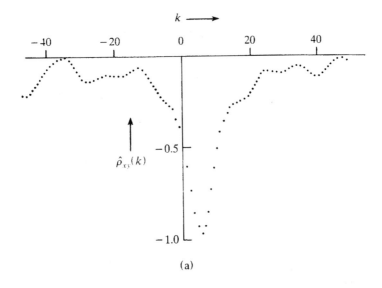

(a)

Cross Correlations and Confidence Bands

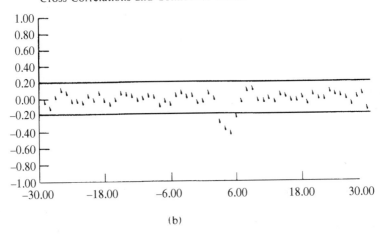

(b)

**FIGURE 9.6**   Estimates of the NCCF for the gas furnace signals: (a) Direct estimate. (b) Estimate after prewhitening. [Adapted from Box and Jenkins, fig. 11.4, and Newton, fig. 4.4, with permission.]

through a fluid medium. In many instances, there are multiple pathways for transmission that need to be known because they contribute to confounding or distorting the desired received signal. For example, in audio systems the multiple paths of reflections need to be known to avoid echoes. Figure 9.7 shows a schematic of an experiment to determine multiple pathways. Obviously, it is desired to only have the direct path in auditorium or sound studio. Before

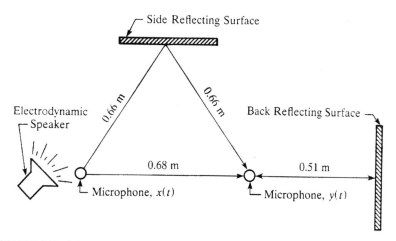

**FIGURE 9.7**   Schematic of setup for multiple-path acoustic experiment. [Adapted from Bendat and Piersol, fig. 6.2, with permission.]

expounding in detail it is necessary to formulate the implementation of the cross correlation.

### 9.3.1   Basic Concepts

Assume a ranging system is designed to generate and transmit a square pulse, $x(t)$, of the amplitude and duration shown in Figure 9.8a. For now assume the environment is ideal; that is, the received pulse has exactly the same shape and amplitude as the transmitted pulse and is only shifted in time, $x(t - \gamma)$. The exact time of arrival, $\gamma = dT$, is obtained by cross correlating a template of the transmitted signal with the received signal, $y(t) = x(t - \tau_d)$, as in Figure 9.8b. Notice that in this situation both the transmitted and received signals are deterministic. Writing this in continuous time gives

$$R_{xy}(\tau) = \frac{1}{W} \int_0^W x(t)y(t + \tau) \, dt$$

$$= \frac{1}{W} \int_0^W x(t)x(t + \tau - \tau_d) \, dt$$

$$= R_{xx}(\tau - \tau_d) \tag{9.18}$$

and is plotted in Figure 9.8c. Note that the integration is confined to the duration of the pulse. This is necessary to standardize the magnitude of $R_{xx}(\tau)$. The

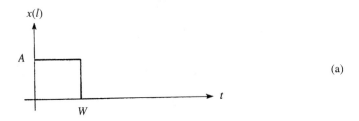

(a)

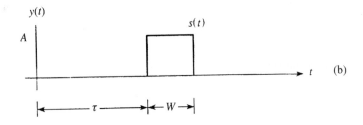

(b)

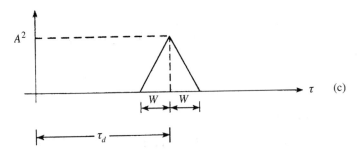

(c)

**FIGURE 9.8** Ideal detection situation: (a) Transmitted pulse. (b) Received pulse. (c) Cross correlation between (a) and (b).

discrete time counterpart for $W = MT$ and $\tau_d = dT$ is

$$R_{xy}(k) = \frac{1}{M} \sum_{n=0}^{M-1} x(n)y(n+k)$$

$$= \frac{1}{M} \sum_{n=0}^{M-1} x(n)x(n+k-d)$$

$$= R_{xx}(k-d), \quad 0 \le k \le N - M \qquad (9.19)$$

The cross correlation peaks at the delay time and its shape is a triangular pulse. The CCF between the transmitted and received signals is the ACF of the pulse shifted by the delay time.

Now proceed to a more realistic situation with a *lossy* medium and a *noisy* environment. The received signal has a reduction in amplitude, loss coefficient $g$, and is contaminated by additive white noise, $\eta(n)$; that is,

$$y(n) = gx(n - d) + \eta(n), \quad |g| \leq 1, \quad \text{Var}[\eta(n)] = \sigma_\eta^2 \qquad (9.20)$$

A random signal is involved and statistical moments must be used:

$$R_{xy}(k) = \text{E}[x(n)y(n + k)] = g\text{E}[x(n)x(n - d + k)] + \text{E}[x(n)\eta(n + k)] \qquad (9.21)$$

Assuming that the signal and noise are uncorrelated,

$$\text{E}[x(n)\eta(n + k)] = R_{x\eta}(k) = m_x m_\eta = 0,$$

and the result is the same as the ideal situation, equation 9.19, except for an attenuation factor. For a multiple-path environment with $q$ paths, the received signal is

$$y(n) = \sum_{i=1}^{q} g_i x(n - d_i) + \eta(n) \qquad (9.22)$$

Its cross correlation with the transmitted signal is

$$R_{xy}(k) = \text{E}[x(n)y(n + k)] = \sum_{i=1}^{q} g_i \text{E}[x(n)x(n - d_i + k)] + \text{E}[x(n)\eta(n + k)]$$

$$= \sum_{i=1}^{q} g_i R_{xx}(k - d_i) \qquad (9.23)$$

and contains multiple peaks of different heights, $A^2 g_i$.

## 9.3.2 Application of Pulse Detection

The measurement of the fetal electrocardiogram (FECG) is important for monitoring the status of the fetus during parturition. The electrodes are placed on the abdomen of the mother, and the resulting measurement is a summation of the FECG and the maternal electrocardiogram (MECG) with very little noise, as shown in Figure 9.9a. Often both ECG waveforms are superimposed, making it impossible to ascertain the nature of the fetus's heart activity. In such cases, cross correlation is used to obtain a recording of the isolated FECG. First the signal is highpass filtered to stabilize the baseline and produce a zero mean signal, as shown in Figure 9.9b, producing $y(n)$. Examining this trace shows that sometimes the FECG and MECG are separate and distinct. Through an interactive program one obtains a template of the MECG. This is used as the reference signal, $x(n) = m(n)$. Then an estimate of $R_{xy}(k)$ is found and will be

$$\hat{R}_{xy}(k) = \sum_{i=1}^{q} g_i \hat{R}_{mm}(k - d_i) + \sum_{j=1}^{r} g_j \hat{R}_{mf}(k - d_j) \qquad (9.24)$$

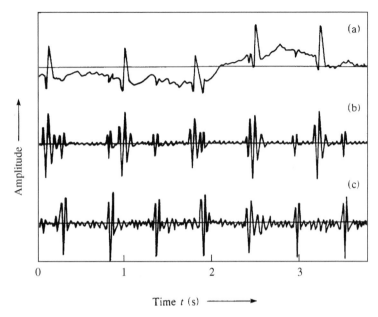

**FIGURE 9.9** Maternal and fetal ECG signals: (a) Original abdominal ECG. (b) Filtered ECG. (c) FECG after MECG is subtracted from plot (b). [Adapted from Nagel, fig. 1, with permission.]

where $\hat{R}_{mm}(k)$ is the sample ACF of the MECG waveform and $\hat{R}_{mf}(k)$ is the sample cross correlation between the maternal and fetal ECGs. Since no loss is involved, $g_i = g_j = 1$. Examination of Figure 9.9b reveals that the maximum absolute value of $m(n)$ is greater than that of $f(n)$; therefore $R_{mm}(0) > \max R_{mf}(k)$. $\hat{R}_{xy}(k)$ is searched for peak values and the time of the peaks, $d_1, \ldots, d_q$, correspond to the times of occurrence of the MECG. At these times the template, $m(n)$, is subtracted from $y(n)$, resulting in the FECG, as shown in Figure 9.9c.

### 9.3.3 Random Signals

In the situation where the goal is to determine the relationship between two random signals, the correlation functions do not involve deterministic signals and are interpreted slightly differently. However, since only stationary signals are being considered, the mathematical manipulations are identical. The only practical difference is the number of points in the summation of equation 9.19. Now for the time delay in a lossy and noisy environment,

$$y(n) = gx(n - d) + \eta(n) \tag{9.25}$$

$$R_{xy}(k) = E[x(n)y(n + k)] = gE[x(n)x(n + k - d)] + E[x(n)\eta(n + k)]$$
$$= g \cdot R_{xx}(k - d) + R_{x\eta}(k) \tag{9.26}$$

The term $R_{x\eta}(k)$ is the cross correlation between the signal and noise. The usual assumption is that $\eta(n)$ is a zero mean white noise that is uncorrelated with $x(n)$. It comprises the effect of measurement and transmission noise. Again, $R_{x\eta}(k) = m_x m_\eta = 0$, and again the cross correlation is simply the autocorrelation function of the reference signal shifted by the time delay and multiplied by the attenuation factor. The peak value is $R_{xy}(d) = g \cdot R_{xx}(0) = g\sigma_x^2$.

Consider the multiple-path acoustic environment shown in Figure 9.7. The reference signal, $x(n)$, is a bandlimited noise process with bandwidth, $B$, of 8 KHz and is sketched in Figure 9.10a. With only the direct path, equation 9.25 represents the measurement situation, Figure 9.10b the received signal, and Figure 9.11a shows $R_{xy}(k)$. The cross correlation function peaks at 2 ms, which is consistent with the fact that the path length is 0.68 m and the speed of sound in air is 340 m/s. With one reflecting surface, the CCF appears as that in Figure 9.11b. Another term, $g_2 R_{xx}(k - d_2)$, is present and peaks at $\tau_2 = d_2 T = 3.9$ ms. This

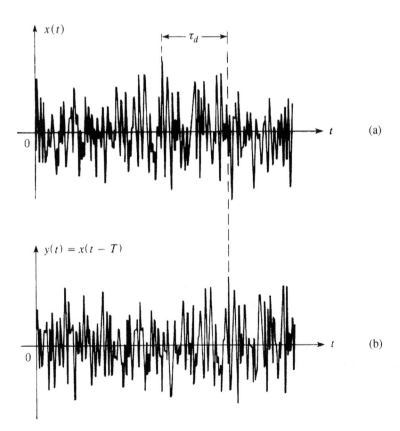

**FIGURE 9.10** Bandlimited acoustic signal: (a) Transmitted signal. (b) Received signal. [Adapted from Coulon, fig. 13.23, with permission.]

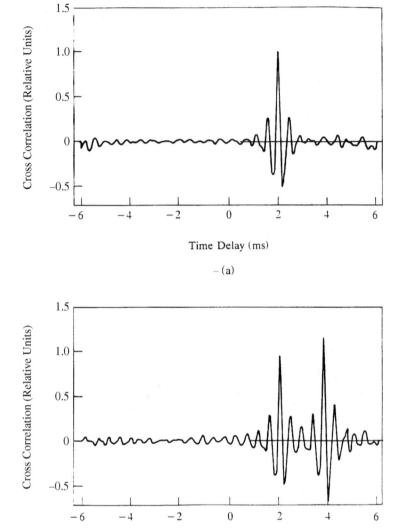

**FIGURE 9.11** Cross correlation functions for multiple-path acoustic experiment with $T = 0.012$ ms: (a) Direct path only. (b) Direct and side reflection path present. [Adapted from Bendat and Piersol, fig. 6.3, with permission.]

means that an additional pathway with a length of 1.32 m exists. The geometry of the situation indicates that this reflection comes from a side wall.

### 9.3.4 Time Difference of Arrival

In ranging systems, arrays of sensors are used to detect the waveform sent from a transmitter, emitter source, or the reflection from a target. The basic hypothesis of the measurement situation is that the waveform is a plane wave and the sensors are close together so that the two received signals are composed of the same waveform with added noise. The difference in the time of arrival is used to estimate the angular position of the source.

A schematic is shown in Figure 9.12 for an acoustic source. The two measured signals are modeled as

$$y(n) = x(n - n_y) + \eta_y(n), \quad \text{sensor 1}$$

and $\hspace{10cm}$ (9.27)

$$z(n) = gx(n - n_z) + \eta_z(n), \quad \text{sensor 2}$$

where $g$ represents a relative attenuation factor. The white-noise processes are uncorrelated with each other and the signal waveform $x(n)$. The cross correlation function for these signals is

$$R_{yz}(k) = gR_{xx}(k + n_y - n_z) \hspace{4cm} (9.28)$$

Thus the CCF will have the shape of the ACF of $x(n)$ and peak at the *time difference of arrival* (*TDOA*).

### 9.3.5 Marine Seismic Signal Analysis

Marine seismic explorations are undertaken to determine the structure of the layered media under the water. Identifying the material attributes and the geometry of the layers are important goals not only in searching for hydrocarbon formations but also for geodetic study. A boat tows the acoustical signal source and an array of hydrophones. The energy source is either an explosion or a high-pressure, short-duration air pulse. A typical acoustic source waveform, $x(t)$, is plotted in Figure 9.13a. The underground media is modeled as a layered media, as shown in Figure 9.13b. The source signal undergoes reflections at the layer boundaries, and the received signal has the form

$$y(t) = \sum_{i=1}^{\infty} g_i x(t - \tau_i) + \eta(t) \hspace{4cm} (9.29)$$

The amplitude coefficients, $g_i$, are related to the reflection coefficients at each layer and the delay times, $\tau_i$, are related to the travel distances to each layer and

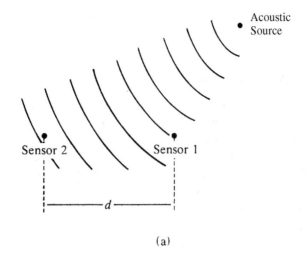

(a)

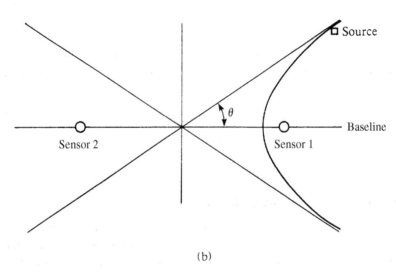

(b)

**FIGURE 9.12**   Determining bearing angle—geometrical arrangement of (a) Acoustic source and two sensors and (b) Bearing angle reference point. [Adapted from Chen, fig. 3, with permission.]

the speed of sound propagation through each layer, $c_i$. The information in the signals is usually within the frequency range from 10 Hz to 10 KHz. The delay times are found by digitizing $x(t)$ and $y(t)$ and estimating their cross correlation function. Once they are found, the amplitude coefficients can be found by dividing the amplitude of the peaks of the CCF by the peak value of the ACF of $x(t)$ (F. El-Hawary, 1988).

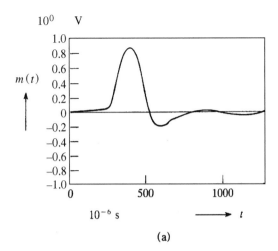

$10^0$  V

$m(t)$

$10^{-6}$ s $\longrightarrow$ t

(a)

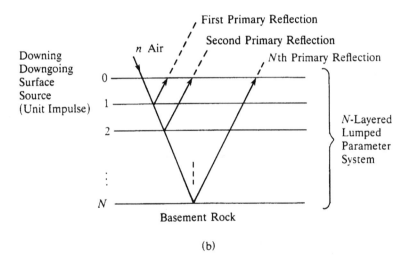

First Primary Reflection

Second Primary Reflection

$n$ Air

Downing
Downgoing
Surface
Source
(Unit Impulse)

$N$th Primary Reflection

$N$-Layered
Lumped
Parameter
System

Basement Rock

(b)

**FIGURE 9.13** Marine seismology: (a) Acoustic source waveform. (b) Schematic of layers of reflection. [Adapted from El-Hawary, fig. 20.1 and Chen, fig. 3, with permission.]

### 9.3.6 Procedure for Estimation

Now we will summarize the steps that must be performed to implement the concepts and procedures for estimating cross correlation functions.

#### Pulse Detection

**Detrending**    Both signals must be examined for trends and detrended, cf. Section 3.5.6.

**Correlation detection**    Select the reference signal or waveform and compute estimates of the cross correlation functions.

### Correlation between Signals

**Detrending**    Both signals must be examined for trends and detrended.

**Alignment and prewhitening**    Compute the cross correlation function estimates. Look for a dominant lag. If one exists, align the two signals and store the lag time for next step. Develop an AR model for the two signals and generate their residual sequences.

**Correlation functions**    Compute the auto and cross correlation function estimates using the residual sequences and shift it using the lag time from the previous step.

**Significance**    Test the magnitudes of the cross correlation function estimates to determine which, if any, are different from zero.

## 9.4   CROSS-SPECTRAL DENSITY FUNCTIONS

### 9.4.1   Definition and Properties

In Chapter 6, we defined the cross-spectral density function in the context of linear discrete time systems with input $x(n)$ and output $y(n)$. In review, the correlation function relationship is

$$R_{yx}(k) = R_x(k)*h(k)$$
$$\text{DTFT}[R_{yx}(k)] = S_{yx}(f) = S_x(f)H(f) \tag{9.30}$$

where $S_{yx}(f)$ is the *cross-spectral density function* (*CSD*). It can be seen from equation 9.30 that the CSD can be used to determine a system's transfer function if the PSD of the input signal is also known. It is also used directly as another methodology for determining signal relationships. The CSD has a magnitude relationship with the PSDs of the component signals; that is,

$$|S_{yx}(f)|^2 \le S_x(f)S_y(f) \tag{9.31}$$

Its proof is left as an exercise. A function called the *magnitude squared coherence function* (*MSC*) is defined as

$$K_{yx}^2(f) = \frac{|S_{yx}(f)|^2}{S_y(f)S_x(f)} = \frac{S_{yx}(f)S_{yx}(f)^*}{S_y(f)S_x(f)} \tag{9.32}$$

With the inequality of equation 9.31, the bound on the MSC is

$$0 \le K_{yx}^2(f) \le 1 \tag{9.33}$$

An important interpretation of the coherence function arises if it is assumed that $y(n)$ is a system output; then $Y(f) = H(f)X(f)$. Using equation 9.30 and system relationships, the coherence is

$$K_{yx}^2(f) = \frac{H(f)S_x(f)H(f)^*S_x(f)^*}{H(f)H(f)^*S_x(f)S_x(f)} = 1 \tag{9.34}$$

Thus if two signals are linearly related their coherence is unity. Therefore coherence becomes a good basis for determining the linear relationship of frequency components.

Because the phase angle is an important component of the CSD, the *complex coherence function* is also defined and is

$$K_{yx}(f) = +\sqrt{K_{yx}^2(f)} \angle S_{yx}(f) \tag{9.35}$$

The importance of the phase angle is that it reflects the time shift between frequency components in the signals $y(n)$ and $x(n)$. The time shift comes forth in a direct manner from the situations concerning ranging in Section 9.3. It was shown that even in a noisy and lossy environment that the correlation functions have the relationship

$$R_{xy}(k) = g \cdot R_{xx}(k - d) \tag{9.26}$$

Taking the DTFT, we have

$$S_{xy}(f) = gS_x(f)e^{-j2\pi fdT} \tag{9.36}$$

and the slope of the phase angle curve is proportional to the time delay, $\tau = dT$. The attenuation factor is easily found by

$$g = \frac{|S_{xy}(f)|}{S_x(f)} \tag{9.37}$$

As with the ordinary DTFT, the CSD has its real part that is an even function and its imaginary part that is an odd function. This can be shown by writing the CCF as a summation of even and odd functions. Create them as

$$\lambda_{yx}(k) = \frac{1}{2}(R_{yx}(k) + R_{yx}(-k))$$
$$\psi_{yx}(k) = \frac{1}{2}(R_{yx}(k) - R_{yx}(-k)) \tag{9.38}$$

and

$$R_{yx}(k) = \lambda_{yx}(k) + \psi_{yx}(k) \tag{9.39}$$

The CSD can then be expressed as

$$S_{yx}(f) = \sum_{k=-\infty}^{\infty} (\lambda_{yx}(k) + \psi_{yx}(k))e^{-j2\pi fkT}$$

$$= \sum_{k=-\infty}^{\infty} \lambda_{yx}(k)e^{-j2\pi fkT} + \sum_{k=-\infty}^{\infty} \psi_{yx}(k)e^{-j2\pi fkT}$$

$$= \Lambda_{yx}(f) + j\Psi_{yx}(f) \tag{9.40}$$

where $\Lambda_{yx}(f) = \Re[S_{yx}(f)]$ and $\Psi_{yx}(f) = \Im[S_{yx}(f)]$. $\Lambda_{yx}(f)$ is called the *cospectrum* and $\Psi_{yx}(f)$ is called the *quadrature spectrum*. As can be anticipated, these form the *cross magnitude*, $|S_{yx}(f)|$, and *cross phase*, $\phi_{yx}(f)$, spectra where

$$|S_{yx}(f)| = \sqrt{\Lambda_{yx}^2(f) + \Psi_{yx}^2(f)}$$

and

$$\phi_{yx}(f) = \tan^{-1} \frac{\Psi_{yx}(f)}{\Lambda_{yx}(f)} \tag{9.41}$$

### 9.4.2 Properties of Cross-Spectral Estimators

#### 9.4.2.1 Definition

There are many different estimators and properties of cross correlation and cross-spectral density functions. The ones being emphasized here are those that are needed to test the independence between two time series and that lay a foundation for further study of system identification. The properties of the estimators for the CSD are derived exactly as the estimators for the PSD in Chapter 7. Simply substitute the signal $y(n)$ for the first of the two $x(n)$ signals in the equations. The estimator for the CSD is

$$\hat{S}_{yx}(f) = T \sum_{k=-M}^{M} \hat{R}_{yx}(k)e^{-j2\pi fkT} \tag{9.42}$$

where the CSD is evaluated at frequencies $f = \pm m/2MT$ with $-M \leq m \leq M$. In terms of the frequency number, this is

$$\hat{S}_{yx}(m) = T \sum_{k=-M}^{M} \hat{R}_{yx}(k)e^{-j\pi mk/M} \tag{9.43}$$

The analog to the periodogram is

$$\hat{S}_{yx}(m) = \frac{1}{NT} Y^*(m)X(m) \tag{9.44}$$

### 9.4.2.2  Mean and Variance for Uncorrelated Signals

The variance of the estimate of the CSD is a function of the ACFs of the component signals. The variance and distributional properties of the CSD of independent signals will be presented because they provide the basis for developing statistical tests for assessing the correlation between the signals in the frequency domain. Using the same spectral approach as in Chapter 7 for deriving the statistical properties of the estimators, the definition of the DFT is

$$\frac{X(m)}{\sqrt{NT}} = A(m) - jB(m)$$

with

$$A(m) = \sqrt{\frac{T}{N}} \sum_{n=0}^{N-1} x(n) \cos\left(\frac{2\pi mn}{N}\right) \tag{9.45}$$

and

$$B(m) = \sqrt{\frac{T}{N}} \sum_{n=0}^{N-1} x(n) \sin\left(\frac{2\pi mn}{N}\right)$$

With the parameters $T$ and $N$ being part of the real and imaginary components, we have

$$\hat{S}_{yx}(m) = \frac{T}{N} Y^*(m)X(m) = (A_y(m) + jB_y(m))(A_x(m) - jB_x(m)) \tag{9.46}$$

Dropping the frequency number for simplicity in this explanation, the sample cospectrum and quadrature spectrum are, respectively,

$$\hat{\Lambda}_{yx}(m) = (A_y A_x + B_y B_x); \qquad \hat{\Psi}_{yx}(m) = (A_x B_y - A_y B_x) \tag{9.47}$$

It is known from Chapter 7 that, for Gaussian random processes, the real and imaginary components of the sample spectra are Gaussian random variables with a zero mean and variance equal to $S_y(m)/2$ or $S_x(m)/2$. If the two processes are uncorrelated, then

$$E[\hat{\Lambda}_{yx}(m)] = E[\hat{\Psi}_{yx}(m)] = 0 \tag{9.48}$$

The variance for the sample cospectrum is

$$\begin{aligned}
\text{Var}[\hat{\Lambda}_{yx}(m)] &= E[A_y^2 A_x^2 + B_y^2 B_x^2 + 2A_y A_x B_y B_x] \\
&= (E[A_y^2]E[A_x^2] + E[B_y^2]E[B_x^2]) \\
&= \left(\frac{S_y(m)}{2} \frac{S_x(m)}{2} + \frac{S_y(m)}{2} \frac{S_x(m)}{2}\right) \\
&= \frac{S_y(m)S_x(m)}{2}
\end{aligned} \tag{9.49}$$

The variance for the sample quadrature spectrum is the same, and the covariance between $\hat{\Lambda}_{yx}(m)$ and $\hat{\Psi}_{yx}(m)$ is zero.

The distribution of the magnitude of the sample CSD estimator is derived through its square.

$$|\hat{S}_{yx}(m)|^2 = \frac{T^2}{N^2} \, Y^*(m)X(m)Y(m)X^*(m) = \hat{S}_y(m)\hat{S}_x(m) \qquad (9.50)$$

Now we introduce the random variable

$$\Gamma^2(m) = \frac{4|\hat{S}_{yx}(m)|^2}{S_y(m)S_x(m)} = \frac{2\hat{S}_y(m)}{S_y(m)} \frac{2\hat{S}_x(m)}{S_x(m)} = UV \qquad (9.51)$$

Knowing that each sample PSD has a chi-square distribution and that they are independent of each other, we obtain

$$\begin{aligned} E[\Gamma^2(m)] &= E[U]E[V] = 2 \cdot 2 = 4 \\ E[\Gamma^4(m)] &= E[U^2]E[V^2] = 8 \cdot 8 = 64 \end{aligned} \qquad (9.52)$$

and

$$\mathrm{Var}[\Gamma^2(m)] = 48$$

Using equations 9.51 and 9.52, the mean and variance of the squared sample CSD are found to be

$$E[|\hat{S}_{yx}(m)|^2] = S_y(m)S_x(m)$$

and

$$\mathrm{Var}[|\hat{S}_{yx}(m)|^2] = 3S_y^2(m)S_x^2(m) \qquad (9.53)$$

The sample phase spectra is

$$\hat{\phi}_{yx}(m) = \tan^{-1}\left(\frac{\hat{\Psi}_{yx}(m)}{\hat{\Lambda}_{yx}(m)}\right) = \tan^{-1}\left(\frac{A_yB_x - A_xB_y}{A_yA_x + B_yB_x}\right) \qquad (9.54)$$

Since the terms $A_i$ and $B_i$ are independent Gaussian random variables ranging from $-\infty$ to $\infty$, the numerator and denominator terms are approximately Gaussian, independent, and possess the same variance. Thus it can be stated that $\hat{\phi}_{yx}(m)$ has an approximately uniform distribution ranging between $-\pi/2$ and $\pi/2$ (Jenkins and Watts, 1968).

### 9.4.2.3   Adjustments for Correlated Signals

It can be seen that for two uncorrelated signals $\hat{S}_{yx}(m)$ is unbiased and inconsistent. Remember that this development is based on a large value of $N$. As is known from the study of PSD estimators, these CSD estimators also have a bias since the CCF is biased (as shown in equation 9.13). The exception is when $y(n)$ and $x(n)$ are uncorrelated. Thus,

$$E[\hat{S}_{yx}(m)] = S_{yx}(m)*W(m) \qquad (9.55)$$

where $W(m)$ again represents the lag spectral window. The same procedures are used to reduce this inherent bias:

a. A lag window must be applied to the sample CCF when using equation 9.43.
b. Data windows must be applied to the acquired signals when using equation 9.44 and the power corrected for process loss.

The CSD is not used in general for the same reason that the CCVF is not used much in the time domain—it is unit-sensitive. The normalized version of the CSD, the MSC, and the phase angle of the CSD are used in practice. Thus we will concentrate on estimating the coherence and phase spectra. Before developing the techniques, we will briefly summarize several applications in the next section.


## 9.5  APPLICATIONS

Noise and vibration problems cause much concern in manufacturing. Sound intensity measurements are used to quantify the radiation of noise from vibrating structures and to help locate the noise generation sources. At one point on a vibrating structure the deflections were measured using accelerometers. The sound intensity was also measured at a place in space close to the point of vibration measurement. The PSD for the sound intensity is plotted in Figure 9.14a. There are definite peaks at 49, 61, and 100 Hz. The negative peak at 49 Hz indicates a 180° phase shift with respect to the other frequencies. The CSD estimate is plotted in Figure 9.14b. There are distinct peaks at 49 and 61 Hz but not at 100 Hz. This means that the surface is contributing to the noise at the lower frequencies but not at 100 Hz. Another part of the structure is creating this latter noise frequency (N. Thrane and S. Gade, 1988).

The study of the noise generated by a jet engine requires knowing whether the noise is coherent or diffuse, and at what distances from the engine any noise may become diffuse. The coherence function will quantify the strength of synchronous noise between two positions. Figure 9.15a shows the configuration for measuring the noise from the exhaust of a jet engine. Measurements were made from both locations in order to average 256 estimates of the needed spectra. Figures 9.15b and 9.15c show the phase and coherence spectra; the bandwidth is 4 Hz. The shape of the phase spectrum is consistent with the distance between the two locations. The MSC shows that the noise is coherent from 0 to 500 Hz and that higher-frequency components are diffuse (Bendat and Piersol, 1980).

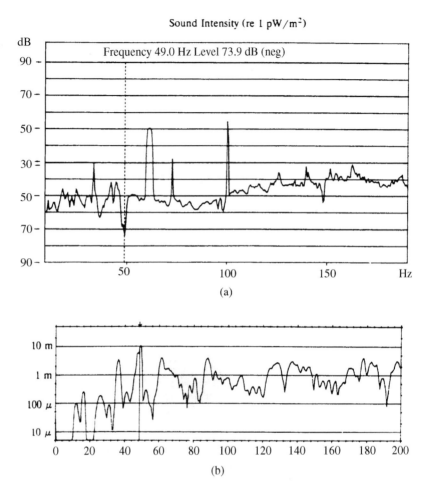

**FIGURE 9.14** Noise measurements from a vibrating structure: (a) PSD of sound intensity. (b) CSD of vibration and sound intensity. [Adapted from Thrane and Gade, figs. 1 and 2, with permission.]

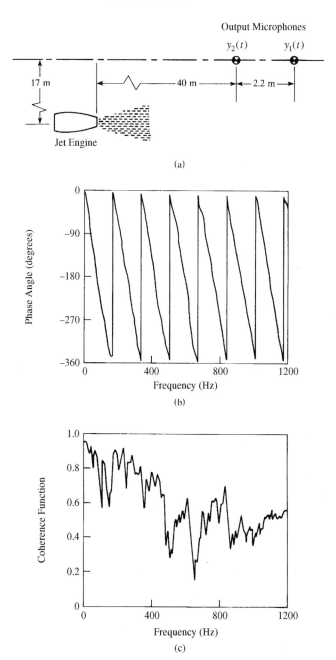

**FIGURE 9.15** Jet exhaust sound measurements at two locations: (a) Measurement configuration. (b) Phase spectra. (c) Coherence spectra. [Adapted from Bendat and Piersol, figs. 7.5 and 7.6, with permission.]

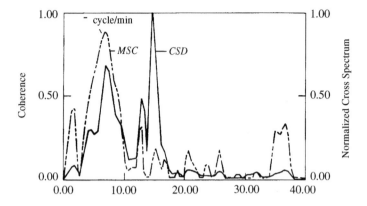

**FIGURE 9.16** Coherence function (MSC) and normalized cross-spectral density function (CSD) for the EGG in the colon. [Adapted from Reddy et al., fig. 10b, with permission.]

Investigating how the colon moves food along the digestive tract involves coherence analysis. The muscle activity of the colon is measured at two sites that are 3 cm apart. The signals are sampled at 5 Hz, and 30 segments of data lasting 51.2 sec are acquired. The CSD and MSC are estimated and plotted in Figure 9.16. The CSD shows peaks at 7 and 15 cycles per minute (c/min) whereas the MSC has a major peak only at 7 c/min. The interpretation of the MSC is that the 7 c/min oscillation is propagated down the colon. The existence of the 15 c/min oscillation in the CSD means that it exists strongly at one of the sites but is not propagated (Reddy et al., 1987).

## 9.6   TESTS FOR CORRELATION BETWEEN TIME SERIES

### 9.6.1   Coherence Estimators

The sample CSD provides the basis for a set of tests complementary to the CCF for testing the amount of correlation between two time series. The coherency, being a normalized magnitude, is a logical test criterion. However, if the sample squared coherency is considered, it provides no information since

$$\hat{K}_y^2(m) = \frac{Y^*(m)X(m)Y(m)X^*(m)}{Y^*(m)Y(m)X(m)X^*(m)} = 1 \tag{9.56}$$

Thus let us examine the effect that spectral smoothing will have on the properties of this estimator. It was seen before that

$$E[\hat{R}_{yx}(k)] = \left(1 - \frac{|k|}{N}\right)R_{yx}(k) \tag{9.57}$$

which resulted in a biased CSD estimator. As in autospectral estimation, a lag window is applied to the CCF estimate and a smoothed estimate of the CSD is produced with the mean being

$$E[\tilde{S}_{yx}(m)] = T \sum_{k=-M}^{M} w(k)\left(1 - \frac{|k|}{N}\right)R_{yx}(k)e^{-j\pi mk/M} \qquad (9.58)$$

Remember that the characteristics of the lag window dominate the effects of the data window, which results in a smoothed version of the theoretical CSD with

$$E[\tilde{S}_{yx}(m)] \approx S_{yx}(m)*W(m) = \tilde{S}_{yx}(m) = \tilde{\Lambda}_{yx}(m) + j\tilde{\Psi}_{yx}(m) \qquad (9.59)$$

Extreme care must be exercised when using this approach. *If the value of M is not large enough* to encompass any significant CCF values, then equation 9.59 is highly and erroneously biased. To remove this effect, one must find the major significant time shift using equation 9.12 and *then align the signals y(n) and x(n) and then perform the CSD estimation.* For PSD estimation is has been found in general that the quality of the estimate was the same whether ensemble averaging, correlational smoothing, or spectral smoothing was used to reduce the variance of the estimate. The same principle is true concerning CSD estimation, the important parameters being the spectral resolution and the degrees of freedom. From this point, the tilde ( ˜ ) symbol is used to represent estimates obtained using any of the approaches to reduce the variance. It seems that the most used smoothing procedure in CSD estimation is the Welch method (WOSA) with 50% overlap.

## 9.6.2 Statistical Properties of Estimators

The coherence spectrum, squared coherence spectrum, and phase spectrum are now estimated using these averaged or smoothed estimates; that is,

$$\tilde{K}_{yx}^2(m) = \frac{\tilde{\Lambda}_{yx}^2(m) + \tilde{\Psi}_{yx}^2(m)}{S_y(m)S_x(m)} \qquad (9.60)$$

$$\tilde{\phi}_{yx}(m) = \tan^{-1}\left(\frac{\tilde{\Psi}_{yx}(m)}{\tilde{\Lambda}_{yx}(m)}\right) \qquad (9.61)$$

As with the autospectral estimation, the window properties dictate the amount of variance reduction in the smoothing process. Recall that in the BT smoothing

$$\text{Var}[\tilde{S}(m)] = \frac{S^2(m)}{N} \sum_{k=-M}^{M} w^2(k) = \frac{S^2(m)}{\text{VR}} \qquad (9.62)$$

where VR is the variance reduction factor. Also recall that for ensemble averaging and spectral smoothing, VR equals the number of spectral values being averaged together. Similar results occur for estimating the cross-spectral density functions.

However, the derivation of their sampling distributions is very complex and voluminous and is reserved for advanced study. Detailed derivations can be found in Jenkins and Watts (1968) and Fuller (1976). Thankfully, however, the estimation techniques are similarly implementable and will be summarized and used.

The variance for the smoothed coherency and squared coherency estimators are

$$\text{Var}[|\tilde{K}_{yx}|] = \frac{1}{2 \text{ VR}} (1 - K_{yx}^2)^2$$

and (9.63)

$$\text{Var}[\tilde{K}_{yx}^2] = \frac{1}{2 \text{ VR}} 4 K_{yx}^2 (1 - K_{yx}^2)^2$$

The variance of the smoothed phase estimator, in radians, is

$$\text{Var}[\tilde{\phi}_{yx}(m)] = \frac{1}{2 \text{ VR}} \left( \frac{1}{K_{yx}^2} - 1 \right) \tag{9.64}$$

The covariance between the smoothed coherence and phase spectral estimators is approximately zero. It is important to notice that these variance expressions are dominated by two terms, VR and MSC. *The important practical reality is that the variance reduction factor is controllable, whereas the coherence spectrum is not—which can defeat any averaging or smoothing efforts.*

Since the CSD is biased by the inherent spectral windows, likewise is the estimate of the MSC. Another source of bias exists because the signal component of one measurement is delayed with respect to the other measurement. It has been shown that

$$E[\tilde{K}_{yx}^2(m)] \approx \left( 1 - \frac{\tau_d}{NT} \right) K_{yx}^2(m) \tag{9.65}$$

where $\tau_d$ is the time shift (Carter, 1988). Simulation studies have demonstrated that this bias can be appreciable. Fortunately, this source of bias can be controlled by *aligning* both signals as the first step in the estimation procedure.

As with any estimators that are intricate, the sampling distributions are quite complicated. This is certainly true for $\tilde{K}_{yx}^2(m)$ and $\tilde{\phi}_{yx}(m)$ (Carter, 1988). The distributions depend on whether or not $K_{yx}^2(m)$ is zero. If $K_{yx}^2(m) = 0$, then $\tilde{\phi}_{yx}(m)$ is uniformly distributed on the interval $(-\pi/2, \pi/2)$ and $\tilde{K}_{yx}^2(m)$ has an $F$ distribution. If $K_{yx}^2(m) \neq 0$, then $\tilde{\phi}_{yx}(m)$ converges to a normal distributed random variable and $\tilde{K}_{yx}^2(m)$ converges to one for a multiple correlation coefficient. Several variable transformations have been developed so that only one test is needed for each estimator. These will be reviewed in the next section. If the number of degrees of freedom is large, the sample coherence, complex coherence, and phase become unbiased estimators with normal distributions and the variances stated above.

### 9.6.3 Confidence Limits

It has been recognized that the variance of the smoothed coherence estimator is identical to the variance of an ordinary correlation coefficient. Hence the Fisher "$z$" transformation can be applied and the estimator becomes

$$\tilde{Z}_{yx}(m) = \tanh^{-1}(|\tilde{K}_{yx}(m)|) = \frac{1}{2} \ln\left(\frac{1 + |\tilde{K}_{yx}(m)|}{1 - |\tilde{K}_{yx}(m)|}\right) \qquad (9.66)$$

The function $\tilde{Z}_{yx}(m)$ is a biased and consistent estimator with

$$E[\tilde{Z}_{yx}(m)] = \tanh^{-1}(|K_{yx}(m)|) + \frac{1}{v - 2}; \qquad \text{bias} = b = \frac{1}{v - 2}$$

$$\text{Var}[\tilde{Z}_{yx}(m)] = \frac{1}{v - 2} \qquad (9.67)$$

where $v$ is the number of degrees of freedom of the variance reduction process used, $v = 2B_e NT$ for a smoothing procedure. This transformation is valid for $v \geq 20$ and $0.3 \leq K_{yx}^2(m) \leq 0.98$. Empirical improvements have been developed that enable statistical testing for the entire range of actual coherence and lower degrees of freedom. A correction factor for bias in the coherence domain is

$$B = \frac{1}{2v}(1 - \tilde{K}_{yx}^2(m)) \qquad (9.68)$$

so that in equation 9.66 the estimate for the coherence function should be replaced by

$$\tilde{K}_{yx}^2(m) \Rightarrow \tilde{K}_{yx}^2(m) - \frac{1}{2v}(1 - \tilde{K}_{yx}^2(m)) \qquad (9.69)$$

The actual bias of the transformed variable does not change. A variance correction (VC) is used in the $z$ domain

$$\text{VC} = 1 - 0.004^{1.6\tilde{K}^2 + 0.22} \qquad (9.70)$$

so that the new variance of $\tilde{Z}_{yx}(m)$ is

$$\text{Var}[\tilde{Z}_{yx}(m)] = \text{VC}\,\frac{1}{v - 2} \qquad (9.71)$$

The confidence intervals are established by assuming that the transformed estimator has a Gaussian distribution. Then for an estimated transformed spectrum, $\tilde{Z}_{yx}(m)$, the confidence limits are

$$\tilde{Z}_{yx}(m) - b \pm z\left(1 - \frac{\alpha}{2}\right)\sqrt{\frac{\text{VC}}{v - 2}} \qquad (9.72)$$

where $z(1 - \alpha/2)$ indicates the $z$ value of the cdf of the $N(0, 1)$ random variable for the probability of $1 - \alpha/2$. The confidence limits for the coherence function are established by making the hyperbolic tangent transformation of the limits in equation 9.72 (Ottnes and Enochson, 1972).

A usable sample distribution of the phase spectrum estimator is difficult to derive because an accurate approximation is unwieldy. A transformation is used to make the resulting variable approximately Gaussian. A tangent transformation is used and

$$\tilde{\theta}_{yx}(m) = \tan(\tilde{\phi}_{yx}(m)) \tag{9.73}$$

which produces a variance

$$\text{Var}[\tilde{\theta}_{yx}(m)] = \sigma_\theta^2 \approx \sec^4\left(\phi_{yx}(m) \cdot \frac{1}{2\,\text{VR}} \cdot \left(\frac{1}{K_{yx}^2(m)} - 1\right)\right) \tag{9.74}$$

The confidence limits for a $(1 - \alpha)$ confidence level are

$$\tilde{\theta}_{yx}(m) \pm z\left(1 - \frac{\alpha}{2}\right)\sigma_\theta \tag{9.75}$$

Because the components of the actual complex coherence spectrum needed in equation 9.74 are unknown, they must be replaced by their estimates. However, since $\tilde{\phi}_{yx}(m)$ and $\phi_{yx}(m)$ are independent, it is expected that when the limits in equation 9.75 are transformed back into angle units they will be independent of the actual phase angle (Jenkins and Watts, 1968).

### EXAMPLE 9.3

The coherence spectra in Figure 9.16 show the synchronous oscillation at 7 c/min of two locations in a colon. The confidence limits for this peak magnitude will be calculated. The signals were acquired at 5 Hz and the spectra were estimated using segment averaging with $K = 30$ and segments lengths being 51.2 seconds long. The frequency number of the peak is $m = 53$ and $\tilde{K}_{yx}^2(m) = 0.9$. The $z$ transformation is

$$\tilde{Z}_{yx}(m) = \frac{1}{2}\ln\left(\frac{1 + 0.95}{1 - 0.95}\right) = 1.81844$$

with $b = \dfrac{1}{v - 2} = \dfrac{1}{60 - 2} = 0.0172$. The bias correction is

$$B = \frac{1}{2v}(1 - \tilde{K}_{yx}^2(m)) = \frac{1}{120}(1 - 0.9) = 0.00083$$

The corrected $z$ value is

$$\tilde{Z}_{yx}(m) = \frac{1}{2}\ln\left(\frac{1 + 0.948}{1 - 0.948}\right) = 1.81407$$

The variance correction factor is

$$VC = 1 - 0.004^{1.6k^2+0.22} = 0.99989$$

and the corrected variance is

$$Var[\tilde{Z}_{yx}(m)] = \frac{0.99989}{58} = 0.01724$$

The 95% confidence limits are

$$\tilde{Z}_{yx}(m) - b \pm z(0.975)\sqrt{0.01724} = 1.79683 \pm 0.25735 = 2.05418, \ 1.53948$$

In the coherence domain, the limits become

$$K_{UL} = \tanh(2.05418) = 0.96766$$
$$K_{LL} = \tanh(1.53948) = 0.91203$$

The magnitude estimated was 0.94868. Thus the estimation procedure is quite good as the limits are close to $\tilde{K}_{yx}(53)$.

## 9.6.4 Procedure for Estimation

To summarize, the following steps must be performed to implement the concepts and procedures for estimating cross magnitude and phase spectra.

### Correlation Approach

**Detrending**     Both signals must be examined for trends and detrended. These trends will contribute to artifactual low-frequency components in all the magnitude spectra.

**Correlation functions**     Compute the auto and cross correlation function estimates. Look for the minimum number of lags needed for smoothing the autocorrelation estimates and conspicuous lags between the two signals. If a dominant lag exists, align the two signals and store the lag time for a later step. Recalculate the cross correlation estimate.

**Spectra**     Estimate the auto and cross spectra and the coherence spectrum for several values of bandwidths and maximum correlation lags. Perform the window closing, that is, look for the minimum lag that causes convergence of the estimates.

**Estimation**     Compensate the phase spectra for alignment and transform the coherence and phase spectra into their testing variables. Calculate the confidence limits and inverse transform the spectra and their associated confidence limits to the coherence and phase spectra. Interpret these spectra according to the hypotheses necessary.

Stability Margin $= \min(s_1, s_2)$

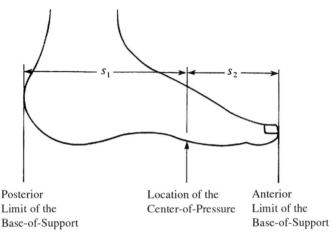

| Posterior | Location of the | Anterior |
| Limit of the | Center-of-Pressure | Limit of the |
| Base-of-Support | | Base-of-Support |

(a)

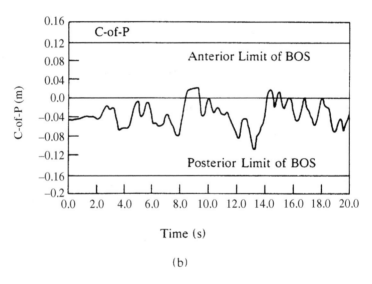

Time (s)

(b)

**FIGURE 9.17**  Investigation of balance and stability: (a) Schematic of foot and position of center of pressure. (b) Sample function of the movement of center of pressure. [Adapted from Maki et al., figs. 1, 4, and 5, with permission.]

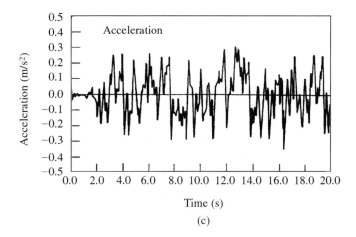

Time (s)

(c)

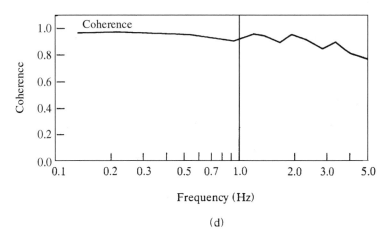

Frequency (Hz)

(d)

**FIGURE 9.17** Investigation of balance and stability: (c) Sample function of the acceleration of the floor. (d) Estimate of the coherence function. [Adapted from Maki et al., figs. 1, 4, and 5, with permission.]

### Direct Spectral Approach

**Detrending**    Perform detrending as described above.

**Spectra**    Estimate the auto and cross spectra using the periodogram approach. Smooth or average as necessary to reduce the variance. If segment averaging is used, be aware that the segments must be long enough to encompass any delayed phenomena. Perhaps a signal alignment will also be necessary.

**Estimation**    Perform the estimation as described above.

## 9.6.5 Application

The study of human balance has become very important because of the effect of drugs and aging on the control of balance. An effective and valid analytic procedure is an important component of this study. The coherence function was used to assess whether the results of a certain experimental procedure could be analyzed using linear relationships.

The rationale is explained using Figure 9.17a. As a person stands, the position of the resulting force on the floor is located within the boundary of the foot; it is called the center of pressure. It will move as the body is perturbed. The body is perturbed by moving the floor slightly and recording the acceleration of the floor and the location of the center of pressure. The floor only moves the foot forward and backward. A sample of the recordings of the movement of the center of pressure and the acceleration are plotted in Figure 9.17b and c. The signals were recorded for 185 s with $T = 0.06$ s. The MSC is estimated with the WOSA method. Each segment contained 256 points and a Hamming window with 50% overlap was used. The resulting $\tilde{K}_{yx}^2(m)$ is plotted in Figure 9.17d. Appreciate that almost all magnitudes of the MSC have a value of 1. Thus a linear systems approach for signal analysis seems valid.

## REFERENCES

J. Bendat and A. Piersol, *Engineering Applications of Correlation and Spectral Analysis*, Wiley, New York, 1980.

J. Bendat and A. Piersol, *Random Data, Analysis and Measurement Procedures*, Wiley, New York, 1986.

G. Box and G. Jenkins, *Time Series Analysis, Forecasting and Control*, Holden-Day, Oakland, CA, 1976.

G. Carter, "Coherence and Time Delay Estimation," in *Signal Processing Handbook*, C. Chen, Ed., Dekker, New York, 1988.

C. Chen, "Sonar Signal Processing," in *Digital Waveform Processing and Recognition*, C. Chen, Ed., CRC Press, Boca Raton, FL, 1982.

F. de Coulon, *Signal Theory and Processing*, Artech House, Dedham, MA, 1986.

El-Hawary, "Marine Geophysical Signal Processing," in *Signal Processing Handbook*, C. Chen, Ed., Dekker, New York, 1988.

W. Fuller, *Introduction to Statistical Time Series*, Wiley, New York, 1976.

G. Jenkins and D. Watts, *Spectral Analysis and Its Applications*, Holden-Day, San Francisco, 1968.

B. Maki, P. Holliday, and G. Fermie, "A Posture Control Model and Balance Test for the Prediction of Relative Postural Stability," *IEEE Trans. Biomed. Eng.* **34**:797–810 (1987).

J. Nagel, "Progresses in Fetal Monitoring by Improved Data Acquisition," *IEEE Eng. in Med. and Biol. Magazine* **3**(3):9–13 (1984).

H. Newton, *TIMESLAB: A Time Series Analysis Laboratory*, Wadsworth & Brooks/Cole, Pacific Grove, CA, 1988.

R. Ottnes and L. Enochson, *Digital Time Series Analysis*, Wiley, New York, 1972.

S. Reddy, S. Collins, and E. Daniel, "Frequency Analysis of Gut EMG," *Crit. Rev. in Biomed. Eng.* **15**(2): (1987).

M. Schwartz and L. Shaw, *Signal Processing: Discrete Spectral Analysis, Detection, and Estimation*, McGraw-Hill, New York, 1975.

M. Silvia, "Time Delay Estimation," in *Handbook of Digital Signal Processing—Engineering Applications*, D. Elliott, Ed., Academic Press, New York, 1987.

"Special Issue on Time Delay Estimation," *IEEE Trans. Acoust., Speech, Signal Proc.*, June (1981).

N. Thrane and S. Gade, "Use of Operational Deflection Shapes for Noise Control of Discrete Tones," *Bruel & Kjaer Tech. Rev.* **1**: (1988).

## EXERCISES

**9.1**  Prove that $C_{yx}^2(k) \leq C_{yy}(0)C_{xx}(0)$.

**9.2**  Derive the inequality $|R_{yx}(k)| \leq \frac{1}{2}(R_{yy}(0) + R_{xx}(0))$. [*Hint*: Start with equation 9.5 and make one minor change, and let $a = 1$.]

**9.3**  Equation 9.17 states the variance of the cross covariance estimator for two independent first-order AR processes. Derive the expression.

**9.4**  What happens to the bias in the estimate of the CCVF after the signals are prewhitened?

**9.5**  Derive the cross correlation function, equation 9.28, for the time difference of arrival situation. What is the TDOA? Knowing the distance between the sensors and the speed of sound in water (1500 m/s), estimate the radial position of the source, $\theta$, as shown in Figure 9.12b.

**9.6**  A single receiver is used to track a source using passive sonar and an autocorrelation technique. The multiple paths of the acoustic signal received are sketched in Figure E9.6. The multiple-path reflections can be used to estimate the depth, $h_s$, of the source. Assume that the source produces pulses of sound with a rectangular waveshape of duration 2 ms. What are the constraints on the path lengths so that the multiple-path reflections produce distinct pulses in the ACF of the received signal?

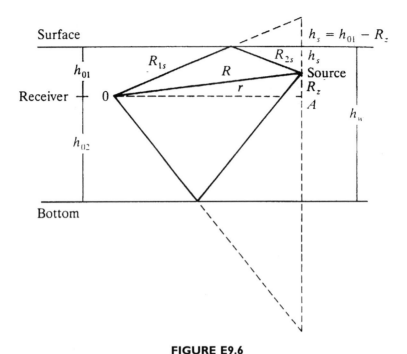

**FIGURE E9.6**

**9.7** Prove the relationship $|S_{yx}(m)|^2 \leq S_{yy}(m)S_{xx}(m)$. Start with the relationship

$$\left( \frac{Y^*(m)}{\sqrt{S_{yy}(m)}} - \frac{X(m)}{\sqrt{S_{xx}(m)}} \right)^2 \geq 0$$

and take expectations.

**9.8** Prove that the variance of the sample quadrature spectrum for two independent white-noise signals is $\sigma_y^2 \sigma_x^2 / 2$.

**9.9** Prove that the covariance between the sample cospectrum and quadrature spectrum for any uncorrelated signals is zero. Use the same information used in the derivation of the variance of the sample cospectrum.

**9.10** For the coherence function in Figure 9.17, find the confidence limits at 1.0 Hz.

## Computer Exercises

**9.11** The bivariate signals relating average temperatures and birthrate are listed in Appendix 9.2.

   a. Estimate the normalized cross covariance function between these two signals.

b. Determine AR models for these signals. What are their parameters?

c. Prewhiten these signals and estimate the NCCF between the noise signals.

d. What does $\hat{\rho}_{yx}(k)$ indicate? Is there any difference between this function and that in part (a)?

**9.12** This is an exercise to practice detecting the delay of a random signal in a noisy environment.

a. Generate 100 points of a Gaussian AR(1) process with a variance of 10; $a(1) = 0.5$. This is $x(n)$.

b. Form three other signals by

   1. delaying $x(n)$ by 10 time units and attenuating it by a factor of 2,

   2. adding uniform white noise to the process in part (b1) with variances of 1, 2, and 5. That is $y_i(n) = 0.5x(n - 10) + \eta_i(n), i = 1, 2, 3$.

c. Estimate the cross correlation function between $x(n)$ and each of the $y_i(n)$. How do they differ and why?

# APPENDIX 9.1   BIVARIATE GAS FURNACE DATA*

| n | $X_n$ | $Y_n$ | n | $X_n$ | $Y_n$ | n | $X_n$ | $Y_n$ |
|---|-------|-------|---|-------|-------|---|-------|-------|
| 1 | −0.109 | 53.8 | 51 | 1.608 | 46.9 | 101 | −0.288 | 51.0 |
| 2 | 0.000 | 53.6 | 52 | 1.905 | 47.8 | 102 | −0.153 | 51.8 |
| 3 | 0.178 | 53.5 | 53 | 2.023 | 48.2 | 103 | −0.109 | 52.4 |
| 4 | 0.339 | 53.5 | 54 | 1.815 | 48.3 | 104 | −0.187 | 53.0 |
| 5 | 0.373 | 53.4 | 55 | 0.535 | 47.9 | 105 | −0.255 | 53.4 |
| 6 | 0.441 | 53.1 | 56 | 0.122 | 47.2 | 106 | −0.229 | 53.6 |
| 7 | 0.461 | 52.7 | 57 | 0.009 | 47.2 | 107 | −0.007 | 53.7 |
| 8 | 0.348 | 52.4 | 58 | 0.164 | 48.1 | 108 | 0.254 | 53.8 |
| 9 | 0.127 | 52.2 | 59 | 0.671 | 49.4 | 109 | 0.330 | 53.8 |
| 10 | −0.180 | 52.0 | 60 | 1.019 | 50.6 | 110 | 0.102 | 53.8 |
| 11 | −0.588 | 52.0 | 61 | 1.146 | 51.5 | 111 | −0.423 | 53.3 |
| 12 | −1.055 | 52.4 | 62 | 1.155 | 51.6 | 112 | −1.139 | 53.0 |
| 13 | −1.421 | 53.0 | 63 | 1.112 | 51.2 | 113 | −2.275 | 52.9 |
| 14 | −1.520 | 54.0 | 64 | 1.121 | 50.5 | 114 | −2.594 | 53.4 |
| 15 | −1.302 | 54.9 | 65 | 1.223 | 50.1 | 115 | −2.716 | 54.6 |
| 16 | −0.814 | 56.0 | 66 | 1.257 | 49.8 | 116 | −2.510 | 56.4 |
| 17 | −0.475 | 56.8 | 67 | 1.157 | 49.6 | 117 | −1.790 | 58.0 |
| 18 | −0.193 | 56.8 | 68 | 0.913 | 49.4 | 118 | −1.346 | 59.4 |
| 19 | 0.088 | 56.4 | 69 | 0.620 | 49.3 | 119 | −1.081 | 60.2 |
| 20 | 0.435 | 55.7 | 70 | 0.255 | 49.2 | 120 | −0.910 | 60.0 |
| 21 | 0.771 | 55.0 | 71 | −0.280 | 49.3 | 121 | −0.876 | 59.4 |
| 22 | 0.866 | 54.3 | 72 | −1.080 | 49.7 | 122 | −0.885 | 58.4 |
| 23 | 0.875 | 53.2 | 73 | −1.551 | 50.3 | 123 | −0.800 | 57.6 |
| 24 | 0.891 | 52.3 | 74 | −1.799 | 51.3 | 124 | −0.544 | 56.9 |
| 25 | 0.987 | 51.6 | 75 | −1.825 | 52.8 | 125 | −0.416 | 56.4 |
| 26 | 1.263 | 51.2 | 76 | −1.456 | 54.4 | 126 | −0.271 | 56.0 |
| 27 | 1.775 | 50.8 | 77 | −0.944 | 56.0 | 127 | 0.000 | 55.7 |
| 28 | 1.976 | 50.5 | 78 | −0.570 | 56.9 | 128 | 0.403 | 55.3 |
| 29 | 1.934 | 50.0 | 79 | −0.431 | 57.5 | 129 | 0.841 | 55.0 |
| 30 | 1.866 | 49.2 | 80 | −0.577 | 57.3 | 130 | 1.285 | 54.4 |
| 31 | 1.832 | 48.4 | 81 | −0.960 | 56.6 | 131 | 1.607 | 53.7 |
| 32 | 1.767 | 47.9 | 82 | −1.616 | 56.0 | 132 | 1.746 | 52.8 |
| 33 | 1.608 | 47.6 | 83 | −1.875 | 55.4 | 133 | 1.683 | 51.6 |
| 34 | 1.265 | 47.5 | 84 | −1.891 | 55.4 | 134 | 1.485 | 50.6 |
| 35 | 0.790 | 47.5 | 85 | −1.746 | 56.4 | 135 | 0.993 | 49.4 |
| 36 | 0.360 | 47.6 | 86 | −1.474 | 57.2 | 136 | 0.648 | 48.8 |
| 37 | 0.115 | 48.1 | 87 | −1.201 | 58.0 | 137 | 0.577 | 48.5 |
| 38 | 0.088 | 49.0 | 88 | −0.927 | 58.4 | 138 | 0.577 | 48.7 |
| 39 | 0.331 | 50.0 | 89 | −0.524 | 58.4 | 139 | 0.632 | 49.2 |
| 40 | 0.645 | 51.1 | 90 | 0.040 | 58.1 | 140 | 0.747 | 49.8 |
| 41 | 0.960 | 51.8 | 91 | 0.788 | 57.7 | 141 | 0.900 | 50.4 |
| 42 | 1.409 | 51.9 | 92 | 0.943 | 57.0 | 142 | 0.993 | 50.7 |
| 43 | 2.670 | 51.7 | 93 | 0.930 | 56.0 | 143 | 0.968 | 50.9 |
| 44 | 2.834 | 51.2 | 94 | 1.006 | 54.7 | 144 | 0.790 | 50.7 |
| 45 | 2.812 | 50.0 | 95 | 1.137 | 53.2 | 145 | 0.399 | 50.5 |
| 46 | 2.483 | 48.3 | 96 | 1.198 | 52.1 | 146 | −0.161 | 50.4 |
| 47 | 1.929 | 47.0 | 97 | 1.054 | 51.6 | 147 | −0.553 | 50.2 |
| 48 | 1.485 | 45.8 | 98 | 0.595 | 51.0 | 148 | −0.603 | 50.4 |
| 49 | 1.214 | 45.6 | 99 | −0.080 | 50.5 | 149 | −0.424 | 51.2 |
| 50 | 1.239 | 46.0 | 100 | −0.314 | 50.4 | 150 | −0.194 | 52.3 |

*X: 0.60–0.04 input gas rate in cu. ft/min Y: % $CO_2$ in outlet gas
Sampling interval 9 seconds N = 296 pairs of data points

## Series J Continued

| n | $X_n$ | $Y_n$ | n | $X_n$ | $Y_n$ | n | $X_n$ | $Y_n$ |
|---|---|---|---|---|---|---|---|---|
| 151 | −0.049 | 53.2 | 201 | −2.473 | 55.6 | 251 | 0.185 | 56.3 |
| 152 | 0.060 | 53.9 | 202 | −2.330 | 58.0 | 252 | 0.662 | 56.4 |
| 153 | 0.161 | 54.1 | 203 | −2.053 | 59.5 | 253 | 0.709 | 56.4 |
| 154 | 0.301 | 54.0 | 204 | −1.739 | 60.0 | 254 | 0.605 | 56.0 |
| 155 | 0.517 | 53.6 | 205 | −1.261 | 60.4 | 255 | 0.501 | 55.2 |
| 156 | 0.566 | 53.2 | 206 | −0.569 | 60.5 | 256 | 0.603 | 54.0 |
| 157 | 0.560 | 53.0 | 207 | −0.137 | 60.2 | 257 | 0.943 | 53.0 |
| 158 | 0.573 | 52.8 | 208 | −0.024 | 59.7 | 258 | 1.223 | 52.0 |
| 159 | 0.592 | 52.3 | 209 | −0.050 | 59.0 | 259 | 1.249 | 51.6 |
| 160 | 0.671 | 51.9 | 210 | −0.135 | 57.6 | 260 | 0.824 | 51.6 |
| 161 | 0.933 | 51.6 | 211 | −0.276 | 56.4 | 261 | 0.102 | 51.1 |
| 162 | 1.337 | 51.6 | 212 | −0.534 | 55.2 | 262 | 0.025 | 50.4 |
| 163 | 1.460 | 51.4 | 213 | −0.871 | 54.5 | 263 | 0.382 | 50.0 |
| 164 | 1.353 | 51.2 | 214 | −1.243 | 54.1 | 264 | 0.922 | 50.0 |
| 165 | 0.772 | 50.7 | 215 | −1.439 | 54.1 | 265 | 1.032 | 52.0 |
| 166 | 0.218 | 50.0 | 216 | −1.422 | 54.4 | 266 | 0.866 | 54.0 |
| 167 | −0.237 | 49.4 | 217 | −1.175 | 55.5 | 267 | 0.527 | 55.1 |
| 168 | −0.714 | 49.3 | 218 | −0.813 | 56.2 | 268 | 0.093 | 54.5 |
| 169 | −1.099 | 49.7 | 219 | −0.634 | 57.0 | 269 | −0.458 | 52.8 |
| 170 | −1.269 | 50.6 | 220 | −0.582 | 57.3 | 270 | −0.748 | 51.4 |
| 171 | −1.175 | 51.8 | 221 | −0.625 | 57.4 | 271 | −0.947 | 50.8 |
| 172 | −0.676 | 53.0 | 222 | −0.713 | 57.0 | 272 | −1.029 | 51.2 |
| 173 | 0.033 | 54.0 | 223 | −0.848 | 56.4 | 273 | −0.928 | 52.0 |
| 174 | 0.556 | 55.3 | 224 | −1.039 | 55.9 | 274 | −0.645 | 52.8 |
| 175 | 0.643. | 55.9 | 225 | −1.346 | 55.5 | 275 | −0.424 | 53.8 |
| 176 | 0.484 | 55.9 | 226 | −1.628 | 55.3 | 276 | −0.276 | 54.5 |
| 177 | 0.109 | 54.6 | 227 | −1.619 | 55.2 | 277 | −0.158 | 54.9 |
| 178 | −0.310 | 53.5 | 228 | −1.149 | 55.4 | 278 | −0.033 | 54.9 |
| 179 | −0.697 | 52.4 | 229 | −0.488 | 56.0 | 279 | 0.102 | 54.8 |
| 180 | −1.047 | 52.1 | 230 | −0.160 | 56.5 | 280 | 0.251 | 54.4 |
| 181 | −1.218 | 52.3 | 231 | −0.007 | 57.1 | 281 | 0.280 | 53.7 |
| 182 | −1.183 | 53.0 | 232 | −0.092 | 57.3 | 282 | 0.000 | 53.3 |
| 183 | −0.873 | 53.8 | 233 | −0.620 | 56.8 | 283 | −0.493 | 52.8 |
| 184 | −0.336 | 54.6 | 234 | −1.086 | 55.6 | 284 | −0.759 | 52.6 |
| 185 | 0.063 | 55.4 | 235 | −1.525 | 55.0 | 285 | −0.824 | 52.6 |
| 186 | 0.084 | 55.9 | 236 | −1.858 | 54.1 | 286 | −0.740 | 53.0 |
| 187 | 0.000 | 55.9 | 237 | −2.029 | 54.3 | 287 | −0.528 | 54.3 |
| 188 | 0.001 | 55.2 | 238 | −2.024 | 55.3 | 288 | −0.204 | 56.0 |
| 189 | 0.209 | 54.4 | 239 | −1.961 | 56.4 | 289 | 0.034 | 57.0 |
| 190 | 0.556 | 53.7 | 240 | −1.952 | 57.2 | 290 | 0.204 | 58.0 |
| 191 | 0.782 | 53.6 | 241 | −1.794 | 57.8 | 291 | 0.253 | 58.6 |
| 192 | 0.858 | 53.6 | 242 | −1.302 | 58.3 | 292 | 0.195 | 58.5 |
| 193 | 0.918 | 53.2 | 243 | −1.030 | 58.6 | 293 | 0.131 | 58.3 |
| 194 | 0.862 | 52.5 | 244 | −0.918 | 58.8 | 294 | 0.017 | 57.8 |
| 195 | 0.416 | 52.0 | 245 | −0.798 | 58.8 | 295 | −0.182 | 57.3 |
| 196 | −0.336 | 51.4 | 246 | −0.867 | 58.6 | 296 | −0.262 | 57.0 |
| 197 | −0.959 | 51.0 | 247 | −1.047 | 58.0 | | | |
| 198 | −1.813 | 50.9 | 248 | −1.123 | 57.4 | | | |
| 199 | −2.378 | 52.4 | 249 | −0.876 | 57.0 | | | |
| 200 | −2.499 | 53.5 | 250 | −0.395 | 56.4 | | | |

## APPENDIX 9.2   BIVARIATE TEMPERATURE AND BIRTH DATA IN NEW YORK CITY

Monthly Temperatures 1946–59*

| | | | | | | | |
|---|---|---|---|---|---|---|---|
| 11.506 | 11.022 | 14.405 | 14.442 | 16.524 | 17.918 | 18.959 | 18.309 |
| 18.160 | 16.691 | 14.480 | 17.862 | 12.082 | 10.558 | 12.138 | 14.442 |
| 16.152 | 17.714 | 19.015 | 19.164 | 17.900 | 16.933 | 13.364 | 11.468 |
| 10.000 | 10.985 | 12.993 | 14.480 | 16.134 | 17.862 | 19.238 | 19.089 |
| 18.067 | 15.576 | 14.926 | 12.398 | 12.435 | 12.416 | 13.067 | 15.093 |
| 16.766 | 18.680 | 19.796 | 19.424 | 17.398 | 16.877 | 13.866 | 12.658 |
| 13.048 | 11.301 | 12.063 | 14.164 | 16.041 | 18.067 | 19.108 | 18.736 |
| 17.212 | 16.394 | 13.922 | 11.691 | 11.970 | 11.970 | 12.900 | 14.888 |
| 16.747 | 17.955 | 19.201 | 18.903 | 17.714 | 16.097 | 13.234 | 12.435 |
| 11.952 | 11.970 | 12.639 | 15.204 | 16.283 | 18.699 | 19.851 | 18.959 |
| 18.030 | 15.316 | 14.126 | 12.323 | 12.230 | 12.230 | 13.234 | 14.647 |
| 16.729 | 18.494 | 19.312 | 19.164 | 18.011 | 16.338 | 14.368 | 12.751 |
| 11.022 | 12.658 | 12.844 | 15.037 | 16.190 | 18.253 | 19.182 | 18.513 |
| 17.621 | 16.506 | 13.717 | 11.840 | 10.855 | 11.636 | 12.825 | 15.019 |
| 16.952 | 17.937 | 20.000 | 19.535 | 17.770 | 16.283 | 13.327 | 10.818 |
| 11.264 | 12.026 | 12.230 | 14.126 | 15.818 | 18.420 | 18.699 | 18.903 |
| 17.175 | 16.041 | 13.866 | 12.844 | 10.613 | 12.082 | 12.955 | 15.000 |
| 16.747 | 18.792 | 19.312 | 18.680 | 18.067 | 15.632 | 14.349 | 12.732 |
| 11.152 | 10.372 | 12.639 | 14.981 | 16.097 | 17.602 | 19.331 | 18.978 |
| 17.695 | 15.539 | 14.182 | 10.781 | 11.059 | 11.152 | 12.621 | 14.981 |
| 17.100 | 18.160 | 19.015 | 19.331 | 18.346 | 16.115 | 13.532 | 12.398 |

Monthly Births 1946–1959*

| | | | | | | | |
|---|---|---|---|---|---|---|---|
| 26.663 | 23.598 | 26.931 | 24.740 | 25.806 | 24.364 | 24.477 | 23.901 |
| 23.175 | 23.227 | 21.672 | 21.870 | 21.439 | 21.089 | 23.709 | 21.669 |
| 21.752 | 20.761 | 23.479 | 23.824 | 23.105 | 23.110 | 21.759 | 22.073 |
| 21.937 | 20.035 | 23.590 | 21.672 | 22.222 | 22.123 | 23.950 | 23.504 |
| 22.238 | 23.142 | 21.059 | 21.573 | 21.548 | 20.000 | 22.424 | 20.615 |
| 21.761 | 22.874 | 24.104 | 23.748 | 23.262 | 22.907 | 21.519 | 22.025 |
| 22.604 | 20.894 | 24.677 | 23.673 | 25.320 | 23.583 | 24.671 | 24.454 |
| 24.122 | 24.252 | 22.084 | 22.991 | 23.287 | 23.049 | 25.076 | 24.037 |
| 24.430 | 24.667 | 26.451 | 25.618 | 25.014 | 25.110 | 22.964 | 23.981 |
| 23.798 | 22.270 | 24.775 | 22.646 | 23.988 | 24.737 | 26.276 | 25.816 |
| 25.210 | 25.199 | 23.162 | 24.707 | 24.364 | 22.644 | 25.565 | 24.062 |
| 25.431 | 24.635 | 27.009 | 26.606 | 26.268 | 26.462 | 25.246 | 25.180 |
| 24.657 | 23.304 | 26.982 | 26.199 | 27.210 | 26.122 | 26.706 | 26.878 |
| 26.152 | 26.379 | 24.712 | 25.688 | 24.990 | 24.239 | 26.721 | 23.475 |
| 24.767 | 26.219 | 28.361 | 28.599 | 27.914 | 27.784 | 25.693 | 26.881 |
| 26.217 | 24.218 | 27.914 | 26.975 | 28.527 | 27.139 | 28.982 | 28.169 |
| 28.056 | 29.136 | 26.291 | 26.987 | 26.589 | 24.848 | 27.**543** | 26.896 |
| 28.878 | 27.390 | 28.065 | 28.141 | 29.048 | 28.484 | 26.**634** | 27.735 |
| 27.132 | 24.924 | 28.963 | 26.589 | 27.931 | 28.009 | 29.**229** | 28.759 |
| 28.405 | 27.945 | 25.912 | 26.619 | 26.076 | 25.286 | 27.**660** | 25.951 |
| 26.398 | 25.565 | 28.865 | 30.000 | 29.261 | 29.012 | 26.**992** | 27.897 |

*read across rows, N = 168 pairs

# INDEX

**381**